Eco-Friendly Energy Processes and Technologies for Achieving Sustainable Development

Mir Sayed Shah Danish
University of the Ryukyus, Japan

Tomonobu Shah Senjyu
University of the Ryukyus, Japan

IGI Global
PUBLISHER of TIMELY KNOWLEDGE

A volume in the Advances in Environmental Engineering and Green Technologies (AEEGT) Book Series

Published in the United States of America by
IGI Global
Engineering Science Reference (an imprint of IGI Global)
701 E. Chocolate Avenue
Hershey PA, USA 17033
Tel: 717-533-8845
Fax: 717-533-8661
E-mail: cust@igi-global.com
Web site: http://www.igi-global.com

Library of Congress Cataloging-in-Publication Data

Names: Danish, Mir Sayed Shah, 1987- editor.
Title: Eco-friendly energy processes and technologies for achieving sustainable development / Mir Sayed Shah Danish, editor.
Description: Hershey, PA : Engineering Science Reference, 2021. | Includes bibliographical references and index. | Summary: "This book identifies sustainability pillars such as environmental, technical, social, institutional, and economic disciplines and explores the longevity of these disciplines through a resource-oriented approach"-- Provided by publisher.
Identifiers: LCCN 2020011727 (print) | LCCN 2020011728 (ebook) | ISBN 9781799849155 (hardcover) | ISBN 9781799856306 (paperback) | ISBN 9781799849162 (ebook)
Subjects: LCSH: Sustainable development. | Renewable energy sources. | Sustainable agriculture.
Classification: LCC HC79.E5 E2166 2021 (print) | LCC HC79.E5 (ebook) | DDC 338.9/27--dc23
LC record available at https://lccn.loc.gov/2020011727
LC ebook record available at https://lccn.loc.gov/2020011728

This book is published in the IGI Global book series Advances in Environmental Engineering and Green Technologies (AEEGT) (ISSN: 2326-9162; eISSN: 2326-9170)

British Cataloguing in Publication Data
A Cataloguing in Publication record for this book is available from the British Library.

Advances in Environmental Engineering and Green Technologies (AEEGT) Book Series

Sang-Bing Tsai
Zhongshan Institute, University of Electronic Science and Technology of China, China & Wuyi University, China
Ming-Lang Tseng
Lunghwa University of Science and Technology, Taiwan
Yuchi Wang
University of Electronic Science and Technology of China Zhongshan Institute, China

ISSN:2326-9162
EISSN:2326-9170

Mission

Growing awareness and an increased focus on environmental issues such as climate change, energy use, and loss of non-renewable resources have brought about a greater need for research that provides potential solutions to these problems. Research in environmental science and engineering continues to play a vital role in uncovering new opportunities for a "green" future.

The **Advances in Environmental Engineering and Green Technologies (AEEGT)** book series is a mouthpiece for research in all aspects of environmental science, earth science, and green initiatives. This series supports the ongoing research in this field through publishing books that discuss topics within environmental engineering or that deal with the interdisciplinary field of green technologies.

Coverage

- Green Technology
- Sustainable Communities
- Industrial Waste Management and Minimization
- Alternative Power Sources
- Air Quality
- Electric Vehicles
- Renewable Energy
- Radioactive Waste Treatment
- Water Supply and Treatment
- Pollution Management

IGI Global is currently accepting manuscripts for publication within this series. To submit a proposal for a volume in this series, please contact our Acquisition Editors at Acquisitions@igi-global.com or visit: http://www.igi-global.com/publish/.

The Advances in Environmental Engineering and Green Technologies (AEEGT) Book Series (ISSN 2326-9162) is published by IGI Global, 701 E. Chocolate Avenue, Hershey, PA 17033-1240, USA, www.igi-global.com. This series is composed of titles available for purchase individually; each title is edited to be contextually exclusive from any other title within the series. For pricing and ordering information please visit http://www.igi-global.com/book-series/advances-environmental-engineering-green-technologies/73679. Postmaster: Send all address changes to above address.

Titles in this Series

For a list of additional titles in this series, please visit:
http://www.igi-global.com/book-series/advances-environmental-engineering-green-technologies/73679

Energy-Efficient Underwater Wireless Communications and Networking
Nitin Goyal (Chitkara University Institute of Engineering and Technology, Chitkara University, Punjab, India) Luxmi Sapra (Chitkara University Institute of Engineering and Technology, Chitkara University, Punjab, India) and Jasminder Kaur Sandhu (Chitkara University Institute of Engineering and Technology, Chitkara University, Punjab, India)
Engineering Science Reference • © 2021 • 339pp • H/C (ISBN: 9781799836407) • US $195.00

Impacts of Climate Change on Agriculture and Aquaculture
Ahmed Karmaoui (Southern Center for Culture and Sciences, Morocco) Kirby Barrick (University of Florida, USA) Michael Reed (University of Kentucky, USA) and Mirza Barjees Baig (King Saud University, Saudi Arabia)
Engineering Science Reference • © 2021 • 333pp • H/C (ISBN: 9781799833437) • US $195.00

Development Challenges and Strategies in Global Arid Regions
Umar Garba Benna (Benna Associates, Nigeria)
Engineering Science Reference • © 2021 • 300pp • H/C (ISBN: 9781799837930) • US $195.00

Molecular Plant Breeding and Genome Editing Tools for Crop Improvement
Pradip Chandra Deka (Sir Padampat Singhania University, India)
Engineering Science Reference • © 2021 • 489pp • H/C (ISBN: 9781799843122) • US $195.00

Recent Advancements in Bioremediation of Metal Contaminants
Satarupa Dey (Shyampur Siddheswari Mahavidyalaya, India) and Biswaranjan Acharya (School of Computer Engineering, KIIT University (Deemed), India)
Engineering Science Reference • © 2021 • 363pp • H/C (ISBN: 9781799848882) • US $195.00

Global Environmental Impacts on Food Security, Nutrition, and Human Health
Hicham Chatoui (Private University of Marrakech, (UPM), Morocco & Cady Ayyad University, Morocco) Mohamed Merzouki (Faculty of Science and Technology, Sultan Moulay Slimane University, Beni Mellal, Morocco) Hanane Moummou (Private University of Marrakech (UPM), Marrakech, Morroco) Mounir Tilaoui (Private University of Marrakech (UPM), Sultan Moulay Slimane University, Beni Mellal, Morocco) and Nabila Saadaoui (Private University of Marrakech (UPM), Marrakech, Morroco)
Engineering Science Reference • © 2020 • 300pp • H/C (ISBN: 9781799837480) • US $195.00

IGI Global
PUBLISHER of TIMELY KNOWLEDGE

701 East Chocolate Avenue, Hershey, PA 17033, USA
Tel: 717-533-8845 x100 • Fax: 717-533-8661
E-Mail: cust@igi-global.com • www.igi-global.com

Table of Contents

Detailed Table of Contents

This chapter has the objective to analyze the urban ecology, the biological diversity or biodiversity, and their adaptive cycle as the fundamentals of green economic growth. The analysis begins questioning the implications that some assumptions of urban ecology and biodiversity, such as the socio-ecosystems, resilience, ecosystem services, and adaptive cycle have on the creation of green economic growth. A series of different dimensions of resilience are proposed as subsystems that contribute to the general resilience of a system. The method used is the analytical based on a review of the conceptual and theoretical literature. This analysis concludes that the connectivity of processes and functions of urban ecology and biodiversity are relevant to the creation of green economic growth in terms of green economic value.

With the ever-increasing population, there is exploitation of our environment to meet the growing needs of the inhabitants. Along with pollution, climate change, global warming, etc., one of the major problems associated with overpopulation is the generation of a large amount of non-biodegradable solid wastes along with various other industrial effluents. Plastic, being inexpensive and durable, have found their way to be produced in large amounts for almost every purpose. This results in the accumulation of plastic products around the world causing plastic pollution. During the past few centuries, the idea of "green chemistry" has gained maximum attention, and therefore, efforts have been made by the researchers to reduce pollution with the help of some environmentally friendly solution. In this concern, bioplastics and biofilms have gained remarkable attention over the last two decades. The major advantages of bioplastics are their biodegradability and the reduced emission of carbon dioxide in their synthesis.

Energy is an important issue for achieving sustainable development. Sustainable Development Goals emphasize renewable energy sources, green energy technologies, and energy efficiency for sustainable world. In addition, energy consumption patterns can influence energy efficiency as much as green energy policies. In this point, sustainable consumption patterns should be adapted into energy consumption. 3R principle can be an alternative model for sustainable energy consumption. This study aims to explore how household can adapt 3R principle in energy consumption patterns in Turkey. Turkey aims to achieve sustainable development goals and especially works to achieve energy security and efficiency in the long term. In this point, the contribution of households will play a critical role for energy efficiency and security in a country. The study analyzed open access data from TURKSTAT (Turkish Statistical

Kenya, just like other developing countries of the world, is energy deficient. Power supply is characterized by flaws including inaccessibility by marginalized rural communities, high cost of electricity, power outages, and over-reliance on biomass and hydropower sources. Despite these trends, the country has great potential to produce green energy including wind, biofuels, solar, and geothermal power in excess of the entire country's demand. This chapter explores the status of each of these green energy sources and their production potential using in Kenya. Findings show that the production capacity for all of the sources is high though limited tapping and harnessing of these energy resources is done. This could be attributable to lack of exploitation technology and expertise, huge capital investments involved and the uncertainty on the effects of these renewable energy sources to food and water security. For optimal exploitation, Kenya must invest in the energy sector hugely to enhance technical know-how on its development and cater for the associated financial opportunity costs.

Today, public spaces have changed depending on the differentiating human needs. The purpose of the study is to bring together a variety of users and to discuss a life in which various actions such as gathering and meeting take place in cities by taking an example based on the concept of ergonomics based on the human factor. In the study, the relationship between human behavior and physical environment, air-conditioned bus stops in Gümüşhane city, which is one of the places offered for different user groups, was examined. Arrangements for the use of the physical environment for people in many cities have not yet reached the desired level. Ergonomic evaluation of the effectiveness of all human-oriented studies will increase the importance and value of the work done. The more psychological and emotional needs of people provide, the more people will have access to a society with increased satisfaction.

Water is the sole reason for the existence of life and prime nutrient to sustain it. So, it becomes extremely important to preserve wastewater to utilize it efficiently for various purposes. The methods and techniques that have the least investment capital, qualitative output, and environment-friendly content are desired. Such unaccustomed methods are our hope for a cleaner, safer, and brighter future. This chapter revolves around the methods employed to brawl the threats posed to the viability of water at earth from emerging contaminants and the efforts made by mankind to prevent it. Nanotechnology has potent attributes for water decontamination. Several nanotech-based devices, methods, and treatments have been used in

Preface

A rapid change in technology and lifestyle have led nations with a dramatic increase in energy demand, which has faced the globe with the challenge of ensuring sustainable and clean energy production. Growing energy demand has been the main causer of environmental pollution. With today's cutting-edge renewable technologies must propound sustainability culture in energy systems' planning and design, to transforming the world with the 2030 Sustainable Development Goals (SDGs). Efficient exploitation of renewable energy sources and technologies for residential, commercial, industrial, and agricultural sectors offers the opportunity to diminish energy dependence, ensure efficiency and reliability, reduce pollutant emissions, and buoyant economy. This book aims to cover importance of the state-of-the-art and findings of researchers in the field of eco-friendly energy processes and its technologies exploitation on a broad theme of sustainable development.

The Eco-Friendly Energy Processes and Technologies for Achieving Sustainable Development is part of a peer-review series that establishes interdisciplinary coverage in terms of sustainable development. This book argues sustainability pillars (environmental, technical and technological, social, institutional, and economic disciplinaries), and explores the adaption of these pillars for long-run sustainability.

The Eco-Friendly Energy Processes and Technologies for Achieving Sustainable Development is part of a series of peer-reviewed books that emanate recent researches based on an exhaustive topic of the day "Sustainable Development Goals (SDGs)".

The target audience of this book will be composed of professionals and researchers working in the field of sustainability outreach (technical (energy resources and technologies), institutional (energy policy and strategy), environmental, social (socio-friendly energy), economical (socio-economic)) in various disciplines. Moreover, this book as series of sustainability outreach covers all proposed viable options for societies transition in the 21st century for a well-being life to eradicate poverty, mitigate climate change, promote lifelong learning opportunities, empower sociality, deploy affordable energy, sustain economic growth, offer innovation, reduce inequality, and finally to ensure global sustainability.

ORGANIZATION OF THE BOOK

The book is organized into 13 chapters. Brief introductory information and main points of each of the chapters follows:

Chapter 1 analyzes the system of green resilience eco-urban land uses oriented in urban social-ecological systems. It reviews and analyses the relevant literature in green social-ecosystem resilience concepts and presents a discussion in relation to sustainable development and ecological sustainability. It further discusses and gives an in-depth overview of the urban social ecosystems as a working structural and functional unit, describes decision support tools that could be applied to sustainable green land uses and development, and offers some strategies for engaging in urban ecosystems, ecological sustainability, and adaptive development. It is concluded that urban land use through the innovative pro-environmental solutions can, in a natural way, support the system of green resilience eco-oriented urban land uses in urban ecosystems and serve to improve the quality of life in the city.

Chapter 2 deals with the concept of sustainable development that looks to balance different and often competing needs against an awareness of the environmental, social and economic limitations we face as a society. Most of the present development activities are being conceived and implemented without fully considering the wider or future impacts. A manifestation of such kind of development approach is quite evident in the form of large-scale financial crises caused by irresponsible banking, changes in the global climate resulting from our dependence on fossil fuel-based energy sources. The longer we pursue unsustainable development, the more frequent and severe its consequences are likely to become, which is why we need to take action now.

Chapter 3 analyzes the urban ecology, biological diversity or biodiversity, and their adaptive cycle as the fundamentals of green economic growth. The analysis begins questioning the implications that some assumptions of urban ecology and biodiversity, such as the socio-ecosystems, resilience, ecosystem services, and adaptive cycle have on the creation of green economic growth. A series of different dimensions of resilience are proposed as subsystems that contribute to the general resilience of a system. The method used is the analytical based on a review of the conceptual and theoretical literature. This analysis concludes that the connectivity of processes and functions of urban ecology and biodiversity are relevant to the creation of green economic growth in terms of green economic value.

Chapter 4 discusses overpopulation and environmental-related problem. Along with pollution, climate changes, global warming, etc., one of the major problems associated with overpopulation is the generation of a large amount of non-biodegradable solid wastes along with various other industrial effluents. Plastic, being inexpensive and durable has found its way to be produced in large amounts for almost every purpose. This results in the accumulation of plastic products around the world causing plastic pollution. During the past few centuries, the idea of "Green Chemistry" has gained maximum attention and therefore, efforts have been made by the researchers to reduce pollution with the help of some environment-friendly solution. In this concern, bioplastics and biofilms have gained remarkable attention from the last two decades. The major advantages of bioplastics are its biodegradability and the reduced emission of carbon dioxide in its synthesis.

Chapter 5 emphasizes renewable energy sources, green energy technologies, and energy efficiency as a tool for sustainable development. In addition, energy consumption patterns can influence energy efficiency as much as green energy policies. At this point, sustainable consumption patterns should be adapted to energy consumption. 3R principle can be an alternative model for sustainable energy consumption. This study aims to explore how households can adapt 3R principle in energy consumption patterns in Turkey. Turkey aims to achieve sustainable development goals and especially works to achieve energy security and efficiency in the long term.

Chapter 6 reviewed the Algeria energy sector and presents alternatives for nonrenewable resources to ensure global sustainability and mitigate climate changes. Since the emergence of the industrial revolution, the use of energy resources has increased considerably, particularly nonrenewable (coal, oil, and gas), so these resources are no longer sufficient to cover the different energy needs, and this has become a challenge to the energy independence of many governments now. This has caused the world to scramble for other ways to satisfy these needs. In which the results of scientific research and development involved from alternative use of old energy sources and named Green economy, and many think about the future of energy despite the barriers that still hinder the general application of this type of economy. Energy consumption in Algeria is focused almost entirely on fossil fuels, hydrocarbons, and gas in particular. In recent years, Algeria is deciding to move into green economy is the unique solution for meeting future energy requirements and helping reduce environmental risks.

Chapter 7 explores the existing literature related to the causality-testing techniques used by several researchers and academicians in the field of sustainable development. This study sums up with a wide variety of existing causality-testing techniques in the literature (both conventional and latest/improved ones). Still, researchers are making efforts for further studies and strive to continually look for examining these causal linkages in a better way. Although several variables have been used in previous studies to explore the causality relationships for sustainable development, still it seems there is no limit to the number of such explanatory variables. Therefore, causality testing for sustainable development has a broader scope always for further research advancements in this area.

Chapter 8 analyses the implications and interrelationships between a sustainable urban agro ecology and the food system. The beginning assumption of this analytical review considers that sustainable urban agro ecology has positive implications in the development of a sustainable urban food system. The analysis is based on the theoretical and empirical literature review confronted with common spatial-functional observations of urban development and configurations. The analysis concludes that the sustainable urban food system based on agro ecology is growing as an alternative movement towards the building and maintenance of a fairer and healthier urban sustainable environmental development. Keywords: Agro ecology, food systems, sustainable, urban.

Chapter 9 explores the status of each of these green energy sources and their production potential using in Kenya. Findings show that the production capacity for all of the sources is high though limited tapping and harnessing of these energy resources are done. This could be attributable to lack of exploitation technology and expertise, substantial capital investments involved and the uncertainty on the effects of these renewable energy sources on food and water security. For optimal exploitation, Kenya must invest in the energy sector hugely to enhance technical know-how on its development and cater for the associated financial opportunity costs.

Chapter 10 brings together a variety of users and to discuss a life in which various actions such as gathering and meeting take place in cities by taking an example based on the concept of ergonomics based on the human factor. In the study, the relationship between human behavior and physical environment, air-conditioned bus stops in Gümüşhane city, which is one of the places offered for different user groups, was examined. Arrangements for the use of the physical environment for people in many cities have not yet reached the desired level. Ergonomic evaluation of the effectiveness of all human-oriented studies will increase the importance and value of the work done. The more psychological and emotional needs of people provide, the more people will have access to a society with increased satisfaction.

Chapter 11 revolves around the methods employed to brawl the threats posed to the viability of water at earth from emerging contaminants and the efforts made by mankind to prevent it. Nanotechnology has potent attributes for water decontamination. Several nanotech-based devices, methods, and treatments have been used in recent times for wastewater treatment such as nanomembranes, nano photocatalysts, nano-cellulose, nano-absorbent and nanocomposite materials. The chapter discusses these contemporary developments in nano-based technologies and the various advancements which have been recently made in conventional techniques used for wastewater treatment technologies.

Chapter 12 explores the system of green resilience eco-urban land uses oriented in urban social-ecological systems. It reviews and analyses the relevant literature in green social-ecosystem resilience concept and presents a discussion in relation to sustainable development and ecological sustainability. It further discusses and gives an in-depth overview of the urban social ecosystems as a working structural and functional unit, describes decision support tools that could be applied to sustainable green land uses and development, and offers some strategies for engaging in urban ecosystems, ecological sustainability, and adaptive development.

Chapter 13 conducts the feasibility of Visual Computing (VC) and Machine Learning (ML) method applying in textile recycle industry for efficiently manages post-consumer textile waste. It includes an image-based VC technology for supporting textile waste reuse and resale, and a material identification system for sorting textile materials by using Near Infrared (NIR) / Hyperspectral Spectroscopy technology to support efficiently recycling to reuse the textile fiber will be evaluated. The process involved collecting and validating reference samples, apply ML technique to auto recognize the garment type and features applying visual technology. Afterward, the sorted garments would be measured and pre-treated by NIR / Hyperspectral spectrum and building up the parameters for spectral patterns calculation for recycling process to recover the fiber.

Mir Sayed Shah Danish
University of the Ryukyus, Japan

Tomonobu Senjyu
University of the Ryukyus, Japan

Chapter 1
System of Green Resilience Eco-Oriented Land Uses in Urban Socio-Ecosystems

José G. Vargas-Hernández
https://orcid.org/0000-0003-0938-4197
University Center for Economic and Managerial Sciences, University of Guadalajara, Mexico

Mir Sayed Shah Danish
University of the Ryukyus, Japan

ABSTRACT

This chapter analyzes the system of green resilience eco-urban land uses oriented in urban social-ecological systems. It reviews and analyses the relevant literature in green social-ecosystem resilience concepts and presents a discussion in relation to sustainable development and ecological sustainability. It further discusses and gives an in-depth overview of the urban social ecosystems as a working structural and functional unit, describes decision support tools that could be applied to sustainable green land uses and development, and offers some strategies for engaging in urban ecosystems, ecological sustainability, and adaptive development. It is concluded that urban land use through the innovative pro-environmental solutions can, in a natural way, support the system of green resilience eco-oriented urban land uses in urban eco-systems and serve to improve the quality of life in the city.

INTRODUCTION

Human development has had a profound imprint on nature and coevolving ecosystems which had resulted in complex, economic-socio-ecological challenges for sustainability and future development. Sustainability is a set of goals combining social equity, economic viability and ecological integrity (Curwell, Deakin, & Symes, 2005; Jenks & Jones, 2010). Humankind alters the dynamics of ecosystems through anthropologic activities that change the atmosphere, climate, land surface, forest, sea, and waters. Humankind depends on healthy and resilient social-ecological systems and sustainable environments.

DOI: 10.4018/978-1-7998-4915-5.ch001

Human wellbeing, economic growth, and social development are dependent on the interrelationships between and within the regions and environmental sustainability (Arrow et al., 1995; Folke et al., 1998). Uncertainty, diversity, and variability of a social-ecosystem increases their inevitability, reducing their capacity to cope with disturbance and change within functional groups in the adaptive capacity of ecosystems (Folke et al. 2002; Jackson et al. 2001; Scheffer et al. 2001).

Cities have been portrayed as predominantly monumental static architectural structures of increasing ecological complexity that change over the long term. Contextual and conceptual factors of urban change can be assembled into a framework of ecological urban design. Disturbances change the resilient capacity of nature to supply ecosystem services, degrading socioecological systems and leading to social and economic vulnerability. Inhabitants of urban areas pose a high impact on the ecosystems services with trade and consumption, claiming support in waste absorption, carbon emissions, residential water use, wood for industrial purposes (Folke et al., 1997, Grimm et al., 2008). Adverse human impacts on the social-ecosystem and the biosphere can be minimized through the use of resilient and sustainability environmental approaches such as environmental resources management, conservation biology, environmental and ecological economics, bio economics, green technology, etc.

According to Levin (1999), fragility is inherent to the ecosystems services on which human's dependent and not nature. A global health control of the ecosystems, published in 2005 by The United Nations Millennium Ecosystem Assessment (MA) observed that technological advancement is the main cause of the Earth's ecosystem services degradation and are used unsustainably. The growing eco-deterioration is an impediment to combating poverty (Millennium Ecosystem Assessment (MA), 2005). If the ecosystem and natural resources are controlled by few people, they do it for short-term economic gain.

Sustainability is the capacity of a system or process to be preserved, enhanced, upheld, or maintained. Sustainability is the capacity of a biological systems and processes to endure disturbances and remain vigorously diverse. Sustainability is the systematic combination of environmental science and sustainable development (Lynn et al., 2014).

Sustainability relies on development from interconnection of ecology, culture, politics, and economics domains (James, et al., 2015). Ecology refers to the paradigm, knowledge, methods and procedures of contemporary ecological science (Kolasa & Pickett, 2005). Ecology is a multidimensional and complex concept that requires an interdisciplinary framework of analysis for its application in urban spaces (Pickett & Cadenasso, 2002). The dimensions of ecology are the notion, the models and the metaphors used to communicate assumptions, values and experiences (Cadenasso et al., 2006a).

Sustainability is a social challenge that has an impact on economic efficiency, production, distribution, urbanization, transportation, lifestyles and ethical consumerism. Promoting sustainable and resilient lifestyles in sustainable urban social-ecosystems to conserve ecological resources ranges from among other activities from green spaces, sustainable architecture, green housing and building, ecovillages, green business and green economic sectors, sustainable urban agriculture and farming, green technologies, renewable energies and designing reversible and flexible systems (Zhang and Babovic, 2012). The specific types of sustainability are sustainable architecture or ecological economics, sustainable technology, and agriculture (Costanza and Patten, 1995).

Sustainability includes concerns on global climate change, ozone depletion, biodiversity, wild life conservation, deforestation, genetically engineering, urbanizations, etc. A sustainable environmental and ecological system is identified in resilient social ecosystems with the capacity to continue functioning despite facing natural and anthropogenic disturbances, but anticipating and preparing ahead for them.

GREEN SOCIAL-ECOSYSTEM RESILIENCE

Human beings have had a profound imprint on nature and on dealing with the resulting challenges. One of the most challenging events is the global warming phenomenon which increases the average temperature of organisms while speeding up their metabolism. An ecosystem is a structural and functional unit of the biosphere and the complex interrelationships with human activities that threaten its sustainability. Resilience has its source of thinking in ecology (Holling, 1973; Wu & Wu, 2013), a concept that can be applied in urban environments and cities (Musacchio, 2008). Resilience deals with the opportunities arising from ecosystemic changes in its structures, processes, emergence of great developments and coping with disturbances.

The resilience approach provides a conceptual and theoretical framework for interdisciplinary collaboration with ecological economics, sustainable development and governance (Lambin, 2005). Resilience is a concept and model framework used to operationalizing normative sustainability (Childers, Pickett, Grove, Ogden, & Whitmer, 2014). Resilience is a non-normative conceptual model for the achievement of economic, social and environmental goals disaggregating process (Walker, Holling, Carpenter, & Kinzig, 2004) into capacities for adaptation (Yohe & Tol, 2002).

Resilience is a dynamic process interfering with symbiotic relationships within and between the social ecosystem and its environment. Resilience is a fundamental capacity that individuals, groups, organizations, and ecosystems have that respond to significant disruptive changes. Resilience is essential to sustainable development in dynamic equilibrium with the natural disturbances to the social – ecosystem. Resilience refers to the ecosystem's ability to respond productively to significant disruptive change and adapt successfully to external variables that threaten its existence.

Resilience is the capacity of the social-ecological system to reorganize itself after any disturbance. Social-ecosystem resilience lies in the continuous cycling movement of adaptation and transformation while maintaining system integrity and viability. Resilience involves the reaction of the socio-ecological system under disruption and destruction with the capacity to recovering and developing in a state of uncertainty, discontinuity, and emergency.

Human-induced erosion of resilience has consequences on shifts of the social ecosystems. Response diversity in sustaining ecosystem states, functional groups and ecosystem services is linked when facing disturbances and environmental change. Response diversity among species within a functional group is a component of resilience. Response diversity interacts operating across spatial and temporal scales. High response diversity increases the insuring buffer for resilience and the effectiveness of ecosystem management policies and actions.

Moderate perturbations and disturbance events on the stability of the social ecosystems either may be absorbed by the resilient ecosystems through reorganization or it may bring small changes in resilience supported by response diversity (Deutsch et al. 2003). For example, response diversity maintains resilience in lakes stressed by toxic chemicals and acidification (Carpenter and Cottingham 1997).

The social system structure establishes an equilibrium in its interactions with the ecosystems by accepting a specific level of stability. Sustainable development involves local and global efforts to meet core human needs without distorting the ecosystem (Kates et al., 2005; IISD, 2009; EurActiv, 2004). Sustainable development implies proactive and responsible decision-making with innovation processes to maintain social ecological systems equilibrium between resilience, economic efficiency, social inclusion and equality, political justice and vibrant cultural values aimed to ensure a livelily and desirable ecosystem on the long term for all the species (Liam et al., 2013).

Resilience is an approach for cogitating and critiquing of social–ecological systems with policy implications for sustainable development (Folke et al., 2002). Resilience of social-ecological systems integrates self-organization, learning, adaptation and persistence. Social-ecological resilience measures the degree of the system self-organization and coping with disturbance and still maintaining its inertia, attraction and capacities for learning and adaptation (Carpenter et al. 2001).

Sustainability is a normative social goal promoted by using the mechanisms of ecological resilience. The ecological resilience includes the patch ecology or landscape, the notion of metacity and experimental modeling and design.

The ecological resilience is the capacity of a site in urban systems to adjust and control of interactions with external vulnerabilities, disruptions and shocks making more sustainable the urban component of the ecosystem, given its nature. It is different from engineering resilience which has the capacity of returning to the previous existing state before perturbation. Resilience sustains the urban ecosystem states under uncertain, variable and complex environments. Response diversity is critical to resilience during ecosystem reorganization.

The terms urban and city are used as the totality of densely settled regions, and urban systems have spatial and functional contexts combine social, biological, built and geophysical components (Graham & Marvin, 2001; Seto et al., 2010; Naveh, 2000; Pickett & Grove, 2009). Some social dimensions of urban systems are economics of production and consumption, the economic and political power, social inclusion, identity, equality and change, social justice and vulnerability, the nature of livelihood and lifestyle (Dow, 2000; Grove et al., 2006; Machlis, Force, & Burch, 1997).

Response diversity does not necessarily support that high species biodiversity entails high ecosystem resilience and less vulnerable to environmental change. Migration of people to the large cities have resulted in a reduction in the ecosystems biodiversity (MA, 2005; Sala et al., 2000). Biodiversity is a critical factor of social-ecosystems while cultural diversity and common property systems are related with the resilience building of urban socio-ecological-systems. Urbanization processes have led to higher levels of cultural diversity in cities (Zanoni and Janssens, 2009).

Urban space influences environmental values of urban populations (Miller, 2005; Tidball et al., 2010). The urban space and territory are changing rapidly and becoming globally interconnected across contrasting types of land and facing new environmental, demographic and social threats and pressures. Urban green commons (UGCs) have potential to facilitate civic participation and cultural integration, to manage urban land and biological diversity in urban spaces and to promote urban resilience building (Colding and Barthel, 2012). Urban resilience in the context of an urban system is an ecological approach to analyze the changes, disturbances, vulnerabilities and mutability of city development and their relationships with the climate change.

Urban transformation benefits from the shift of an urban ecology framework from earlier approaches such as the metabolic urban energy budgers advance to the city resilience approach with components of these hybrid systems such as the biophysical-social structures and processes (Cadenasso et al., 2006b; Cadenasso & Pickett, 2013; Cadenasso & Pickett, 2008; Pickett, Cadenasso, & Grove, 2004).

The vulnerability of an ecosystem state is related with other organisms in a functional group. Contraction of spatial resilience increases the disturbance to catastrophic levels (Nyström and Folke 2001). Ecosystem resilience has a cross-scale response diversity, so biodiversity has an important role in formulating policy on ecosystem resilience for sustained economic and socioecological development. Managing for resilience copes with uncertainty in the biosphere shaped by human action (Folke et al. 2002).

The preservation of congenital features leads to robustness of the ecosystem despite behavior alterations of its component parts and environment. The resilience of ecosystems provides backups to replace failures in the species through the diversification and redundancy of biodiversity with overlapping ecological functions in natural ecosystems. Ecological functions across a range of scales can have diverse disturbances. Ecological disruption occurs across a limited range of spatial and time scales allowing undisturbed scales to persist and ecological functions to operate.

Response diversity is relevant to resilience because the diversity of species contributes to ecosystem functions of renewal and reorganizations in environmental changes. Response diversity contributes to resilience when planning, management and restoration of a social-ecosystem. Response diversity provides adaptive capacity in uncertain and complex systems and human-dominated environments.

The property of response diversity is a variability in responses of species to environmental change that is critical to social ecosystem resilience. Response diversity is the set of reactions among species contributing to the same ecosystem function to environmental change. Species and populations diversity within functional groups maintain ecological redundancy in ecosystem services (Luck et al. 2003). Resilience equated with species should consider that species loss as non-random in relation to response diversity and ecosystem functioning. However, response diversity in ecosystem resilience is related to ecosystem disturbances and environmental changes. Biological diversity is essential in social ecosystem resilience and in sustainable ecosystem change (Peterson et al. 1998). Biological diversity enhances the social ecosystem resilience required to secure the production of ecosystem services.

Resilience can interfere and be in conflict with other beneficial social objectives, such as economic efficiency because the costs of redundancy. So economic efficiency reduces resilience. Resilience of social ecosystems requires more adaptability to stress and still maintains the stability domain from extrinsic disturbances, that is, to find a solution of the conflict between stability and resilience for sustainable development in terms of complex system cycles.

Resilience is concerned with the management of sustainable interactions between human-developed systems and natural ecosystems. Social resilience linked to social change is the communities' capacity to cope with extrinsic disturbances to their social infrastructure, such as political turbulence, socio-economic reforms, and environmental variability (Adger 2000; Anderies et al., 2004).

Ecosystem resilience is the capacity to accommodate and adapt to disturbances and the buffer capacity of persistence. Ecosystem resilience is the amount of disturbance a system can absorb and still may remain within the same state and domain of attraction (Holling 1973, 1996). Ecosystem resilience is the capacity degree of self-organization, learning and adaptation (Carpenter et al. 2001), encompassing its ability subject to disturbance and change to reorganize and renew itself. The dynamic changes of the ecosystem resilience are related to the diversity of species and the functional groups (Walker 1992, 1997; Norberg et al. 2001).

Social-ecosystem resilience is an essential factor to uncertain and complex systems for sustainable natural resources and ecosystem services (Gunderson and Holling 2002). Any resilience disequilibrium between sustainability and development change leads to collapse of the ecosystems. Thus, it is extremely difficult to transform a resilient ecosystem to a more congenial one (Scheffer et al., 2001; Gunderson and Holling, 2002; Walker et al., 2004). Congeniality with nature to prevent degradation is doing things nature's way.

Sustainable planning urban ecosystems framework draws from urban ecology, green resilience, green infrastructure, multifunctional and sustainable landscapes planning, etc., with the full purpose of creating and developing healthy and sustainable economic, social–ecological urban systems. Urban ecology is rooted in landscape ecology and has been integrating some principles from physical and atmospheric sciences, soil, hydrology, social sciences, etc. (Sukopp, 1990; Collins et al., 2000; McDonnell & Hahs, 2009; Pickett, Burch, Dalton, & Foresman, 1997; Redman, Grove, & Kuby, 2000). Green resilience deals with the coping of the planet with anthropogenic disturbances and still remaining viable for future generations. Ecological resilience is the ecosystem´s capacity to cope with disturbances and retaining its structure and functions.

Resilience concerns the capacities to absorb the uncertainty resulting from shocks and conflicts to maintain function and for renewal, growth, development and re-organization (Gunderson and Holling, 2002; Berkes et al., 2003). The concept of green resilience may be from a narrow to more broad interpretation of the socio-economic and ecological context. An ecological unit is the functioning of components and their relationships and interactions with each other forming a complex and dynamic whole.

The accountability of natural resources in an ecosystem occurs through the adaptation process after an external disturbance. Adaptive processes contribute to the adaptive cycle in urban socio-ecological systems. Sustainability measurement requires transparent accountability between the resources depleted and those used to replace them in the ecosystem. Sustainable and green resiliency of an ecosystem can be measured at events' junctures with naturally occurring regenerative forces interacting with energy released into disturbances. These regenerative forces are the solar energy, water, soil, atmosphere, vegetation, and biomass (Ben, 2013). Green resilience is an institutional capacity to cope and deal with stresses and conflicts arising from climate change, unforeseen contingencies, unsustainable development to live and other emerging environmental issues within the ecosphere.

The old perspective of resilience assumed stable and static equilibrium to build socio-economic and ecological adaptation to nature (Berkes et al., 2003, Smit and Wandel, 2006). Resilience has some tools based on structured scenarios and adaptive management to build in complex and uncertain systems. These tools increase capacity to build ecological resilience. Natural resilience sources of complex adaptive systems are more than preservation of ecosystems and resistance to change. Traditional and dominant theoretical perspectives have been assuming a stable and resilient environment where nature and its flows of natural resources are in static equilibrium after stressors are controlled or removed (Holling, 1973).

Green resilience sources are in intertwining relationships in complex adaptive systems of dynamic changes in the ecosystem. Dynamic adaptive capacity of the ecosystem provided by resilience interplays between development change and sustainability (Smit and Wandel 2006). The institutional capacity of the social system to cope with change in the ecosystem by anticipating and combating disturbances leads to adaptive development integrating resilience in the interactions in ecological sustainable development. Adaptive development in the social ecosystems has as elements the ability of ecological assessment of actual events and taking corrective action.

Resilience is the system´s ability to adapt to any kind of disturbance and self-re-organize while undergoing transformation processes retaining its initial forms, roles, identity, and feedback characteristics (Walker et al., 2004). Any disturbance of a strong resilient socio-economic and ecological system may have the potential to create new opportunities for innovation and advancement. However, when the system is weak, any disturbance may be disastrous (Adger, 2006). A sustainable and resilient social-ecosystem relies on nature in interaction with small dependence on human inputs because the anthropogenic interferences may interrupt the provision of inputs.

Sustainable development should reduce social and ecological damage on both a local and a global scale through the operation of fairness and social inclusion, protection of the environment, and economic efficiency. Green resilience-building in complex, uncertain and unpredictable urban ecosystems is supported by structured scenarios and active adaptive management for sustainable development. Green resilience management enhances sustainable development in changing complex environments where the future is uncertain and unpredictable (Walker et al., 2004; Adger et al., 2005). Strategies in any system of urban green resilience focus on green infrastructure to transform and adapt resources to face future challenges such as climate change and food insecurity.

The social ecosystem is currently threatened by anthropogenic activities at the brink of collapse that requires prompt action. In this context of social-ecological systems, management and flexible collaboration aimed to develop policy frameworks as a basis to build adaptive capacity is required. Nature should be strengthened to stimulate development through the interaction and interdependence with humankind and to enhance resilience in social-ecosystems.

The phenomenon of resilience in sustainable development may lead to some policy recommendations to improve the interrelationships between the economic efficiency and the social-ecosystems and biosphere, and to develop some flexibility and innovative relationships of collaboration to achieve sustainability and its operationalization in the context of socio-ecological resilience. Other policy recommendations should focus on the development of indicators to measure any change of the level of resilience and to signal and monitor uncertainties of social – ecosystem variables and to manage diversity.

URBAN SOCIAL-ECOSYSTEMS

In 1930, the term ecosystem was coined by Roy Clapham to describe the biological and physical environmental components as a unit where all are related to each other. The ecosystem was described as the interaction between the living creatures or bioconenosis and the environment where they live, or biotope (Tansley, 1935). According to the Convention on the Biological Diversity (1992), an ecosystem is a dynamic complex formed by plants, animals and micro-organism communities and their non-living environment interacting as a functional unit.

The biosphere is the largest ecosystem. The biosphere interacts with and exchanges matter and energy with the lithosphere, hydrosphere and atmosphere. The level of the ecosystem is above the ecological community corresponding to the biomes and the biosphere, also considered as the regional ecosystems. Ecosystems are complex and adaptive systems characterized by historical dependency and non-linear dynamics with multiple basins of attraction (Levin 1999). The ecosystem refers to the level of biological organizations formed by living organisms interacting with each other and in a symbiotic relationship with their environment. An ecosystem is capable of harnessing solar energy during photosynthesis and converted into carbon dioxide and other inorganic chemicals functional to organic life.

Ecosystem processes are seasonally dynamic responding to the solar input driving the largest biogeochemical carbon cycle of the global cycles (Odum, 1971). An ecosystem-level process can be represented by a biogeochemical cycle of exchange of matter between organisms and their environment (Golley, 1993). Energy transfer and matter cycling processes are essential in determining ecosystem structures and functions of a wide diversity of species, as well as in defining the interactions between organisms and their environment (Golley, 1993). The living things are continuously competing to reproduce and survive with one another. When dominant species decline, other subdominant species in the same functional group survive (Elmqvist et al. 2001).

Sustained social ecosystem capacity copes with and adapts to environmental changes in the context of multiple-equilibrium systems and human-dominated environments (Folke et al. 1996; Norberg et al. 2001; Luck et al. 2003). Environmental change with low response diversity may result in extinct or ecologically insignificant functional groups contributing to ecosystem services. Sustaining a social ecosystem from compounded perturbations requires functional groups of species available for renewal and reorganization (Lundberg and Moberg 2003).

Response diversity management sustains and enhances the flow of ecosystem services subject to disturbances and operating within and across special and temporal scales. The impact of disturbances on species loss has implications for social ecosystem and the sustained flow of ecosystem services (Zimov et al. 1995). The persistence of functional groups of species prevents shifts to ecosystem states to sustain the ecosystem services. Species loss may entail low rates of ecosystems processes. Low or absent response diversity redirects ecosystem development into a different pathway. Stress-sensitive species populations decline and response diversity deteriorates while less sensitive species have minimal changes in ecological processes (Schindler 1990; Frost et al. 1995, Rudd et al. 1988).

An ecosystem is an open system requiring a flow of energy and matter between diverse organisms and the environment driving biogeochemical processes. A functioning system has living and non-living components interacting with each other. Any ecosystem has as components the abiotic or living organisms and abiotic which refers to the non-living matter of the environment. The biotic components are the living forms that inhabit the ecosystem and have a biogeochemical energy cycle. The abiotic are the nonorganic components of the environment that determine the life forms that can thrive. Thus, the living organisms are in continuous interactions with the environmental biotic and abiotic components (Golley, 1993). An ecosystem consists of all natural living things interacting each other and with all non-living factors of the environment (Christopherson 1997).

An ecosystem is made up of a set of components including the relationships between living and non-living factors, a biotic element or some source of a single living plant with some energy and an abiotic environment with elements of the Earth system. Energetic processes in ecosystems are formed by thropic levels defined by the role of organisms in its flow of energy. Ecosystems fall between the extremes of complexity (Odum, 1971). The term ecosystem refers to the bionic community of living organisms and biocoenosis always in continuous interaction with their environment or biotope functioning in a space as a unit.

Urbanization is rapidly expanding in cities, which are becoming the frontier for landscape ecology in urban ecosystem science. Urban design and planning take into account the coordination of resources, capacities and efforts to be implemented in comprehensive agendas across the city. Landscape ecology comprises the concepts of landscape sustainability and ecosystem services in changing landscapes (Wu & Qu, 2013). Landscape sustainability is the capacity to consistently provide long-term, landscape-specific ecosystem services that are essential for maintaining and improving human well-being (Wu 2013).

Cities that are committed to urban planning and implementing of ecosystem services supporting greening projects for transforming urban green infrastructure are likely to benefit through value creation for all the population. A community organization focused on the urban environment attains social ecologically sustainable urban development. Modernist model of urban development is portrayed as a life cycle of cities moving from commodity exchange and industrialization, through an ecological version that mitigates all the ills of the urban ecosystem. Urban development emerges in high levels of biodiversity areas with high ecosystem productivity (Ricketts and Imhoff, 2003; Hansen et al., 2004; Ljungqvist et al., 2010).

The movement of people into more densely urban settlements can lessen pressure on urban and suburban sprawl and more remotely located ecosystems (Colding, 2011). Biodiversity at suburbs hold more natural and semi-natural land with an increase in the semi-rural urban fringe (Blair, 2001; McKinney, 2002; Sukopp et al., 1979 Colding et al., 2006).

Urban green infrastructure is a strategically planned and managed urban network of natural lands, working landscapes, and urban open green spaces that provide a range of diverse benefits. Ecosystem services derive benefits to humans (Millennium Ecosystem Assessment 2005). The social-ecosystems capacity altered to buffer disturbances has an impact on the flow of the services requires for the well-being. Landscape services derive some benefits to humans working out as a structure–function-value chain to create landscape development (Termorshuizen Opdam 2009).

Urban ecosystem is a potential solution to future ecological and environmental issues created for the loss of natural landscapes. When the natural resources and the ecosystem are controlled by a few privileged people, the capacity of adaptive development is reduced, resilience is diminished and there is more disequilibrium. The balance between natural and urban ecosystems and the surrounding regional landscape is limited. Urban ecosystems services are relevant for urban green spaces such as biodiversity, microclimate control carbon sequestration, soil infiltration, food production, recreation, etc.

Multifunctional landscape frameworks are applied to agroecosystems and for planning urban ecosystems. Multifunctional landscapes provide beneficial production, ecological, and cultural functions, considering the needs and preferences of users and owners (Otte et al. 2007; Lovell et al. 2010). A commitment to multifunctional urban green infrastructure pays off through the value of ecosystem services for urban population. Optimization of urban green space functions might be challenged by using a multifunctional landscape framework for sustainable urban design and planning of green infrastructure. Multifunctional green infrastructure provides ecosystem services and biodiversity that benefit the environment and society.

The relationship between biological diversity and the ecosystems functions is that biodiversity provides renewal and reorganization capacities for social-ecosystems change (Loreau et al. 2001; Kinzig et al. 2002). Organization and functioning of a social ecosystem can be conservative in relation to the patterns of species loss despite the diversity compositions and considerable species turnover (Schindler 1990; Kalin Arroyo et al. 1995; Forys and Allen 2002; Brown et al. 2001, Havlicek and Carpenter 2001). Otherwise, social ecosystems may change when some species are lost or when there is an invasion of other new species (Estes and Duggins 1995; Terborgh et al. 2001; Vitousek and Walker 1989).

Multifunctional urban planning involves participation of diverse stakeholders in decisions on green infrastructure planning and implementation of solutions. Urban ecosystem services are supplied through participatory planning processes focused on multifunctional green resilience infrastructure developed to contribute to sustainable ecological, social and economic welfare of the city. Ecosystem services transform urban green infrastructure through the integration of empowerment and creative process into small-scale greening projects.

Community organizations on nature conservation focus on the protection and conservation of urban social ecosystems from dysfunctional anthropogenic activities in order to adapt the development and achieve economic efficiency. Community greening is a community-based effort to transform underutilized sites and areas into valuable green spaces such as community gardens (Tidball and Krasny 2009).

Urban green space is an underdeveloped piece of land located within the territorial context of a city and open to the public. Urban green spaces are under pressure to provide production, economic, social, cultural and ecological functions. The functions provided by the ecosystems services benefit humans while simultaneously meeting the needs and preferences of local population. Urban green spaces play a critical role in conserving biodiversity, mitigating the impacts of climate change, sequestering carbon, improving microclimate, protecting water resources conserving biodiversity, protecting water resources, suppling fresh food, etc.

ECO-ORIENTED URBAN LAND USES

Sustainable urbanization has received more attention since metropolitan areas are engaged in initiatives aimed at upgrading land use regulations, infrastructure, urban form and ecosystem services, etc., in an effort to protect urban biodiversity and create a better urban resilient socio-ecosystem, to develop a sustainable urban environment with more social justice and economic growth while improving competitiveness and attractiveness of the city (de Jong et al. 2015).

The land use must have a priority linked to nature and biodiversity to support the urban sustainable urban development in a territorial context, and networks of urban open spaces unite nature and gardening to become a green resilient city under a concept of nature in the city and biodiversity (Quincerot & Weil, 2009, p. 175; Daune & Mongé, 2011). Land uses can be appropriate to develop a network of green open spaces for gardening penetrating urban core areas (Quincerot & Weil, 2009). A multifunctional role of land uses can provide urban corridors of gardening spaces to preserve nature, agriculture, farming, to perform educational and recreational activities helping to protect biodiversity and the natural environment while improving resilient socio-ecosystem services in the context of increasing housing, building and inner-city densification.

Urban land is a natural resource-constraint threatening the prospect to meet the needs of increasing urban population, economic prosperity, and socio-political stability, and influencing policy decisions. Resource depletion and pollution are two related global issues to land-use change.

The accelerated conversion and change of land use from arable or forested lands to urban land use threatens the biodiversity, species and natural resources, which historically had provided the sustenance, valuable goods and services, climate regulation, water recharger, nutrient recycler, agricultural products, timber, seafood, waste assimilations, etc. All of these benefits are in danger to be lost with the increasing urban land use. To respond to the growing challenges posed by the conversion of land uses from agricultural to urban, the local government should formulate and implement urban development strategies aimed to ensure more eco-efficient benefits in terms of economic growth and productivity that are more socially equitable and inclusive and environmental sustainable.

Pressures are arising from climate change are related on the increasing scarcity of land for urban use and other natural resources, such as water, natural nutrients, etc., at a time when population is continuing to rise and becoming an important concern and relevant issue for the policy agenda. Resilient ecological land use orientation is only one of the governance challenges, structures and processes of the urban social eco-systems converging with energy, fuel, climate, water, food, etc. These challenges incorporate and integrate the economic, social and environmental perspectives and initiatives of diverse stakeholders to invest in natural resources and capital projects in the generation and development of the green resilience of the city.

Urban land use is inextricably an interrelated function of the resilient socio-ecosystem influenced by the behavior of individuals, groups, organizations and communities targeted at specific and limited functions at a time, such as biomass and water passage and accumulation. Urban land use has some strong relationships with the urban land market, local government regulations, management practices, technology innovations, etc., and is influenced by the complexity of societal processes but ignoring the systemic consequences and effects on the resilient socio-ecosystem services. A sustainable urban transport system using renewable resources and energy, operating efficiently and affordable, minimizes the use of land and limits emissions, waste and noise within the ability of the city to absorb them.

A contemporary ecological movement advocates the principles of sustainable land use involved in urban planning although this movement is not always engaged in specific project to demonstrate the principles (Seana, Johnson & Peters, 1999). Beyond the integration of land use, urban planning intentions and initiatives are formulated that negotiate spaces for less formalized land uses initiatives and practices. Land use issues requires bringing multi-disciplined and cross-departmental approaches that have an impact on decisions with effects on economic efficiency and urban growth, employment, social inclusion and equality, health, sustainable environment, etc. Urban zoning, neighborhood plans, and the spatial-temporal patterns of inequality are some other factors that may have some implications and impacts of land use.

Urban land use can be supported by public actions to promote social dimensions of urban public spaces as part of collective re-appropriated spaces for diversified uses of unused urban vacant land around housing areas. Recovered unused land is often used for agriculture, farming and gardening projects on a temporary basis created on constructible land and adapting to the demands of residents. These public actions for more integral city development are aimed to foster conviviality, proximity and social cohesion between residents (Canton de Genève, 2013) enhancing more urbanite social interaction and cohesion while improving quality of life.

Urban planners and landscape designers can coordinate activities across different fields and urban functions. Multi-scale analysis of land use compositions is a useful element in land use planning and policy design for urban sustainability. Multi-functional analysis of land use may be conducted by a participatory design process of visioning, defining, relating the elements, elaborating the site plan, and the implementation process. Multiple land use scenarios simulate the future changes in composition of their functions under consideration such as self-sufficiency in vegetable production. Changes in land use may lead to inaccurate counts of land abandonment and vacant land (Bowman and Pagano 2004).

Population trends and urban land use and vacancy patterns have to consider their impacts on vacant land and structures. In densely populates areas, urban farming, agriculture and gardening offers an alternative land use for integrating multiple functions. However, current urban land use practices in urban sustainable planning prohibit agricultural and farming activities in urban spaces while there is a significant loss of farmland nearby (Pothukuchi & Kaufman, 1999; 2000).

Urban land use for agriculture, farming and gardening have become attractive because of their potential to meet multiple needs, supplying fresh and healthy food in neighborhoods with limited access, articulating the linkages between the built environment, food systems and health while offering opportunities to residents for employment, recreation and education (Hodgson et al., 2011; Redwood, 2011).

However, agriculture and farming as a forms of land use are not well accepted by local and city governments because of perception of health risks. Urban agriculture and farming practices are socially divisive when the land could be dedicated to other scarce economic activities. Land use requires enhanced security on physical limitations that makes the urban space unbuildable and the urban parcels underutilized (Pagano and Bowman 2000) although may have a potential productive quality.

Urban agriculture, farming and gardening increasingly have become necessary urban land use functions supported by sustainable urban planning and designing to have an influence on city life and forms and publicly accepted benefits in terms of multicultural traditions, recreational trends and poverty alleviation

(Hough, 1995, p. 230). Gardens in Montreal, the city that has the largest municipal urban agriculture program in Canada, have their land use designation (Cosgrove, 2001).

Sustainable urban planning seeks to create attractive land-use combinations to meet the needs of residents and environmental challenges, including urban farming agriculture and gardening. Urban planning seeks to create and incorporate to sustainable city development more attractive land use combinations to provide recreational, educational, tourism, cultural, agricultural and farming, gardening, urban forestry, aquaculture, health services, commerce, (van de Berg and van Veenhuizen, 2005; Deelstra et all, 2006).

The urban land-use planning system contributes to economic growth, social development and environmental sustainability. Vancouver's Food Action Plan (City of Vancouver, 2003) supported land use decisions to serve as a public resource to support the city's sustainability commitments. State regulations, like in Oregon, states that a land use planning system requires that each city have to set an urban growth boundary to show the limits and control "sagebrush subdivisions, coastal condo-mania, and the ravenous rampage of suburbia" but protecting the farm and forest lands (McCall, 1973).

The awareness of sustainable urban land-use planning reinforces the development and maintenance of a database of urban development on issues of urban growth or decline, rundown and waste resources (Carley, 1995; Curwell and Cooper, 1998; Department of the Environment, Transport and the Regions (DETR), 1998). Decision-making about land use urban development and parallel policy processes support urban sustainable planning and policy making based on an inventory of natural resources. Urban land-use planning for commercial farming in most of the cities requires permission and to follow zoning and building regulations, water sources, fire and energy codes, waste disposal, etc.

High land prices and exposure to pollution and contamination of water, air and soil by industrial and commercial activities and traffic affect the land use for housing, education, recreation, agriculture, farming, gardening, etc. hampering investments and leading to poor land use security (Mougeot 2000).

Urban land use planning and design authorities for a sustainable environment does not necessarily consider relevant the natural resources and land provision to support urban food production, except for urban planning policies supporting gardening. Urban land use policies and rules may negatively affect local food production and distribution (City of Portland, 2007). Innovation land use types includes vulnerable groups to benefit (Clement 2013, p. 76). The innovation productive land use as landscape typology is designated for local food production and with innovation ecological resilience; these open spaces require decommissioning of urban services to encourage residents to seek employment in agriculture.

Urban planning and food production policies developed by local authorities may connect the issues of land-use planning and sustainable development. Urban land-use and comprehensive planning of some cities have already included urban farming and agriculture providing a food policy planning and adjusting zoning and permit processes with infrastructure, materials, knowledge, and other resources (Mukherji and Morales 2010). The land-use planning system fills the middle layer vacuum. Although land availability and use are not a statutory requirement for sustainable urban planning, they are at the center of the development of urban food production.

Buildings and infrastructure or built environment put pressure on urban land use. Urban vacant, residential yard and industrial rooftop spaces are considered different land uses. Some land use maps do not distinguish between parking lots and other vacant spaces. A good example has been use of vacant abandon land used to plant flowers, ornamental plants, herbs and vegetables, next to government building, as reported by Hui (2011).

Land use with the purpose of reuse of wastes and disposal may offer another benefit in reducing transportation and city management in the long-term. Land use management calls for more spontaneous dynamics of the soil as a consequence that cities struggle to manage the opposing external and internal frontiers of land (Berger 2007). That is a struggle between the city expansion on lands of the urban periphery and the unused spaces within the internal frontier and urban core as land uses move in more elastic cities (Rusk 1993).

Land-use planning has an important role in urban policy making functions to deliver sustainable development, as for example in urban gardening that contributes to accrue economic benefits by preventing and enhancing health-care (Hynes, 1996, p. 161). Urban land use policies and regulations are deemed as methods for increasing and decreasing urban land vacancies. Management and design of vacant land uses contribute ecologically to sustainable cities by creating webs of sustainable urban spaces to provide ecosystem services. Long-term vacancy requires more oriented regeneration policies (National Land Use Database, 2000). Vacant buildings require recording reliable and comprehensive information.

Innovative and integrated sustainable initiatives and strategies in urban land-use planning can contribute to green resilient economic growth, social development issues such as inclusion and equality and environmental sustainability. The implementation of the strategies is relevant to create a typology of community land use types based on vacancy characteristics, incomes of residents in the area and vacancy market rates in order to convert these vacant spaces into urban green resilience spaces with their various inherent functions (Akers, 2013; Detroit Works Project, 2014, p. 51). Vacant land restoration should be an interdisciplinary approach combining economic, socio ecosystem concerns from a holistic urban land use perspective.

CONCLUSION

Human beings are living in a time of qualitative changes in many areas of urban life. These changes affect many levels of city management. The time has come to redefine the city's functional structure. The functional and spatial division in place in many countries seems to be fossilized and ineffective. In the face of enormous civilization changes, if only related to the possibilities of introducing innovative solutions to cities, it turns out that many existing regulations constitute a barrier and block their rational development. Particularly noteworthy is the urban land use that through the innovative pro-environmental solutions can, in a natural way, support the system of green resilience eco-oriented urban land uses in urban eco-systems and serve to improve the quality of life in the city.

Urban land is a scarce natural and real-estate resource which can be used in a more sustainable manner. Ultimately, at the policy level, new forms of value are ascribed to public and private urban land property in order to join other urban land use priorities such as housing, commercial areas, urban green areas and open spaces, roads and other infrastructure. Efficiency of urban land use can be achieved through different means and variables, among others: energy consumption, water and other natural resources, waste management, etc.

Urban land use has different applications for multiple scales, from the entire city, a neighborhood and communities to the small vacant plots, houses, buildings, etc. Urban land use should be evaluated based on a landscape multifunctionality framework, accounting for the contributions and functions, services and benefits that can be provided to the urban socio ecosystem functions, biodiversity, ecological services, cultural functions, economic efficiency, social inclusion and equality, the benefits to the communities and neighborhoods and the society at large.

To determine the current characteristics of land uses, it is necessary to conduct some base data analysis from primary and secondary sources. This approach allows us to broaden the vision by considering the multifunctional forms of land use of a specific site and defining the various components by comparing data with other cities. Then, land uses are identified for specific functions such as education, recreation, etc. Land use components and their dimensions and locations are proposed, this whole process serves to identify the needs and expectations of the urban community regarding land uses, spatial design and site plans.

City authorities and local governments should adopt innovative urban planning and sustainable development strategies based on the use of land, resource protection practices, environmental and ecological services, social inclusion and egalitarian issues, economic growth and efficiency, etc. Strategic orientations of initiatives can seriously obstruct zoning plans and sustainable urban planning to designate urban agriculture, farming and gardening as a land uses. Land use planning and landscape design should reconcile demands of socio economic activities and urban growth.

Urban sustainable planning systems can encourage the use of urban land for more sustainable development context oriented such as the land use for urban food production. Sustainable urban planning does not consider that the food system is its turf, pointing the responsibilities to land use regulations and built environment. Temporal urban land-use in dense cores with limited open green space may formalize some urban gardening and farming initiatives to provide more qualitative resilient ecosystem services that may contribute to develop a more resilient city.

Urban land use sustainable planning must reconcile the demands posed by all the socioeconomic activities and the urban growth assured by an effective management to create value, health and environmental sustainability. However, the high value of competing land in relation to use needs in cities, some land uses, such as urban agriculture, farming and gardening may not seem to be alternatives for urban settings. Land use sustainable planning integrates urban agriculture, farming and gardening into planning and policy processes. Urban farming, agriculture and gardening offer a new frontier for sustainable land use planners and sustainable landscape designers to become involved in the urban sustainable development and transformation of urban spaces.

Agricultural and farm landscapes are considered by urban land use planners and landscape designers as areas for future sustainable urban development and environment instead to integrate new other production functions. The urban agriculture and farming land uses are inherently multifunctional, offering public benefits beyond the provision of sustainability and commodity outputs. Vacant urban land in cities is increasing in city centers where the land use densities are declining due to manufacturing activities declines after have been reached their limits on employment creation and population density.

Multiple urban land use scenarios are a tool that helps to allocate vacant urban spaces to vegetable production including residential green resilient areas, gardens, rooftops, etc. Land use control and building regulations constrain and limit farming and gardening in urban areas.

REFERENCES

Adger, W. N. (2000). Social and ecological resilience: Are they related? *Progress in Human Geography*, *24*(3), 347–364. doi:10.1191/030913200701540465

Adger, W. N. (2006). Vulnerability. *Global Environmental Change, 16*(3), 268–281. doi:10.1016/j.gloenvcha.2006.02.006

Adger, W. N., Hughes, T. P., Folke, C., Carpenter, S. R., & Rockstrom, J. (2005). Social– ecological resilience to coastal disasters. *Science, 309*(5737), 1036–1039. doi:10.1126cience.1112122 PMID:16099974

Akers, J. (2013). Making markets: Think tank legislation and private property in Detroit. *Urban Geography, 34*(8), 1070–1095. doi:10.1080/02723638.2013.814272

Anderies, J. M., Janssen, M. A., & Ostrom, E. (2004). A framework to analyze the robustness of social-ecological systems from an institutional perspective. *Ecology and Society, 9*(1), 18. http://www. ecologyandsociety.org/vol9/iss1/art18/

Arrow, K., Bolin, B., Costanza, R., Dasgupta, P., Folke, C., Holling, C. S., Jansson, B. O., Levin, S., Mäler, K. G., Perrings, C., & Pimentel, D. (1995). Economic growth, carrying capacity and the environment. *Science, 268*(5210), 520–521. doi:10.1126cience.268.5210.520 PMID:17756719

Berger, A. (2007). *Drosscape: Wasting Land Urban America.* Princeton Architectural Press.

Berkes, F., Colding, J., & Folke, C. (Eds.). (2003). *Navigating Social–Ecological Systems: Building Resilience for Complexity and Change.* Cambridge University Press.

Blair, R. B. (2001). Birds and butterflies along urban gradients in two ecoregions of theU.S. In J. L. Lockwood & M. L. McKinney (Eds.), *Biotic Homogenization* (pp. 459–486). Kluwer. doi:10.1007/978-1-4615-1261-5_3

Brian Walker and David Salt. (2012). *Resilience Practice: Building Capacity to Absorb Disturbance and Maintain Function.* Island Press.

Brown, J. H., Ernest, S. K. M., Parody, J. M., & Haskell, J. P. (2001). Regulation of diversity: Maintenance of species richness in changing environments. *Oecologia, 126*(3), 321–332. doi:10.1007004420000536 PMID:28547444

Cadenasso, M. L., & Pickett, S. T. A. (2008). Urbanprinciples for ecological landscape design and management: Scientiðc fundamentals. *Cities and the Environment, 1*(2), 4. doi:10.15365/cate.1242008

Cadenasso, M. L., & Pickett, S. T. A. (2013). Three tides: The development and state of the art of urban ecological studies. In S. T. A. Pickett, M. L. Cadenasso, & B. P. McGrath (Eds.), *Resilience in urban ecology and design: Linking theory and practice for sustainable cities* (pp. 29–46). Springer. doi:10.1007/978-94-007-5341-9_2

Cadenasso, M. L., Pickett, S. T. A., & Grove, J. M. (2006a). Dimensions of ecosystem complexity: Heterogeneity, connectivity, and history. *Ecological Complexity, 3*(1), 1–12. doi:10.1016/j.ecocom.2005.07.002

Cadenasso, M. L., Pickett, S. T. A., & Grove, J. M. (2006b). Integrative approaches to investigating human-natural systems: The Baltimore EcosystemStudy. *Nature Sciences Sociétés, 14*(1), 1–14. doi:10.1051/nss:2006002

Canton de Genève. (2013). Plan directeur cantonal 2030. A14: Promouvoir de nouvelles formes de jardins fa-miliaux et encourager la création de plantages. Geneva, Switzerland: République et Canton de Genève.

Carley, M. (1995). The bigger picture: Organising for sustainable urban regeneration. *Town and Country Planning*, *64*(9), 236–239.

Carpenter, S. R., & Cottingham, K. L. (1997). Resilience and restoration of lakes. *Conservation Ecology*, *1*(1), 2. doi:10.5751/ES-00020-010102

Carpenter, S. R., Walker, B., Anderies, J. M., & Abel, N. (2001). From metaphor to measurement: Resilience of what to what? *Ecosystems (New York, N.Y.)*, *4*(8), 765–781. doi:10.100710021-001-0045-9

Childers, D. L., Pickett, S. T. A., Grove, J. M., Ogden, L., & Whitmer, A. (2014). Advancing urban sustainability theory and action: Challenges and opportunities. *Landscape and Urban Planning*, *125*, 320–328. doi:10.1016/j.landurbplan.2014.01.022

Christopherson, R. W. (1997). Geosystems: An Introduction to Physical Geography (3rd ed.). Upper Saddle River, NJ: Prentice Hall Inc.

City of Portland. (2007). *ENN-1.01: Food Policy Council (Binding city policy)*. Retrieved September 24, 2007, from http://www.portlandonline .com/auditor/index.cfm?a=ihci&c=checj

City of Vancouver. (2003). *Action plan for creating a just and sustainable food system for the City of Vancouver (RTS Number: 3755)*. Author.

Clement, D. (2013). *The spatial injustice of crisis-driven neoliberal urban restructuring in Detroit (Unpublished master's thesis)*. Miami, FL: University of Miami; Retrieved from http://scholarlyrepository.miami.edu/oa_theses/406/

Colding, J. (2011). The role of ecosystem services in contemporary urban planning. In Urban Ecology. Patterns, Processes and Applications. Oxford University Press. doi:10.1093/acprof:oso/9780199563562.003.0028

Colding, J., & Barthel, S. (2012). *The potential of Urban Green Commons in the resilience building of cities*. Elsevier B.V.

Colding, J., Lundberg, J., & Folke, C. (2006). Incorporating green-area user groups in urbanecosystem management. *Ambio*, *35*(5), 237–244. doi:10.1579/05-A-098R.1 PMID:16989508

Collins, J. P., Kinzig, A., Grimm, N. B., Fagan, W. F., Hope, D., Wu, J., & Borer, E. T. (2000). A new urban ecology. *American Scientist*, *88*(5), 416–425. doi:10.1511/2000.5.416

Convention on the Biological Diversity. (1992). The Earth Summit. Held in Rio de Janeiro on 5 June 1992 and entered into force on 29 December 1993. United Nations, General Assembly.

Cosgrove, S. (2001). *Montréal's community gardening program*. Retrieved July 29, 2007, from http://www.cityfarmer.org/Montreal13.html

Costanza, R., & Patten, B. C. (1995). Defining and predicting sustainability. *Ecological Economics*, *15*(3), 193–196. doi:10.1016/0921-8009(95)00048-8

Curwell, S., & Cooper, I. (1998). The implications of urban sustainability. *Building Research and Information*, *26*(1), 17–28. doi:10.1080/096132198370074

Curwell, S., Deakin, M., & Symes, M. (Eds.). (2005). *Sustainable urbandevelopment, volume1: The frameworkand protocols for environmental assessment*. Routledge. doi:10.4324/9780203299913

Daune, L., & Mongé, N. (2011). L'agriculture urbaine, un fondement dans le projet de territoire. *Urbia—Les Cahiers du développement urbain durable, 12*, 85-106.

de Jong, M., Joss, S., Schraven, D., Zhan, Ch., & Weijnen, M. (2015). Sustainable–Smart–Resilient–Low Carbon–Eco–Knowledge Cities; Making Sense of a Multitude of Concepts Promoting Sustainable Urbanization. *Journal of Cleaner Production, 109*(1), 25–38. doi:10.1016/j.jclepro.2015.02.004

Deelstra, T., Boyd, D., & van den Biggelaar, M. (2006). *Multifunctional land use, promoting urban agriculture in Europe*. Academic Press.

DETR. (1998). *Sustainable Development: Opportunities for Change*. Department of the Environment, Transport and the Regions. Consultation Paper on a Revised UK Strategy. DETR.

Detroit Works Project. (2014). *Detroit future city: Detroit strategic framework plan*. Retrieved from http://detroitfuturecity.com/wp-content/uploads/2014/12/DFC_Full_2nd.pdf

Deutsch, L., Folke, C., & Skanberg, K. (2003). The critical natural capital of ecosystem performance as insurance for human well-being. *Ecological Economics, 44*(2-3), 205–217. doi:10.1016/S0921-8009(02)00274-4

Dow, K. (2000). Social dimensions of gradients in urban ecosystems. *Urban Ecosystems, 4*(4), 255–275. doi:10.1023/A:1015767231137

Elmqvist, T., Wall, M., & Berggren, A. L. (2001). Tropical forest reorganization after cyclone and fire disturbance in Samoa: remnant trees as biological legacies. *Cons Ecol, 5*, 10. http://www.consecol.org/vol5/iss2/art10

Estes, J. A., & Duggins, D. O. (1995). Sea otters and kelp forests in Alaska: Generality and variation in a community ecological paradigm. *Ecological Monographs, 65*(1), 75–100. doi:10.2307/2937159

EurActiv. (2004). *Sustainable Development: Introduction*. Author.

Falk, B. (2013). The resilient farm and homestead. Chelsea Green Publishing.

Folke, C., Carpenter, S. R., Elmqvist, T., Gunderson, L., Holling, C. S., & Walker, B. (2002). Resilience and sustainable development: Building adaptive capacity in a world of transformations. *Ambio, 31*(5), 437–440. doi:10.1579/0044-7447-31.5.437 PMID:12374053

Folke, C., Holling, C. S., & Perrings, C. (1996). Biological diversity, ecosystems and the human scale. *Ecological Applications, 6*(4), 1018–1024. doi:10.2307/2269584

Folke, C., Jansson, A., Larsson, J., & Costanza, R. (1997). Ecosystem appropriation by cities. *Ambio, 26*, 167–172.

Folke, C., Pritchard, L., Berkes, F., Colding, J., & Svedin, U. (1998). The Problem of Fit Between Ecosystems and Institutions. International Human Dimensions Programme (IHDP). *IHDP Working Paper No 2*. www.uni-bonn.de/IHDP/public.htm

Forys, E. A., & Allen, C. R. (2002). Functional group change within and across scales following invasions and extinctions in the Everglades ecosystem. *Ecosystems (New York, N.Y.)*, *5*(4), 339–347. doi:10.100710021-001-0078-0

Frost, T. M., Carpenter, S. R., Ives, A. R., & Kratz, T. K. (1995). Species compensation and complementarity in ecosystem function. In C. G. Jones & J. H. Lawton (Eds.), *Linking species and ecosystems*. Chapman & Hall. doi:10.1007/978-1-4615-1773-3_22

Golley, F. B. (1993). *A History of the Ecosystem Concept in Ecology: More Than the Sum of the Parts*. Yale University Press.

Graham, S., & Marvin, S. (2001). *Splintering urbanism: Networked infrastructures, technological mobilities and the urban condition*. Routledge.

Grimm, N. B., Faeth, S. H., Golubiewski, N. E., Redman, C. L., Wu, J., Bai, X., & Briggs, J. M. (2008). Global change and the ecology of cities. *Science*, *319*(5864), 756–760. doi:10.1126cience.1150195 PMID:18258902

Grove, J. M. (2009). Cities: Managing densely settled social–ecological systems. In F. S. Chapin III, G. P. Koðnas, & C. Folke (Eds.), *Principles of ecosystem stewardship: Resilience-based natural resource management in a changing world* (pp. 281–294). doi:10.1007/978-0-387-73033-2_13

Gunderson, L. H., & Holling, C. S. (Eds.). (2002). *Panarchy: Understanding Transformations in Human and Systems*. Island Press.

Hansen, A. J., Defries, R., & Turner, W. (2004). Land use change and biodiversity: a syn-thesis of rates and consequences during the period of satellite imagery. In G. Gutman & C. Justice (Eds.), *Land Change Science: Observing, Monitoring, andUnderstanding Trajectories of Change on the Earth's Surface* (pp. 277–299). Springer Verlag.

Havlicek, T. D., & Carpenter, S. R. (2001). Pelagic species size distributions in lakes: Are they discontinuous? *Limnology and Oceanography*, *46*(5), 1021–1033. doi:10.4319/lo.2001.46.5.1021

Hodgson, K., Caton Campbell, M., & Bailkey, M. (2011). Urban agriculture: growing healthy, sustainable places. Washington, DC: American Planning Association.

Holling, C. S. (1973). Resilience and stability of ecological systems. *Annual Review of Ecology and Systematics*, *4*(1), 1–23. doi:10.1146/annurev.es.04.110173.000245

Holling, C. S. (1996). Engineering resilience versus ecological resilience. In P. C. Schulze (Ed.), *Engineering within ecological constraints*. National Academy Press.

Hough, M. (1995). *Cities and Natural Process*. Routledge. doi:10.4324/9780203292433

Hui, S. C. M. (2011). Green roof urban farming for buildings in high-density urban cities. *Invited paper for the Hainan China World Green Roof Conference 2011*.

Hynes, H. P. (1996). *A Patch of Eden: America's Inner-city Gardeners*. Chelsea Green Publishing Company.

IISD (2009). *What is Sustainable Development?* International Institute for Sustainable Development.

Jackson, J. B. C., Kirby, M. X., Berger, W. H., Bjorndal, K. A., Botsford, L. W., Bourque, B. J., Bradbury, R. H., Cooke, R., Erlandson, J., Estes, J. A., Hughes, T. P., Kidwell, S., Lange, C. B., Lenihan, H. S., Pandolfi, J. M., Peterson, C. H., Steneck, R. S., Tegner, M. J., & Warner, R. R. (2001). Historical overfishing and the recent collapse of coastal ecosystems. *Science*, *293*(5530), 629–638. doi:10.1126cience.1059199 PMID:11474098

James, P., Magee, L., Scerri, A., & Steger, M. B. (2015). *Urban Sustainability in Theory and Practice*. Routledge.

Jenks, M., & Jones, C. (Eds.). (2010). Dimensions of the sustainable city. New York: Springer.

Kahle, L. R., & Gurel-Atay, E. (Eds.). (2014). *Communicating Sustainability for the Green Economy*. M.E. Sharpe.

Kalin Arroyo, M. T., Zedler, P. H., & Fox, M. D. (Eds.). (1995). *Ecology and biogeography of Mediterranean ecosystems in Chile, California, and Australia*. Springer-Verlag. doi:10.1007/978-1-4612-2490-7

Kates, R., Parris, T., & Leiserowitz, A. (2005). What is Sustainable Development? Goals, Indicators, Values, and practice. *Environment*, *47*(3), 8–21. doi:10.1080/00139157.2005.10524444

Kinzig, A. P., Pacala, S. W., & Tilman, D. (Eds.). (2002). *The functional consequences of biodiversity*. Princeton University Press.

Kolasa, J., & Pickett, S. T. A. (2005). Changing academic perspectives of ecology: A view from within. In M. J. Mappin & E. A. Johnson (Eds.), *Environmental education and advocacy* (pp. 50–71). Cambridge University Press.

Lambin, E. F. (2005). Conditions for sustainability of human-environment systems: Information, motivation, and capacity. *Global Environmental Change*, *15*(3), 177–180. doi:10.1016/j.gloenvcha.2005.06.002

Levin, S. (1999). *Fragile dominion: complexity and the commons*. Perseus Books.

Liam, M., Andy, S., Paul, J., Thom, J. A., Padgham, L., Hickmott, S., Deng, H., & Cahill, F. (2013). Reframing social sustainability reporting: Towards an engaged approach. *Environment, Development and Sustainability*, *15*(1), 225–243. doi:10.100710668-012-9384-2

Ljungqvist, J., Barthel, S., Finnveden, G., & Sörlin, S. (2010). The Urban Anthropocene: Lessons for Sustainability from the Environmental History of Constantinople. In P. Sinclair, F. Herschend, C. Isendahl, & G. Nordquist (Eds.), *The Urban Mind, cultural and environmental dynamics Studies in Global Archaeol-ogy* (Vol. 15, pp. 368–394). University Press.

Loreau, M., Naeem, S., & Inchausti, P. (2001). Biodiversity and ecosystem functioning: Current knowledge and future challenges. *Science*, *294*(5543), 804–808. doi:10.1126cience.1064088 PMID:11679658

Lovell, S. T., DeSantis, S., Nathan, C. A., Olson, M. B., Mendez, V. E., Kominami, H. C., Erickson, D. L., Morris, K. S., & Morris, W. B. (2010). Integrating agroecology and landscape multifunc-tionality in Vermont: An evolving framework to evaluate the design of agroecosystems. *Agricultural Systems*, *103*(5), 327–341. doi:10.1016/j.agsy.2010.03.003

Luck, G. W., Daily, G. C., & Ehrlich, P. R. (2003). Population diversity and ecosystem services. *Trends in Ecology & Evolution*, *18*(7), 331–336. doi:10.1016/S0169-5347(03)00100-9

Lundberg, J., & Moberg, F. (2003). Mobile link organisms and ecosystem functioning: Implications for ecosystem resilience and management. *Ecosystems (New York, N.Y.)*, *6*(1), 87–98. doi:10.100710021-002-0150-4

Machlis, G. E., Force, J. E., & Burch, W. R. Jr. (1997). The human ecosystem. 1. The human ecosystem as an organizing concept in ecosystem management. *Society & Natural Resources*, *10*(4), 347–367. doi:10.1080/08941929709381034

McCall, T. (1973). *Legislative address, Governor Tom McCall, Oregon, (1973).* Retrieved July 20, 2008, from http://arcweb.sos.state.or.us/governors/McCall/legis1973.html

McDonnell, M. J., & Hahs, A. (2009). Comparative ecology of cities and towns: Past, present and future. In M. J. McDonnell, A. Hahs, & J. Breuste (Eds.), *Ecology of cities and towns: A comparative approach* (pp. 71–89). Cambridge University Press. doi:10.1017/CBO9780511609763.006

McKinney, M. L. (2002). Urbanization, biodiversity, and conservation. *Bioscience*, *52*(10), 883–890. doi:10.1641/0006-3568(2002)052[0883:UBAC]2.0.CO;2

Millennium Ecosystem Assessment (MA). (2005). *Synthesis Island Press.* Available on the Internet http://www.MAweb.org

Miller, J. R. (2005). Biodiversity conservation and the extinction of experience. *Trends in Ecology & Evolution*, *20*(8), 430–434. doi:10.1016/j.tree.2005.05.013 PMID:16701413

Mukherji, N., & Morales, A. (2010). *Zoning for urban Agriculture*. Zoning Practice 3.

Musacchio, L. R. (2008). Metropolitan landscape ecology: Using translational research to increase sustainability, resilience, and regeneration. *Landscape Journal*, *27*(1), 1–8. doi:10.3368/lj.27.1.1

National Land Use Database. (2000) *Previously Developed Land (PDL): Vacant Sites and Derelict Sites, and Other Previously Developed Land and Buildings that May be Available for Redevelopment, Data Specification*. NLUD – a partnership Rethinking urban capacity.

Naveh, Z. (2000). The total human ecosystem: Integrating ecology and economics. *Bioscience*, *50*, 357–361.

Nelson, D. R., Adger, W. N., & Brown, K. (2007). Adaptation to environmental change: Contributions of a resilience framework. *Annual Review of Environment and Resources*, *32*, 395–419.

Nyström, M., & Folke, C. (2001). Spatial resilience of coral reefs. *Ecosystems (New York, N.Y.)*, *4*(5), 406–417. doi:10.100710021-001-0019-y

O' Bowman, A. M., & Pagano, M.A. (2004). Terra Incognita: Vacant Land and UrbanStrategies. Georgetown University Press.

Odum, E. P. (1971). *Fundamentals of Ecology* (3rd ed.). W. B. Saunders.

Otte, A., Simmering, D., & Wolters, V. (2007). Biodiversity at the landscape level: Recent concepts and perspectives for multifunctional land use. *Landscape Ecology, 22*(5), 639–642. doi:10.100710980-007-9094-6

Pagano, M. A., & Bowman, A. O'M. (2000). Land in Cities: An Urban Resource. Washington, DC: Brookings Institution, Center on Urban and Metropolitan Policy.

Peterson, G., Allen, C.R., & Holling, C.S. (1998). Ecological resilience, biodiversity, and scale. *Ecosystems, 1*, 6–18.

Pickett, S. T. A., Burch, W. R. Jr, Dalton, S., & Foresman, T. (1997). Integrated urban ecosystem research: Themes, needs, and applications. *Urban Ecosystems, 1*(4), 183–184. doi:10.1023/A:1018579628818

Pickett, S. T. A., & Cadenasso, M. L. (2002). Ecosystem as a multidimensional concept: Meaning, model and metaphor. *Ecosystems (New York, N.Y.), 5*(1), 1–10. doi:10.100710021-001-0051-y

Pickett, S. T. A., Cadenasso, M. L., & Grove, J. M. (2004). Resilient cities: Meaning, models, and metaphor for integrating the ecological, socio-economic, and planning realms. *Landscape and Urban Planning, 69*(4), 369–384. doi:10.1016/j.landurbplan.2003.10.035

Pickett, S. T. A., Groffman, P. M., Brush, G. S., Galvin, M. F., Grove, J. M., ... Troy, A. R. (2008). Exchanges across land-water-scape boundaries in urban systems: Strategies for reducing nitrate pollution. *Annals of the New York Academy of Sciences, 1134*(1), 213–232. doi:10.1196/annals.1439.012 PMID:18566096

Pothukuchi, K., & Kaufman, J. (1999). Placing the food system on the urban agenda: The role of municipal institutions in food systems planning. *Agriculture and Human Values, 16*(2), 213–244. doi:10.1023/A:1007558805953

Pothukuchi, K., & Kaufman, J. (2000). The food system: A stranger to the planning field. *Journal of the American Planning Association, 66*(2), 112–124. doi:10.1080/01944360008976093

Quincerot, R., & Weil, M. (2009). *Genève 2020: Plan di-recteur communal de la Ville de Genève*. Ville de Genève.

Redman, C., Grove, J. M., & Kuby, L. (2000). *Toward a uniðed understanding of human ecosystems: Integrating social sciences into long-term ecological research*. White Paper of the Long-Term Ecological Research. Albuquerque, NM: LTER Network.

Redwood, M., & Bahri, A. (2010). *Wastewater Irrigation and Health: Assessing and Mitigation Risks in Low-income Countries*. Earthscan/IDRC/IWMI.

Ricketts, T., & Imhoff, M. (2003). Biodiversity, urban areas, and agriculture: Locating prior-ity ecoregions for conservation. *Conservation Ecology, 8*(2), 1. doi:10.5751/ES-00593-080201

Rudd, J. W. M., Kelly, C. A., Schindler, D. W., & Turner, M. A. (1988). Disruption of the nitrogen cycle in acidified lakes. *Science, 240*(4858), 1515–1517. doi:10.1126cience.240.4858.1515 PMID:17798981

Rusk, D. (1993). *Cities without Suburbs*. The Woodrow.

Sala, O. E., Chapin, F. S. I., Armesto, J. J., Berlow, E., Bloomfield, J., Dirzo, R., Huber-Sanwald, E., Huenneke, L. F., Jackson, R. B., Kinzig, A., Leemans, R., Lodge, D. M., Mooney, H. A., Oesterheld, M., Poff, N. L., Sykes, M. T., Walker, B. H., Walker, M., & Wall, D. H. (2000). Global biodiversity scenarios for the year 2100. *Science*, *287*(5459), 1770–1774. doi:10.1126cience.287.5459.1770 PMID:10710299

Scheffer, M., Carpenter, S. R., Foley, J. A., Folke, C., & Walker, B. H. (2001). Catastrophic shifts in ecosystems. *Nature*, *413*(6856), 591–596. doi:10.1038/35098000 PMID:11595939

Schindler, D. W. (1990). Experimental perturbations of whole lakes as tests of hypotheses concerning ecosystem structure and function. *Oikos*, *57*(1), 25–41. doi:10.2307/3565733

Seana I., Johnson, L. & Peters, K. (1999). Community gardens and sustainable land use planning: A case-study of the Alex Wilson community garden. *Local Environment*, *4*(1), 33-46. Doi:10.1080/13549839908725579

Seto, K. C., De Groot, R., Bringezu, S., Erb, K., Graedel, T. E., Ramankutty, N., & Skole, D. L. (2010). Stocks, ñows, and prospects of land. In T. E. Graedel & E. van der Voet (Eds.), *Linkages of sustainability* (pp. 72–96). MIT Press.

Smit, B., & Wandel, J. (2006). Adaptation, adaptive capacity and vulnerability. *Global Environmental Change*, *16*(3), 282–292. doi:10.1016/j.gloenvcha.2006.03.008

Sukopp, H., Hejny, S., & Kowarik, I. (Eds.). (1990). *Urban ecology: Plants and plant communities in urban environments*. SPB Academic Publishing.

Sukopp, H., Trautmann, W., & Korneck, D. (1979). The soilflora and vegetation of Berlin'swastelands. In I. Laurie (Ed.), *Nature in Cities*. John Wiley.

Tansley, A. G. (1935). The use and abuse of vegetational concepts and terms. *Ecology*, *16*(3), 284–307. doi:10.2307/1930070

Terborgh, J., Lopez, L., & Nunez, P. (2001). Ecological meltdown in predator-free forest fragments. *Science*, *294*(5548), 1923–1926. doi:10.1126cience.1064397 PMID:11729317

Termorshuizen, J. W., & Opdam, P. (2009). Landscape services as a bridge between landscape ecology and sustainable development. *Landscape Ecology*, *24*(8), 1037–1052. doi:10.100710980-008-9314-8

Tidball, G. K., Krasny, E. M., Svendsen, E., Cambell, L., & Helphand, K. (2010). Stewardship, learning and memory in disaster resilience. *Environmental Learning Research*, *16*(5–6), 591–609. doi:10.1080/13504622.2010.505437

Tidball, K. G., & Krasny, M. E. (2009). *From risk to resilience: what role for community greening and civic ecology in cities?* Wageningen Academic Publishers.

van de Berg, L., & van Veenhuizen, R. (2005). Multiple functions of urban Agriculture. *Urban Agriculture Magazine*, 14.

Walker, B. H., Holling, C. S., Carpenter, S. R., & Kinzig, A. P. (2004). Resilience, adaptability and transformability in social–ecological systems. *Ecology and Society, 9*(2), 5. http://www.ecologyandsociety. org/vol9/iss2/art5/

Wu, J., & Wu, T. (2013). Ecological resilience as a foundation for urban design and sustainability. In S. T. A. Pickett, M. L. Cadenasso, & B. McGrath (Eds.), *Resilience in ecology and urban design: Linking theory and practice for sustainable cities* (pp. 211–229). Springer. doi:10.1007/978-94-007-5341-9_10

Yohe, G., & Tol, R. S. J. (2002). Indicators for social and economic coping capacity – moving toward a working deðnition of adaptive capacity. *Global Environmental Change, 12*(1), 25–40. doi:10.1016/S0959-3780(01)00026-7

Zanoni, P., & Janssens, M. (2009). Sustainable DiverCities. In M. Janssens, D. Pinelli, D. C. Reyman, & S. Wallman (Eds.), *Sustainable Cities. Diversity, Economic Growthand Social Cohesion* (pp. 3–25). Edward Elgar.

Zhang, S. X., & Babovic, V. (2012). A real options approach to the design and architecture of water supply systems using innovative water technologies under uncertainty (PDF). *Journal of Hydroinformatics, 488*. www.frontiersinecology.org

Zimov, S. A., Chuprynin, V. I., Oreshko, A. P., Chapin, F. S. III, Reynolds, J. F., & Chapin, M. C. (1995). Steppe–tundra transition: An herbivore-driven biome shift at the end of the Pleistocene. *American Naturalist, 146*(5), 765–794. doi:10.1086/285824

KEY TERMS AND DEFINITIONS

Eco-Urban: Urban environments that are characterized by having a series of characteristics that define them as sustainable and ecological.

Green Resilience: Based on the infrastructure of urban green areas applied to cities that have the capacity to prepare, resist and recover from a crisis.

Land Use: Encompasses the management and modification of the natural environment to convert it into agricultural land: arable fields, grasslands; or human settlements. The term land use is also used to refer to the different land uses in zoning.

Resilience: The ability to overcome critical moments and adapt after experiencing some unusual and unexpected situation. It also indicates a return to normal.

Urban: Belonging to or relating to the city.

Urban Ecology: A discipline whose object of study is the interrelationships between the inhabitants of an urban agglomeration and its multiple interactions with the environment.

Urban Socio-Ecological Systems: Complex urban structures that can be analyzed considering the social subsystem and the ecological subsystem.

Chapter 2
Sustainable Development and Climate Change

Vartika Singh
Amity University, Noida, India

ABSTRACT

The concept of sustainable development can be best interpreted as an approach to development that looks to balance different, and often competing, needs against an awareness of the environmental, social, and economic limitations we face as a society. Most of the present development activities are being conceived and implemented without fully considering the wider or future impacts. A manifestation of such kind of development approach is quite evident in the form of large-scale financial crises caused by irresponsible banking, changes in the global climate resulting from our dependence on fossil fuel-based energy sources. The longer we pursue unsustainable development, the more frequent and severe its consequences are likely to become, which is why we need to take action now. This chapter explores sustainable development and climate change.

INTRODUCTION

The Sustainable Development Goals (SDGs), also known as the Global Goals, were adopted by all United National Member States in 2015 as a universal call to action to end poverty, protect the planet and ensure that all people enjoy peace and prosperity by 2030. The 17 SDGs are integrated- that is, they recognize that action in one area will affect outcomes in others, and that development must balance social, economic and environmental sustainability. Through the pledge to Leave No One Behind, countries have committed to fast-track progress for those furthest behind first. That is why the SDGs designed to bring the world to several life-changing 'zeros', including zero poverty, hunger, AIDS and discrimination against women and girls. Everyone is needed to reach these ambitious targets. The creativity, knowhow, technology and financial resources from all of society is necessary to achieve the SDGs in every context.

DOI: 10.4018/978-1-7998-4915-5.ch002

Historical Background:

- In June 1992, at the Earth Summit in Rio de Janeiro, Brazil, more than 178 countries adopted Agenda 21, a comprehensive plan of action to build a global partnership for sustainable development to improve human lives and protect the environment.
- Member states unanimously adopted the Millennium Declaration at the Millennium Summit in September 2000 at UN Headquarters in New York. The Summit led to the elaboration of eight Millennium Development Goals (MDGs) by 2015.
- The Johannesburg Declaration on Sustainable Development and the Plan of Implementation, adopted at the World Summit on Sustainable Development in South Africa in 2002, reaffirmed the global community's commitments to poverty eradication and the environment, and built on Agenda 21 and the Millennium Declaration by including more emphasis on multilateral partnerships.
- At the United Nations Conference on Sustainable Development (Rio+20) in Rio de Janeiro, Brazil, in June 2012, Member States adopted the outcome document "The Future We Want" in which they decided, inter alia, to launch a process to develop a set of SDGs to build upon the MDGs and to establish the UN High-level Political Forum on Sustainable Development. The Rio+20 outcome also contained other measures for implementing sustainable development, including mandates for future programs of work in development financing, small island developing states and more.
- In 2013, the General Assembly set up a 30-member Open Working Group to develop a proposal on the SDGs.
- In January 2015, the General Assembly began the negotiation process on the post-2015 development agenda. The process culminated in the subsequent adoption of the 2030 Agenda for Sustainable Development, with 17 SDGs at its core, at the UN Sustainable Development Summit in September 2015.
- 2015 was a landmark year for multilateralism and international policy shaping, with the adoption of several major agreements:
    - Sendai Framework for Disaster Risk Reduction (March 2015)
    - Addis Ababa Action Agenda on Financing for Development (July 2015)
    - Transforming our world: the 2030 Agenda for Sustainable Development
    - Summit in New York in September 2015.
    - Paris Agreement on Climate Change (December 2015)
- Now, the annual High-level Political Forum on Sustainable Development serves as the central UN platform for the follow-up and review of the SDGs.

Today, the Division for Sustainable Development (DSDG) in the United Nations Department of Economic and Social Affairs (UNDESA) provides substantive support and capacity-building for the SDGs and their related thematic issues, including water, energy, climate, oceans, urbanization, transport, science and technology, the Global Sustainable Development Report (GSDR), partnerships and Small Island Developing States. DSDG plays a key role in the evaluation of UN system wide implementation of the 2030 Agenda a reality, broad ownership of the SDGs must translate into a strong commitment by all stakeholders to implement the global goals. DSDG aims to help facilitate this engagement.

CONCEPT OF SUSTAINABLE DEVELOPMENT

"Sustainable Development" is development of all in perpetuity. The term sustainability literally means "a capacity to maintain some entity, outcome, or process over time" (Jenkins, 2009). It poses an intended future for human societies where living conditions and resource utilization practices meet human needs without undermining the sustainability of natural resource systems and the environment, to ensure that future generations may also got their needs fulfilled (Beckerman, 1992). Sustainability and development are normally in apposition, where both could have possible counterproductive effects (Sharpley 2000) but economists give emphasis to there is no contradiction among sustainability and development (Lele, 1991). Sustainable development binds together concern for the carrying capacity of natural systems with the social and economic challenges faced by humanity. There is no development without sustainability or sustainability without development (Sachs 2010). Development is a process whose purposes to increase the quality of life and rise the self-sufficient capacity of economies that are technically more complex and depend on global integration (Remeny, 2004, Willis, 2005). Sustainable development also seen as a situation of maintaining a delicate balance between the human needs on one hand, and preserving natural resources and ecosystems on the other hand, on which we and future generations depend.

There have been various developmental phases through which the concept of sustainable development has undergone since its inception. Organizations and institutions across the globe have come together to work immensely and intensely to evolve concept of sustainable development such as principles and objectives of the sustainable development. The concept has experienced different scrutiny and interpretations over the time while being accepted in different areas of human activity, and the definition of sustainable development has become one of the most cited definitions in the literature. In its development, the concept has been adapting to the contemporary requirements of a complex global development environment, but the underlying principles and goals, as well as the problems of their implementation, remained almost unchanged (Klarin, 2018).

From the very beginning of Sustainable Development concept it has undergone several development phases. The term sustainable development was originally introduced in the field of forestry, and it included measures of afforestation and harvesting of interconnected forests which should not undermine the biological renewal of forests (Šimleša, 2003: Bâc, 2008: Črnjar & Črnjar, 2009). Various organizations and Institutions participate in the concept of sustainable development and now they focus on the principles, objective and their implementation. The concept of Sustainable Development has been accepted in different area of human activity. The concept of Sustainable Development had adapting the contemporary requirements of global environment, but when we talking about problems, goal and principles and implementation it is not changed, only some new goals are introduced and same are updated with the new changing word. All these changes are assembling in framework of millennium Development goals 2015. The main aim of all these work is to achieve the sustainable development and survive on earth in well manner. Over the last decades we observe with the development of humanity lots of negative issues are also increased line natural disaster area, climate change, and instability in socio- economic factors. Human action have negative impact on the environment, earth and natural resources. If human behaviour is responsible towards the earth resources, we can fulfil present demand and same resources for future generation. Concept of sustainable is based on socio-economic development and need of future generation.

SUSTAINABLE DEVELOPMENT GOALS AND INDICATORS

Goal 1: No Poverty

"End poverty in all its forms everywhere."

Extreme Poverty has been cut by more than half since 1990. Still, around 1 in 10 people live on less than the target figure of international $1.90 per day. A very low poverty threshold is justified by highlighting the need of those people who are worst off. SDG 1 is to end extreme poverty globally by 2030.That target may not be adequate for human subsistence and basic needs, however (UNDP 2017). It is for this reason that changes relative to higher poverty lines are also commonly tracked. Poverty is more than the lack of income or resources: People live in poverty if they lack basic services such as healthcare, security, and education. They also experience hunger, social discrimination, and exclusion from decision-making processes. One possible alternative metric is the Multidimensional Poverty Index.

Children make up the majority – more than half – of those living in extreme poverty. In 2013, an estimated 385 million children lived on less than US$1.90 per day. Still, these figures are unreliable due to huge gaps in data on the status of children worldwide. On average, 97 percent of countries have insufficient data to determine the state of impoverished children and make projections towards SDG Goal 1, and 63 percent of countries have no data on child poverty at all. Women face potentially life-threatening risks from early pregnancy and frequent pregnancies. This can result in lost hope for an education and for a better income. Poverty affects age groups differently, with the most devastating effects experienced by children. It affects their education, health, nutrition, and security, impacting emotional and spiritual development.

Achieving Goal 1 is hampered by lack of economic growth in the poorest countries of the world, growing inequality, increasingly fragile childhood, and the impacts of climate change.

Facts and Figures

- 736 million people still live in extreme poverty.
- 10 percent of the world's population live in extreme poverty, down from 36 percent in 1990.
- Some 1.3 billion people live in multidimensional poverty.
- Half of all people living in poverty are under 18.
- One person in every 10 is extremely poor.
- 80 percent of people living on less than $1.90 are in South Asia and sub-Saharan Africa.

Goal 2: Zero Hunger

"End hunger, achieve food security and improved nutrition, and promote sustainable agriculture".

Goal 2 states that by 2030 we should end hunger and all forms of malnutrition. This would be accomplished by doubling agricultural productivity and incomes of small-scale food producers (especially women and indigenous peoples), by ensuring sustainable food production systems, and by progressively improving land and soil quality. Agriculture is the single largest employer in the world, providing livelihoods for 40% of the global population. It is the largest source of income for poor rural households. Women make up about 43% of the agricultural labour force in developing countries, and over 50% in parts of Asia and Africa. However, women own only 20% of the land. Other targets deal with maintaining genetic diversity of seeds, increasing access to land, preventing trade restriction and distortions in world agricultural markets to limit extreme food price volatility, eliminating waste with help from the International Food Waste Coalition, and ending malnutrition and undernutrition of children (Fan, Shenggen and Polman, Paul. 2014).

Globally, 1 in 9 people are undernourished, the vast majority of whom live in developing countries. Undernutrition causes wasting or severe wasting of 52 million children worldwide, and contributes to nearly half (45%) of deaths in children under five – 3.1 million children per year. Chronic malnutrition, which affects an estimated 155 million children worldwide, also stunts children's brain and physical development and puts them at further risk of death, disease, and lack of success as adults (Kinyoki, D.; et al. 2020). As of 2017, only 26 of 202 UN member countries are on track to meet the SDG target to eliminate undernourishment and malnourishment, while 20 percent have made no progress at all and nearly 70 percent have no or insufficient data to determine their progress. A report by the International Food Policy Research Institute (IFPRI) of 2013 stated that the emphasis of the SDGs should not be on ending poverty by 2030, but on eliminating hunger and under-nutrition by 2025. The assertion is based on an analysis of experiences in China, Vietnam, Brazil, and Thailand. Three pathways to achieve this were identified: 1) agriculture-led; 2) social protection- and nutrition- intervention-led; or 3) a combination of both of these approaches.

A study published in Nature concluded that it is unlikely there will be an end to malnutrition by 2030.

Facts and Figures:

- The number of undernourished people reached 821 million in 2017.
- In 2017 Asia accounted for nearly two thirds, 63 percent, of the world's hungry.
- Nearly 151 million children under five, 22 percent, were still stunted in 2017.
- More than 1 in 8 adults is obese.
- 1 in 3 women of reproductive age is anaemic.
- 26 percent of workers are employed in agriculture.

Goal 3: Good Health and Well-Being for People

"Ensure healthy lives and promote well-being for all at all ages."

Significant strides have been made in increasing life expectancy and reducing some of the common killers associated with child and maternal mortality. Between 2000 and 2016, the worldwide under-five mortality rate decreased by 47 percent (from 78 deaths per 1,000 live births to 41 deaths per 1,000 live births). Still, the number of children dying under age five is extremely high: 5.6 million in 2016 alone. New borns account for a growing number of these deaths, and poorer children are at the greatest risk of under-5 mortality due to a number of factors. SDG Goal 3 aims to reduce under-five mortality to at least as low as 25 per 1,000 live births (UNDP 2017). But if current trends continue, more than 60 countries will miss the SDG neonatal mortality target for 2030. About half of these countries would not reach the target even by 2050. Goal 3 also aims to reduce maternal mortality to less than 70 deaths per 100,000 live births. Though the maternal mortality ratio declined by 37 percent between 2000 and 2015, there were approximately 303,000 maternal deaths worldwide in 2015, most from preventable causes. In

2015, maternal health conditions were also the leading cause of death among girls aged 15–19. Data for girls of greatest concern – those aged between 10-14 - is currently unavailable. Key strategies for meeting SDG Goal 3 will be to reduce adolescent pregnancy (which is strongly linked to gender equality), provide better data for all women and girls, and achieve universal coverage of skilled birth attendants.

Similarly, progress has been made on increasing access to clean water and sanitation and on reducing malaria, TB, polio, and the spread of HIV/AIDS. From 2000-2016, new HIV infections declined by 66 percent for children under 15 and by 45 percent among adolescents aged 15–19. However, current trends mean that 1 out of 4 countries still won't meet the SDG target to end AIDS among children under 5, and 3 out of 4 will not meet the target to end AIDS among adolescents. Additionally, only half of women in developing countries have received the health care they need, and the need for family planning is increasing exponentially as the population grows. While needs are being addressed gradually, more than 225 million women have an unmet need for contraception. Goal 3 aims to achieve universal health coverage, including access to essential medicines and vaccines. It proposes to end the preventable death of new-borns and children under 5 and to end epidemics such as AIDS, tuberculosis, malaria, and water-borne diseases, for example. 2016 rates for the third dose of the pertussis vaccine (DTP3) and the first dose of the measles vaccine (MCV1) reached 86 percent and 85 percent, respectively. Yet about 20 million children did not receive DTP3 and about 21 million did not receive MCV1. Around 2 in 5 countries will need to accelerate progress in order to reach SDG targets for immunization.

Attention to health and well-being also includes targets related to the prevention and treatment of substance abuse, deaths and injuries from traffic accidents and from hazardous chemicals and air, water and soil pollution and contamination.

Goal 4: Quality Education

"Ensure inclusive and equitable quality education and promote lifelong learning opportunities"

Since 2000, there has been enormous progress in achieving the target of universal primary education. The total enrolment rate in developing regions reached 91 percent in 2015, and the worldwide number of children out of school has dropped by almost half. There has also been a dramatic increase in literacy rates, and many more girls are in school than ever before. These are all remarkable successes.

Progress has also been tough in some developing regions due to high levels of poverty, armed conflicts and other emergencies. In Western Asia and North Africa, ongoing armed conflict has seen an increase in the number of children out of school (Local Burden of Disease Educational Attainment Collaborators 2020). This is a worrying trend. While Sub-Saharan Africa made the greatest progress in primary school enrolment among all developing regions – from 52 percent in 1990, up to 78 percent in 2012 – large disparities still remain. Children from the poorest households are up to four times more likely to be out of school than those of the richest households. Disparities between rural and urban areas also remain high. Achieving inclusive and quality education for all reaffirms the belief that education is one of the most powerful and proven vehicles for sustainable development. This goal ensures that all girls and boys complete free primary and secondary schooling by 2030. It also aims to provide equal access to affordable vocational training, to eliminate gender and wealth disparities, and achieve universal access to a quality higher education

Facts and Figures:

- At least 400 million people have no basic healthcare, and 40 percent lack social protection.
- More than 1.6 billion people live in fragile settings where protracted crises, combined with weak national capacity to deliver basic health services, present a significant challenge to global health.
- By the end of 2017, 21.7 million people living with HIV were receiving antiretroviral therapy. Yet more than 15 million people are still waiting for treatment.
- Every 2 seconds someone aged 30 to 70 years dies prematurely from no communicable diseases - cardiovascular disease, chronic respiratory disease, diabetes or cancer.
- 7 million people die every year from exposure to fine particles in polluted air.
- More than one of every three women have experienced either physical or sexual violence at some point in their life resulting in both short- and long-term consequences for their physical, mental, and sexual and reproductive health.

Goal 5: Gender Equality

"Achieve gender equality and empower all women and girls"

Ending all discrimination against women and girls is not only a basic human right, it's crucial for sustainable future; it's proven that empowering women and girls helps economic growth and development. UNDP has made gender equality central to its work and we've seen remarkable progress in the past 20 years. There are more girls in school now compared to 15 years ago, and most regions have reached gender parity in primary education.

But although there are more women than ever in the labour market, there are still large inequalities in some regions, with women systematically denied the same work rights as men (UNSD 2018). Sexual violence and exploitation, the unequal division of unpaid care and domestic work, and discrimination in public office all remain huge barriers. Climate change and disasters continue to have a disproportionate effect on women and children, as do conflict and migration. It is vital to give women equal rights land and property, sexual and reproductive health, and to technology and the internet (Firzli, Nicolas 2018). Today there are more women in public office than ever before, but encouraging more women leaders will help achieve greater gender equality.

Facts and Figures

- Women earn only 77 cents for every dollar that men get for the same work.
- 35 percent of women have experienced physical and/or sexual violence.
- Women represent just 13 percent of agricultural landholders.
- Almost 750 million women and girls alive today were married before their 18th birthday.
- Two thirds of developing countries have achieved gender parity in primary education.
- Only 24 percent of national parliamentarians were women as of November 2018, a small increase from 11.3 percent in 1995.

Goal 6: Clean Water And Sanitation

"ensure availability and sustainable management of water and sanitation for all"

Water scarcity affects more than 40 percent of people, an alarming figure that is projected to rise as temperatures do. Although 2.1 billion people have improved water sanitation since 1990, dwindling drinking water supplies are affecting every continent. More and more countries are experiencing water stress, and increasing drought and desertification is already worsening these trends (WHO and UNICEF 2017). By 2050, it is projected that at least one in four people will suffer recurring water shortages.

Safe and affordable drinking water for all by 2030 requires we invest in adequate infrastructure, provide sanitation facilities, and encourage hygiene. Protecting and restoring water-related ecosystems is essential. Ensuring universal safe and affordable drinking water involves reaching over 800 million people who lack basic services and improving accessibility and safety of services for over two billion. In 2015, 4.5 billion people lacked safely managed sanitation services (with adequately disposed or treated excreta) and 2.3 billion lacked even basic sanitations (Kellogg, Diane M. 2017).

Facts and Figures

- 71 percent of the global population, 5.2 billion people, had safely-managed drinking water in 2015, but 844 million people still lacked even basic drinking water.
- 39 percent of the global population, 2.9 billion people, had safe sanitation in 2015, but 2.3 billion people still lacked basic sanitation. 892 million people practiced open defecation.
- 80 percent of wastewater goes into waterways without adequate treatment.
- Water stress affects more than 2 billion people, with this figure projected to increase.
- 80 percent of countries have laid the foundations for integrated water resources management.
- The world has lost 70 percent of its natural wetlands over the last century.

Goal 7: Affordable and Clean Energy

Between 2000 and 2016, the number of people with electricity increased from 78 to 87 percent, and the numbers without electricity dipped to just below one billion. Yet as the population continues to grow, so will the demand for cheap energy, and an economy reliant on fossil fuels is creating drastic changes to our climate. Investing in solar, wind and thermal power, improving energy productivity, and ensuring energy for all is vital if we are to achieve SDG 7 by 2030. Expanding infrastructure and upgrading technology to provide clean and more efficient energy in all countries will encourage growth and help the environment (UNDP 2015).

Facts and Figures

- One in 7 people still lacks electricity, and most of them live in rural areas of the developing world.
- Energy is the main contributor to climate change, it produces around 60 percent of greenhouse gases.
- More efficient energy standards could reduce building and industry electricity consumption by 14 percent.
- More than 40 percent of the world's population—3 billion—rely on polluting and unhealthy fuels for cooking.
- As of 2015, more than 20 percent of power was generated through renewable sources.
- The renewable energy sector employed a record 10.3 million people in 2017.

Goal 8: Decent Work and Economic Growth

"promote sustained, inclusive and sustainable economic growth, full and productive environment and decent work for all"

Over the past 25 years the number of workers living in extreme poverty has declined dramatically, despite the lasting impact of the 2008 economic crisis and global recession. In developing countries, the middle class now makes up more than 34 percent of total employment – a number that has almost tripled between 1991 and 2015.

However, as the global economy continues to recover we are seeing slower growth, widening inequalities, and not enough jobs to keep up with a growing labour force. According to the International Labour Organization, more than 204 million people were unemployed in 2015. The SDGs promote sustained economic growth, higher levels of productivity and technological innovation (Firzli, M. Nicolas J., 2016). Encouraging entrepreneurship and job creation are key to this, as are effective measures to eradicate forced labour, slavery and human trafficking. With these targets in mind, the goal is to achieve full and productive employment, and decent work, for all women and men by 2030.

Facts and Figures:

- An estimated 172 million people worldwide were without work in 2018 - an unemployment rate of 5 percent.
- As a result of an expanding labour force, the number of unemployed is projected to increase by 1 million every year and reach 174 million by 2020.
- Some 700 million workers lived in extreme or moderate poverty in 2018, with less than US$3.20 per day.
- Women's participation in the labour force stood at 48 per cent in 2018, compared with 75 percent for men. Around 3 in 5 of the 3.5 billion people in the labour force in 2018 were men.
- Overall, 2 billion workers were in informal employment in 2016, accounting for 61 per cent of the world's workforce.
- Many more women than men are underutilized in the labour force—85 million compared to 55 million.

Goal 9: Industry, Innovation and Infrastructure

"build resilient infrastructure, promote inclusive and sustainable industrialisation and foster innovation"

Investment in infrastructure and innovation are crucial drivers of economic growth and development. With over half the world population now living in cities, mass transport and renewable energy are becoming ever more important, as are the growth of new industries and information and communication technologies.

Technological progress is also key to finding lasting solutions to both economic and environmental challenges, such as providing new jobs and promoting energy efficiency. Promoting sustainable industries, and investing in scientific research and innovation, are all important ways to facilitate sustainable development. More than 4 billion people still do not have access to the Internet, and 90 percent are from the developing world (*Sustainable Development Knowledge Platform*, 2017). Bridging this digital divide is crucial to ensure equal access to information and knowledge, as well as foster innovation and entrepreneurship.

Facts and Figures

- Worldwide, 2.3 billion people lack access to basic sanitation.
- In some low-income African countries, infrastructure constraints cut businesses' productivity by around 40 percent.
- 2.6 billion people in developing countries do not have access to constant electricity.
- More than 4 billion people still do not have access to the Internet; 90 percent of them are in the developing world.
- The renewable energy sectors currently employ more than 2.3 million people; the number could reach 20 million by 2030.
- In developing countries, barely 30 percent of agricultural products undergo industrial processing, compared to 98 percent high-income countries.

Goal 10: Reduced Inequalitiest

"reduced inequalities within and among countries" HESE widen is partiers

Income inequality is on the rise—the richest 10 percent have up to 40 percent of global income whereas the poorest 10 percent earn only between 2 to 7 percent. If we take into account population growth inequality in developing countries, inequality has increased by 11 percent. Income inequality has increased in nearly everywhere in recent decades, but at different speeds. It's lowest in Europe and highest in the Middle East. These widening disparities require sound policies to empower lower income earners, and promote economic inclusion of all regardless of sex, race or ethnicity. Income inequality requires global solutions. This involves improving the regulation and monitoring of financial markets and institutions, encouraging development assistance and foreign direct investment to regions where the need is greatest (*Sustainable Development Knowledge Platform*, 2017). Facilitating the safe migration and mobility of people is also key to bridging the widening divide.

Facts and Figures

- In 2016, 22 percent of global income was received by the top 1 percent compared with 10 percent of income for the bottom 50 percent.
- In 1980, the top one percent had 16 percent of global income. The bottom 50 percent had 8 percent of income.
- Economic inequality is largely driven by the unequal ownership of capital. Since 1980, very large transfers of public to private wealth occurred in nearly all countries. The global wealth share of the top 1 percent was 33 percent in 2016.
- Under "business as usual", the top 1 percent global wealth will reach 39 percent by 2050.
- Women spend, on average, twice as much time on unpaid housework as men.
- Women have as much access to financial services as men in just 60 percent of the countries assessed and to land ownership in just 42 percent of the countries assessed.

Goal 11: Sustainable Cities and Communites

"make cities and human settlements inclusive, safe, resilient and sustainable"

More than half of us live in cities. By 2050, two-thirds of all humanity 6.5 billion people will be urban. Sustainable development cannot be achieved without significantly transforming the way we build and manage our urban spaces. The rapid growth of cities a result of rising populations and increasing migration has led to a boom in mega-cities, especially in the developing world, and slums are becoming a more significant feature of urban life (Sustainable Development Goals, 2017). Making cities sustainable means creating career and business opportunities, safe and affordable housing, and building resilient societies and economies. It involves investment in public transport, creating green public spaces, and improving urban planning and management in participatory and inclusive ways.

Facts and Figures

- In 2018, 4.2 billion people, 55 percent of the world's population, lived in cities. By 2050, the urban population is expected to reach 6.5 billion.
- Cities occupy just 3 percent of the Earth's land but account for 60 to 80 percent of energy consumption and at least 70 percent of carbon emissions.
- 828 million people are estimated to live in slums, and the number is rising.
- In 1990, there were 10 cities with 10 million people or more; by 2014, the number of mega-cities rose to 28, and was expected to reach 33 by 2018. In the future, 9 out of 10 mega-cities will be in the developing world.
- In the coming decades, 90 percent of urban expansion will be in the developing world.
- The economic role of cities is significant. They generate about 80 percent of the global GDP.

Goal 12: Responsible Consumption and Production

"ensure sustainable consumption and production patterns"

Achieving economic growth and sustainable development requires that we urgently reduce our ecological footprint by changing the way we produce and consume goods and resources. Agriculture is the biggest user of water worldwide, and irrigation now claims close to 70 percent of all freshwater for human use (UNDP, 2018). The efficient management of our shared natural resources, and the way we dispose of toxic waste and pollutants, are important targets to achieve this goal. Encouraging industries, businesses and consumers to recycle and reduce waste is equally important, as is supporting developing countries to move towards more sustainable patterns of consumption by 2030.

A large share of the world population is still consuming far too little to meet even their basic needs. Halving the per capita of global food waste at the retailer and consumer levels is also important for creating more efficient production and supply chains. This can help with food security, and shift us towards a more resource efficient economy.

Facts and Figures

- 1.3 billion tonnes of food is wasted every year, while almost 2 billion people go hungry or undernourished.
- The food sector accounts for around 22 percent of total greenhouse gas emissions, largely from the conversion of forests into farmland.
- Globally, 2 billion people are overweight or obese.

- Only 3 percent of the world's water is fresh (drinkable), and humans are using it faster than nature can replenish it.
- If people everywhere switched to energy efficient lightbulbs, the world would save US$120 billion annually.
- One-fifth of the world's final energy consumption in 2013 was from renewable sources.

Goal 13: Climate Action

"take urgent action to combat climate change and its impacts"

There is no country that is not experiencing the drastic effects of climate change. Greenhouse gas emissions are more than 50 percent higher than in 1990. Global warming is causing long-lasting changes to our climate system, which threatens irreversible consequences if we do not act. The annual average economic losses from climate-related disasters are in the hundreds of billions of dollars. This is not to mention the human impact of geo- physical disasters, which are 91 percent climate-related, and which between 1998 and 2017 killed 1.3 million people, and left 4.4 billion injured. The goal aims to mobilize US$100 billion annually by 2020 to address the needs of developing countries to both adapt to climate change and invest in low-carbon development.

Supporting vulnerable regions will directly contribute not only to Goal 13 but also to the other SDGs. These actions must also go hand in hand with efforts to integrate disaster risk measures, sustainable natural resource management, and human security into national development strategies (Ansuategi, A; Greño, P; Houlden, V; et al., 2015). It is still possible, with strong political will, increased investment, and using existing technology, to limit the increase in global mean temperature to two degrees Celsius above pre-industrial levels, aiming at 1.5°C, but this requires urgent and ambitious collective action.

Facts and Figures

- As of 2017 humans are estimated to have caused approximately 1.0°C of global warming above pre-industrial levels.
- Sea levels have risen by about 20 cm (8 inches) since 1880 and are projected to rise another 30–122 cm (1 to 4 feet) by 2100.
- To limit warming to 1.5C, global net CO2 emissions must drop by 45% between 2010 and 2030, and reach net zero around 2050.
- Climate pledges under The Paris Agreement cover only one third of the emissions reductions needed to keep the world below 2°C.
- Bold climate action could trigger at least US$26 trillion in economic benefits by 2030.
- The energy sector alone will create around 18 million more jobs by 2030, focused specifically on sustainable energy.

Goal 14: Life Below Water

"conserve and sustainably use oceans, seas and marine resources for sustainable development"

The world's oceans – their temperature, chemistry, currents and life drive global systems that make the Earth habitable for humankind. How we manage this vital resource is essential for humanity as a whole, and to counterbalance the effects of climate change (Covert, J., 2017). Over three billion people depend on marine and coastal biodiversity for their livelihoods. However, today we are seeing 30 percent of the world's fish stocks overexploited, reaching below the level at which they can produce sustainable yields.

Oceans also absorb about 30 percent of the carbon dioxide produced by humans, and we are seeing a 26 percent rise in ocean acidification since the beginning of the industrial revolution. Marine pollution, an overwhelming majority of which comes from land-based sources, is reaching alarming levels, with an average of 13,000 pieces of plastic litter to be found on every square kilometre of ocean. The SDGs aim to sustainably manage and protect marine and coastal ecosystems from pollution, as well as address the impacts of ocean acidification (Staples, D., & Hermes, R., 2012). Enhancing conservation and the sustainable use of ocean-based resources through international law will also help mitigate some of the challenges facing our oceans.

Facts and Figures

- The ocean covers three quarters of the Earth's surface and represents 99 percent of the living space on the planet by volume.
- The ocean contains nearly 200,000 identified species, but actual numbers may lie in the millions.
- As much as 40 percent of the ocean is heavily affected by pollution, depleted fisheries, loss of coastal habitats and other human activities.
- The ocean absorbs about 30 percent of carbon dioxide produced by humans, buffering the impacts of global warming.
- More than 3 billion people depend on marine and coastal biodiversity for their livelihoods.
- The market value of marine and coastal resources and industries is estimated at US$3 trillion per year, about 5 percent of global GDP.

Goal 15: Life on Land

"protect, restore and promote sustainable use of terrestrial ecosystems. Sustainably manage forests, combat desertification and halt and reverse land degradation and halt biodiversity loss"

Human life depends on the earth as much as the ocean for our sustenance and livelihoods. Plant life provides 80 percent of the human diet, and we rely on agriculture as an important economic resources. Forests cover 30 percent of the Earth's surface, provide vital habitats for millions of species, and important sources for clean air and water, as well as being crucial for combating climate change. Every year, 13 million hectares of forests are lost, while the persistent degradation of drylands has led to the desertification of 3.6 billion hectares, disproportionately affecting poor communities.

While 15 percent of land is protected, biodiversity is still at risk. Nearly 7,000 species of animals and plants have been illegally traded. Wildlife trafficking not only erodes biodiversity, but creates insecurity, fuels conflict, and feeds corruption (Timko, Joleen, 2018). Urgent action must be taken to reduce the loss of natural habitats and biodiversity which are part of our common heritage and support global food and water security, climate change mitigation and adaptation, and peace and security.

Facts and Figures

- Around 1.6 billion people depend on forests for their livelihoods.
- Forests are home to more than 80 percent of all terrestrial species of animals, plants and insects.

- 2.6 billion people depend directly on agriculture for a living.
- Nature-based climate solutions can contribute about a third of CO2 reductions by 2030.
- The value of ecosystems to human livelihoods and well-being is $US125 trillion per year.
- Mountain regions provide 60-80 percent of the Earth's fresh water.

Goal 16: Peace Justice and Strong Institution

"promote peaceful and inclusive societies for sustainable development, provide access to justice for all and build effective, accountable and inclusive institutions at all levels"

We cannot hope for sustainable development without peace, stability, human rights and effective governance, based on the rule of law. Yet our world is increasingly divided. Some regions enjoy peace, security and prosperity, while others fall into seemingly endless cycles of conflict and violence. This is not inevitable and must be addressed. Armed violence and insecurity have a destructive impact on a country's development, affecting economic growth, and often resulting in grievances that last for generations. Sexual violence, crime, exploitation and torture are also prevalent where there is conflict, or no rule of law, and countries must take measures to protect those who are most at risk (SDG 16, 2017)

The SDGs aim to significantly reduce all forms of violence, and work with governments and communities to end conflict and insecurity. Promoting the rule of law and human rights are key to this process, as is reducing the flow of illicit arms and strengthening the participation of developing countries in the institutions of global governance.

Facts and Figures

- By the end of 2017, 68.5 million people had been forcibly displaced as a result of persecution, conflict, violence or human rights violations.
- There are at least 10 million stateless people who have been denied nationality and its related rights.
- Corruption, bribery, theft and tax evasion cost developing countries US$1.26 trillion per year.
- 49 countries lack laws protecting women from domestic violence.
- In 46 countries, women now hold more than 30 percent of seats in at least one chamber of national parliament.
- 1 billion people are legally 'invisible' because they cannot prove who they are. This includes an estimated 625 million children under 14 whose births were never registered.

Goal 17: Partnerships for the Goals

"strengthen the means of implementation and revitalise the global partnership for sustainable development"

The SDGs can only be realized with strong global partnerships and cooperation. Official Development Assistance remained steady but below target, at US$147 billion in 2017. While humanitarian crises brought on by conflict or natural disasters continue to demand more financial resources and aid. Many countries also require Official Development Assistance to encourage growth and trade.

The world is more interconnected than ever. Improving access to technology and knowledge is an important way to share ideas and foster innovation. Coordinating policies to help developing countries manage their debt, as well as promoting investment for the least developed, is vital for sustainable growth and development. The goals aim to enhance North-South and South-South cooperation by supporting national plans to achieve all the targets (Sustainable Development Goals,2017). Promoting international trade, and helping developing countries increase their exports is all part of achieving a universal rules-based and equitable trading system that is fair and open and benefits all.

Facts and Figures

- The UN Conference on Trade and Development (UNCTAD) says achieving SDGs will require US$5 trillion to $7 trillion in annual investment.
- Total official development assistance reached US$147.2 billion in 2017.
- In 2017, international remittances totalled US$613 billion; 76 percent of it went to developing countries.
- In 2016, 6 countries met the international target to keep official development assistance at or above 0.7 percent of gross national income.
- Sustainable and responsible investments represent high-potential sources of capital for SDGs. As of 2016, US$18.2 trillion was invested in this asset class.
- The bond market for sustainable business is growing. In 2018 global green bonds reached US$155.5billion, up 78 percent from previous year.

DIMENSIONS OF SUSTAINABLE DEVELOPMENT

The discussion so far points out that the term sustainability is a process not an endpoint that can be defined or imagined. Sustainability is the ability to sustain things and processes which can in turn sustain. This is deliberately circular and open ended because, that which needs to be sustained is neither fixed nor constant. Also sustainability is not confined to any single discipline. It is, in fact, multi-dimensional. Sustainable development is a concept with social, ecologic, economic, spatial and cultural dimensions. For the Food and Agriculture Organization (FAO) sustainable development consists of the following four dimensions

The United Nations Industrial Development Organization (UNIDO) has identified three main dimensions of Sustainable Development. These two approaches to identify dimensions of sustainable development differ on the ground that whereas FAO approach kept people as an independent dimension of the sustainable development, UNIDO and other approaches consider people at the center of the concept sustainable development. People according to the later approach are inclusive and the centre of the concept. According to them sustainable development has following three dimensions:

Environmental Dimension of Sustainability

The environmental dimension is concerned with the impact of development on living and non-living natural systems, including ecosystems. The Environmental dimension refers to the management of natural resources with a view to ensure that they are available in the future. However, it also includes issues such as the protection of landscapes, habitat, biodiversity, and the quality of drinking water

Figure 1. Dimension of sustainable development

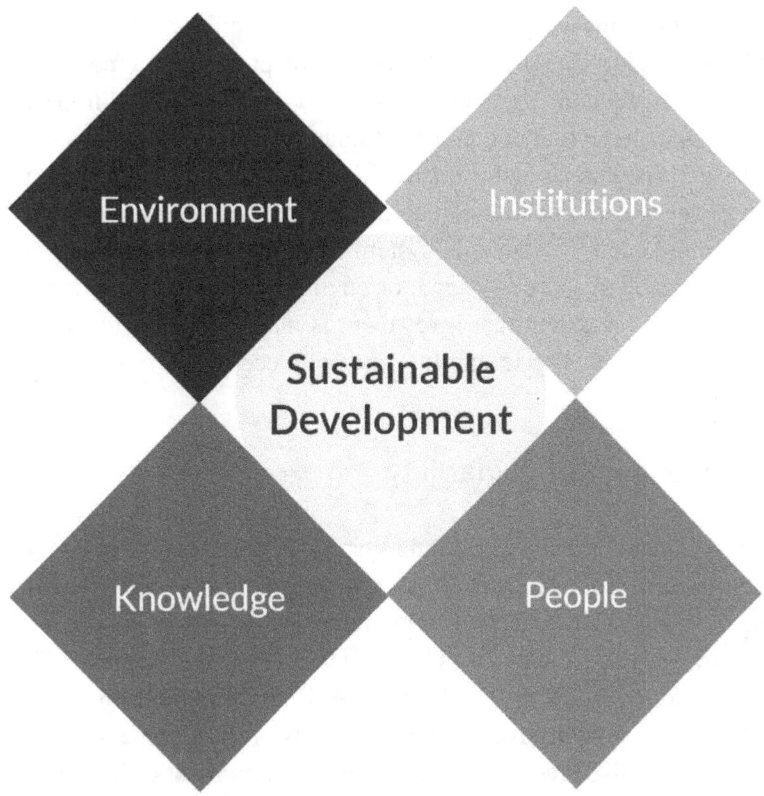

and air. Brundtland (1987) in her famous report argued: "The environment does not exist as a sphere separate from human actions, ambitions, and needs and attempts to defend it in isolation from human concerns have given the very word "environment" a connotation of naivety in some political circles. The word "development" has also been narrowed by some into a very limited focus, along the lines of "what poor nations should do to become richer," and thus again is automatically dismissed by many in the international arena as being a concern of specialists, of those involved in questions of "development assistance." Yet the "environment" is where we live; and "development" is what we all do in attempting to improve our lot within that abode, the two are inseparable. Goodland and Daly (1996) explained environmental sustainability as:

1. **Output Rule:** Waste emissions from a project or action being considered should be kept within the assimilative capacity of the local environment without unacceptable degradation of its future waste absorption capacity or other important services.
2. **Input Rules:**These include:
    a. **Renewable:** Harvest rates of renewable resource inputs must be kept within regenerative capacities of the system that generates them.
    b. **Non-renewable:** Depletion rates of non-renewable resource inputs should be set below the rate at which renewable substitutes are developed by human invention and investment.

Environmental sustainability is essential for the long-term expansion and achievement of Sustainable Development. The natural-environmental constraint to (human) development is the main reason for any concern with sustainability. Agenda 21 identified protection of earth's atmosphere, integrated management of soil, fight against desertification and drought, protection of mountain areas as vulnerable ecosystems, support of a sustainable agriculture and rural development, conservation of biodiversity, protection of ocean and sea, protection of quality and quantity of fresh water resources, environmentally friendly dealing with: toxic chemicals, solid waste and nuclear waste, environmentally friendly disposal of waste, as the categories of ecological sustainability.

Therefore, an environmentally sustainable system must maintain a stable resource base, avoiding over-exploitation of renewable resource systems or environmental sink functions, and depleting non-renewable resources only to the extent that investment is made in adequate substitutes. This includes maintenance of biodiversity, atmospheric stability, and other ecosystem functions not ordinarily classed as economic resources.

Economic Dimensions of Sustainability

The second pillar of sustainability is the economic system. In order to guarantee a healthy environment and stable society, economic systems have to be not only taken into account but also to be seen as a critical tool for reaching the goals of sustainable development. The growth of economies and their structural transformation have always been recognized as being at the core of development. They are still the most important preconditions for the fulfilment of human needs and for any lasting improvements in living conditions. In addition to the quantitative economic aspects of development, an increasing number of qualitative aspects have come to be recognized too. The main argument is that neither economic growth in the aggregate nor growth of income at the personal level is sufficient to guarantee progress of an entire society. Accompanying qualitative changes are needed as well. An economically sustainable system must be able to produce goods and services on a continuing basis, to maintain manageable levels of government and external debt, and to avoid extreme sectoral imbalances which damage agricultural or industrial production (Harris, 2000).

The economic dimension relates to the efficient use of resources, the competitiveness and the viability of all sectors as well as its contributions to the viability of rural areas. The economic dimension looks at the economic performance of development by considering intangible assets such as human capital and how developmental activities can potentially change the economic status of its many stakeholders, through policy change, capacity building and hiring of employees and contractors. Efficient agricultural structures, appropriate technologies as well as the diversification of income sources for farm households are important elements of this dimension (European Commission, 2001). Efficiency of resource use as an important basis for the viability of rural areas cannot be emphasized enough. Concept of economic sustainability is based on possibilities for combining the conflicting notions of a limited natural environment and resources on the one hand and current national economies aimed at unlimited global mass production and mass consumption on the other (Hediger, 1999). Two positions concerning the substitution of different forms of capital are also currently part of the economic discussion:

1. Supporters of a "strong sustainability" defend the position that natural capital resources such as air, land, etc. - cannot be substituted by other forms of capital.

2. Supporters of "weak sustainability" believe that a certain amount of substitution of natural capital is acceptable.

Social Dimensions of Sustainability

Within the sustainability context, 'social sustainability' is the most controversial, and also the youngest; having existed in international politics only since 1992. This component of sustainability has long been neglected but is increasingly making its way to the centre of the international political arena, as was evident at the 2002 World Summit on Sustainable Development (WSSD) in Johannesburg. There is general agreement that the social dimension is one of the three integral pillars of the concept of Sustainable Development (Martin, 2001). However, the concrete definition of this social dimension is less clear-cut. Some interpret it narrowly to mean the interface between environmental and social policies. Yet the social dimension of Sustainable Development is much broader than just the interface between environmental and social policies. According to Martin (2001), it should include other factors that make social policies, and thereby overall economic and societal development, sustainable.

There are different definitions of this broad concept in literature. One conception of socially sustainable development suggests that resources should be used in a way that each succeeding generation can make their own decisions about how best to satisfy their needs. Another conception stresses good social conditions as a goal in its own right, not subordinated to economic growth. In this definition, the challenge for policy makers would be to find win-win policies, i.e., policies that result in social, environmental and economic progress at the same time. On the whole a socially sustainable system must achieve distributional equity, adequate provision of social services including health and education, gender equity, and political accountability and participation (Harris, 2000).Therefore development can be considered as sustainable when there is continuous improvement in living standards, particularly those of the poor and of disadvantaged groups, while at the same time keeping a balance between three Social, Environmental and Economic dimensions of sustainability (Bond and Klonsky, 2006).

CONCLUSION

The Sustainable Development Goals (SDGs), also known as the Global Goals, were adopted by all United Nations Member States in 2015 as a universal call to action to end poverty, protect the planet and ensure that all people enjoy peace and prosperity by 2030. The 17 SDGs are integrated-that is, they recognize that action in one area will affect outcomes in others that development must balance social, economic and environmental sustainability. Through the pledge to Leave No One Behind, countries have committed to fast-track progress for those furthest behind first. That is why the SDGs are designed to bring the world to several life-changing 'zeros', including zero poverty, hunger, AIDS and discrimination against women and girls. Everyone is needed to reach these ambitious targets. The creativity, knowhow, technology and financial resources from all of society is necessary to achieve the SDGs in every context.

REFERENCES

Ansuategi, A., Greño, P., & Houlden, V. (2015). *The impact of climate change on the achievement of the post-2015 sustainable development goals*. CDKN & HR Wallingford.

Bâc, P. D. (2008). A history of the concept of sustainable development: literature review. *Annals of the University of Oradea, Economic Science Series, 17*(2), 576-580. Retrieved November 20, 2015, from http://www.cfr.washington.edu/classes.esrm.458/Paul.%202008.pdf

Beckerman, W. (1992). Economic growth and the environment: Whose growth? Whose environment? *World Development, 20*(4), 481–496. doi:10.1016/0305-750X(92)90038-W

Bond, C. A., & Klonsky, K. (2006). Ecological and economic indicators for sustainability. *SAFS Newsletters, 6*(3). http://safs.ucdavis.edu/newsletter/v06n3/page1.htm

Covert, J. (2017). Planning for the Implementation of SDG-14. *Environmental Policy and Law, 47*(1), 6–8. doi:10.3233/EPL-170003

Črnjar, M. & Črnjar, K. (2009). *Menadžment održivog razvoja*. Opatija: Fakultet za turistički i hotelski menadžment Opatija, Sveučilište u Rijeci.

European Commission. (2001). *A Framework for Indicators for the Economic and Social Dimensions of Sustainable Agriculture and Rural Development*. The Agriculture Directorate-General of the European Union. Available online https://ec.europa.eu/agriculture/publi/reports/sustain/index_en.pdf

Fan, S., & Polman, P. (2014). *An ambitious development goal: Ending hunger and undernutrition by 2025*. Academic Press.

Firzli, M., & Nicolas, J. (2016). *Beyond SDGs: Can Fiduciary Capitalism and Bolder, Better Boards Jumpstart Economic Growth?* Analyse Financière.

Firzli, N. (2018). Greening, Governance and Growth in the Age of Popular Empowerment. *Financial Times*.

Goal 1: No poverty. (2017). *UNDP*.

Goal 12: Responsible consumption, production. (2018). *UNDP*.

Goal 3: Good health and well-being. (2017). *UNDP*.

Goal 7: Affordable and clean energy. (2015). UNDP.

Goodland, R., & Daly, H. (1996). *Environmental Sustainability: Universal and Non-negotiable*. Academic Press.

Harris, J. M. (2000). Basic Principles of Sustainable Development. A Survey of Sustainable Development: Social and Economic Dimensions. Washington, DC: Island Press.

Hediger, W. (1999). Reconciling weak and strong sustainability. *International Journal of Social Economics, 26*(7-9), 1120–1143. doi:10.1108/03068299910245859

Jenkins, W. (2009). *Berkshire encyclopaedia of sustainability: the spirit of sustainability* (1st ed., Vol. 1). Berkshire Publishing Group.

Kellogg, D. M. (2017). *The Global Sanitation Crisis: A Role for Business. In Beyond the bottom line: integrating sustainability into business and management practice.* Greenleaf Publishing.

Kinyoki, D. (2020). Mapping child growth failure across low- and middle-income countries. *Nature*, *577*(7789), 231–234. doi:10.103841586-019-1878-8 PMID:31915393

Lele, S. M. (1991). Sustainable development: A Critical Review. *World Development*, *19*(6), 607–621. doi:10.1016/0305-750X(91)90197-P

Local Burden of Disease Educational Attainment Collaborators. (2020). Mapping disparities in education across low- and middle-income countries. *Nature*, *577*(7789), 235–238. doi:10.103841586-019-1872-1 PMID:31875853

Progress of Goal 10 in 2017. (2017). *Sustainable Development Goal Knowledge Platform.*

Progress of Goal 9 in 2017. (2017). *Sustainable Development Knowledge Platform.*

Remenyi, J. (2004). What is Development? In D. Kingsbury, J. Remenyi, J. McKay, & J. Hunt (Eds.), *Key Issues in Development* (pp. 22–44). Palgrave Macmillan.

Sachs, W. (2010). Environment. In W. Sachs (Ed.), *The Development Dictionary: A guide to knowledge as power* (2nd ed., pp. 24–37). Zed Books.

SDG16 Data Initiative 2017 Global Report. (2017). *SDG16 Report.*

Sharpley, R. (2000). Tourism and Sustainable Development: Exploring the Theoretical Divide. *Journal of Susta Hediger inable. Tourism (Zagreb)*, *8*(1), 1–19. doi:10.1080/09669580008667346

Šimleša, D. (2003). Podržava li biznis održivi razvoj? *Journal for General Social Issues*, *12*(3-4), 403426. Retrieved September 21, 2015, from https://hrcak.srce.hr/19493

Staples, D., & Hermes, R. (2012). Marine biodiversity and resource management – what is the link? *Aquatic Ecosystem Health & Management*, *15*(3), 245–252. doi:10.1080/14634988.2012.709429

Sustainable Development Goal 11. (2017). Sustainable Development Goals.

Sustainable Development Goal 17. (2017). *Sustainable Development Goals.*

Timko, J., Le Billon, P., Zerriffi, H., Honey-Rosés, J., de la Roche, I., Gaston, C., Sunderland, T. C. H., & Kozak, R. A. (2018). A policy nexus approach to forests and the SDGs: Tradeoffs and synergies. *Current Opinion in Environmental Sustainability*, *34*, 7–12. doi:10.1016/j.cosust.2018.06.004

Tomislav, K. (2018). The Concept of Sustainable Development: From its Beginning to the Contemporary Issues. *Zagreb International Review of Economics & Business*, *21*(1), 67–94. doi:10.2478/zireb-2018-0005

United Nations: Gender equality and women's empowerment. (2018). United Nations Sustainable Development.

WHO & UNICEF. (2017). Progress on Drinking Water, Sanitation and Hygiene: 2017 Update and SDG Baselines. Geneva: World Health Organization (WHO) and the United Nations Children's Fund (UNICEF).

Willis, K. (2005). *Theories and Practices of Development*. Routledge.

ADDITIONAL READING

Hanson, K. T., Puplampu, K. P., & Shaw, T. M. (Eds.). (2017). *From Millennium Development Goals to Sustainable Development Goals*. doi:10.4324/9781315228068

Idowu, S. O., Schmidpeter, R., & Zu, L. (Eds.). (2020). *The Future of the UN Sustainable Development Goals*. Samuel O. doi:10.1007/978-3-030-21154-7

Servaes, J. (Ed.). (2017). *Sustainable Development Goals in the Asian Context*. doi:10.1007/978-981-10-2815-1

Chapter 3
Green Economic Growth Based on Urban Ecology and Biodiversity

José G. Vargas-Hernández
https://orcid.org/0000-0003-0938-4197
University Center for Economic and Managerial Sciences, University of Guadalajara, Mexico

ABSTRACT

This chapter has the objective to analyze the urban ecology, the biological diversity or biodiversity, and their adaptive cycle as the fundamentals of green economic growth. The analysis begins questioning the implications that some assumptions of urban ecology and biodiversity, such as the socio-ecosystems, resilience, ecosystem services, and adaptive cycle have on the creation of green economic growth. A series of different dimensions of resilience are proposed as subsystems that contribute to the general resilience of a system. The method used is the analytical based on a review of the conceptual and theoretical literature. This analysis concludes that the connectivity of processes and functions of urban ecology and biodiversity are relevant to the creation of green economic growth in terms of green economic value.

INTRODUCTION

Cities are complex ecological systems dominated by humans (Alberti, 2006). Although human manipulations of nature make cities foundamentally different from other types of natural ecosystems, the city can be considered as an ecosystem open to entry of energy, for consumption and waste generation. For this reason, we can speak about Urban Metabolism, similar to what we do for living organisms (Kennedy, 2011).

Expanding the ecological knowledge of urban regions could facilitate greater integration of nature, representing an investment opportunity and at the same time improving the quality of life of the population. High urban biodiversity can be able to provide many services, including cooling the urban area, reducing urban flood risk, filtering pollutants, supplying food, and providing accessible recreation. To

DOI: 10.4018/978-1-7998-4915-5.ch003

understand and manage the complexity of nature in cities requires knowledge of the dynamics of both ecosystem and social systems (Douglas, 2011, May, 1974).

Although an urban development without control or planning may be the reason for a greater ecological crisis, the phenomenon of demographic concentration may represent a positive element, which offers the possibility of designing lifestyles in compliance with environmental sustainability. As we are about to see, to achieve this goal it is necessary to boost green economic growth, based on the increase of resilience, redundancy and incorporation of adaptive cycles at the urban level. One of the first aspects to consider is the respect or regeneration of ecological nichos, able to allow local species to develop in harmony with human activities.

To accomplish this, there are several ecological parameters to consider in city planning, such as the Biotope Area Factor used in Berlin since 1990, which represents the division between the actually ecological areas for the total area. These green areas should also include the presence of ecological corridors to allow species to move according to their needs.

Building resilient cities means being able to combine the various social, infrastructural and environmental dimensions of resilience. The availability of environmental services in terms of biodiversity present in the urban context, allows the increase of redundacy and therefore of the environmental resilience index, taking advantage of the principle of adaptability of cycles. But all this would not be possible with the simple planning of land use and space in the urban context. It is equally important that this relies on social resilience, that is, on the distinct individual, group, organizational, community and pshycological dimensions.

In the first part of this work, a series of different dimensions of resilience are proposed as subsystems that contribute to the general resilience of a system. Understanding the different dimensions of resilience can help in creating innovative mechanisms for achieving resilience.

In the second part we analyze the relationships between resilience, biodiversity and adaptive cycles. Finally, it comes to establishing a theoretical contribution useful for accompanying the various transition processes towards a green economic growth in agreement with adaptive practices of urban planning and with the business sector.

URBAN ECOLOGY

The evolution sciences and the ecology of large urban systems assumes elasticity and springing back and do not assume deterministic trajectories and endpoints of ecological systems. Social and ecological systems are coupled and interdependent. Review of urban ecology focusing on land-use ecological combinations suggested by ecological premises to promote biodiversity, are based on the synergistically interactions of constituent land uses supporting biodiversity when clusteres and intercepted in an urban matrix (Calkins, 2005). In ecology, meta is a regional landscape community dynamics of population in scattered and patchy locations. An urbanmeta-mosaic comprises landscapes describing matter, organisms, energy and information flows, landscapes of social, evolved or political choices; and the spatial outcomes of the choices and flows.

The ecological value can be increased in high density urban contexts through the strategic implementation at different scales of green spaces interventions and solutions. Ecological indexes such as the Biotope Area Factor employed in Berlin, measured the ecological value as a percentage of permeability and evapo-transporation of the land surface of the urban green spaces including natural and artificial elements (International Energy Agency 2010). Some urban functions are more efficient with the other have a low BAF performance. The lower the urban fabric density the better the potential performance of BAF and the lower urban fabric coverage ratio, the higher the potential BAF.

The concept of resilience in urban ecology has different meanings; it is defined as the capacity of an ecosystem to absorb disturbances and natural disasters, reorganizing the undergoing changes to retain essentially the same function, structure, identity and feedbacks (Berkes et al., 2003; Carpenter and Folke, 2006; Holling, 1973). Resilience is an autonomous adaptation that responds as conditions change (Center for Clean Air Policy, 2011). Slow losses of resilience lead to large changes of the ecosystem when it crosses a threshold subjected to a random event as may be the climate change (Folke et al. 2005, Groffman et al. 2006).

Natural systems recover rapidly (Conway-Cranos, 2012). The strategic building in natural events may change the original levels of the resilience systems leading that the resilience index provides wrong information withouth connecting to a monitoring system of social, psychological, physical, structural and environmental parameters. An expert system is managed by automatic control systems whyle a non expert system is directed to the community resilience to indicate critical conditions. The resilience of a system is present when there is a coping strategy at a particular point in time.

These ecological solutions can be implemented through local private or public funded green strategies such as a municipal ecological networks (Peraboni, 2010). The different types of land-use have an ecological complementation involving different urban green spaces and areas which contribute to sustainable ecosystem services and build resilience in urban ecosystems (MA, 2005). Also, different types of resilience may have different outcomes on different types of urban land-use. Intentional resilience is related to planning ahead using common sense to maximize synergies. Accidental resilience considers that many solutions are not necessarily driven by climate concerns (Clay Nesler Efficient and Resilient Buildings: Vice President, Global Energy and Sustainability, Johnson Controls).

Recent insights on resilience and vulnerability have been developed in the context of socio-ecosystems based on complex systems and uncertainty theories. Distributed socio-ecosystems composed of independent and interactive elements deliver equivalent and better functionality with greater resilience. Resilience is the ability of the system to cope with esternal factors undermining with a system bouncing back. Resilience is defined as the capacity of complex and uncertain systems to cope with internal and external stresses through the implementation of strategies of adaptation and mutation processes and to return or achieve an equilibrium state. The system resilience becomes relevant when the equilibrium state is shifted after disruptions, disasters, pertubations and discontinuities occur.

Nature is resilient but receive impacts from the complex socio ecosystem design. Open systems have some properties that enable them the capacity to manage perturbations and disasters and to evolve and recover. Complex and hierarchically organized systems tend to be more resistant to stress within narrow boundaries, and is vulnerable to small and unforeseen perturbations.

Monitoring and assessments are essential to protect urban natural spaces and resilience. The elements of a complex urban system resilience assessment in a territorial application include the evaluation of all the interacting elements of all the systems in the method defined in the Resilience Index. This resilience index combines the subindexes of social, infrastructural and environmental resilience. The resilience coefficient is critical evolving in conection with both endogenous and exogenous factors of the system. This could be predicted the synthetic index at different levels contributing to the improvement of resilience.

Resilience as the ability to resist disorder characterized by thermodynamic changes, is considered the essence of sustainability. Resilient designing systems support sustainability. Technological improvements in socio-ecosystem design improve the human well-being by enhancing interactions between the nature, the ecosystem services and society, but requires to remove the adverse effects on the biosphere and reduce their adverse impacts and unintended consequiences. The incremental costs of design systems for new construction may be minimal. Environmental issues enhance sustainability and resilience as being global of transboundary (Collins and Kearins, 2010).

Resilience is fostered by maintaining a sustainable supply of ecosystems services to face the threats posed by climate change. Urban ecosystem services cycle and store carbon for regulating climate, help to baffer against disasters, disruptions and disturbances. Urban ecosystem services benefit population offering opportunities for recreation and education besides clean air and water, flood control, etc. Urban vegetation reduces air and noise pollution significantly.

Ontology can provide a frame of reference for resilience (Kuhn, 2003). The concept of resilience is dealing with environmental changes, disasters and disturbances caused by climate change which brings the point to develop vulnerability assessments and information management systems. Ecological resilience is a relevant concept in urban sciences of environment, sustainability and governance, defined as complex system´s ability to cope with external stresses and return to an equilibrium state, using adaptation and mutation strategies (De Lotto, Esopi & Sturla, 2017). Poor governance leads to disfunctional behavior and decline of system conditions. Ecosystem ongoing stress are more vulnerable to change from disasters and disturbances (Gunderson and Holling, 2002; Walker and Salt, 2006). Eco system components allowing to propagate shocks are more vulnerable to systemic risks.

The socio-ecological system has quality inherent resilience. Resilience as a quality is an entity perceived and masured (Masolo et al., 2003). The qualities of resilience are considered essential to enable to take action to prevent the socio ecosystem breakdown. Social - ecological resilience, is the capacity of cross-scale interactions of complex system to respond and absorb disturbances, self-organize, learn and adapt. Resilience operationalizations is irregular because the different definitions. The debate on the meaning of resilience and its operationalization is the result of multiple approaches from ecology, sustianble sciences, climate change, disaster management, etc., in such a wat that the term resilience contains several ambiguities that are addressed ro posit a formal theory of resilience by Daniel (2011) and Daniel and Ortmann (2011) based on ontology engineering principles.

Resilience applied to engineered system design considers it is a process of hierarchical decomposition that have some properties to make it more or less resilient. Resilient engineering system manifests itself as diminishing failure probabilities (O´Rourke 2007). The resilience concepts have been already adopted for academics and practitioners of risk management.

Resilient ecosystems cope with the risks giving confidence and trust to management involved to solve the ecological situations. Green resilience can be considered as the climate adaptation and mitigation synergies. Resilient functional urban ecosystems are essential to provide ecosystem services and goods to benefit environment, structural and socio-economic systems. A resilient urban ecosystem pursue a balanced urban sustainable development model based on social participation planning and management in integration of economic, social and environmental issues, preservation and enhancement of local resources and reduction of environmental impacts.

Ecological land-use complementation theoretical framework draws on the island biogeography theory (MacArthur and Wilson, 1967). Ecological land-use complementation (ELC) is a theoretical framework that merges ecology and the landscape complementation/supplementation concepts developed by Dunning et al. (1992). The ELC approach is used for land management promoting the integration of qualitative attributes of species in biodiversity and urban residents. In heterogeneous patch types such as urban landscapes, the species need to move to obtain the complemented resources to fulfill different life cycles (Pope et al., 2000). A missing process of the landscape or seascape can be reestablished in other dispersed patch withing the range. Dispersal of species in heterogeneous landscapes confers resilience to disturbances (Peterson et al. 1998, Nyström and Folke 2001, Loreau et al. 2003, Cardinale et al. 2004).

Some mechanisms to achieve resilience are sense-making, organization structures, mistakeorientation, structural flexibility and redundancy. Redundancy of species decrese where are sort into niches (Reich et al., 2012), and are critical for functioning under particular environmental conditions (Isbell et al., 2011). Ecosystem services resilience is dependent upon the sets of species existing in an ecosystem and performing similar ecosystem processes (Walker et al. 1999, Elmqvist et al. 2003). Species-level ecological data is required about function and response (Cumming and Child, 2009). In biological ecology, the shift in life history from r-selected features of rapid growth, high investment and disperse nature of offspring to K-selected features on slow growth, greater investment and local dispersal (Gunderson & Holling, 2002).

The socio-ecological system can be characterized as a single equilibrium and as a multiple equilibria (Holling, 1996). A specific socio-ecological system is an instance of a class socio-ecological system. The socio-ecological system is a sub-class of ecological system. The concept of socio-ecological system is a class that exists within resilience and is a type of an ecological system. Thus, the socio-ecological system as a class is a type of the relation with the ecology system or class.

In Ecology, the equilibrium state of an ecosystem is defined as Climax. The term equilibrium or steady state refers to the state where the socio-ecological system does not change unles it is subject to disturbances. The implicit assumption of stability is that there is only one equilibrium or steady state. In the event of existing other states, some safeguards should be applied. An ecological disturbance occurring at defined time and space scales may allow persistence of species, structures and processes at not affected scales (Elmqvist et al. 2003). Ecological processes replicated across a range of scales confers resilience (Peterson et al. 1998).

Systems design poses among the new technical challenges to achieve robustness through resilience, adaptability and diversity are the new concepts for task management, system state indicators are based on energy attributes. Adaptability requires the system change in the face of disturbances to retain essential properties (Folke et al., 2004; Walker et al., 2004).

Ecological resilience is supported by the capacities of the ecological system to absorb disturbance and reorganize while changing to retain essentially the same functions, structures, identities and feedbacks (Walker et al, 2004). Processes and feedbacks and later returning to its previous disturbance state (Walker & Salt, 2006). Constructive positive feedback loops help to identify the vulnerabilities in a crisis and the strengths of the community (IFRC 2004). Qualitative feedback from quantitative ecosystem services and viceversa, as well as human well-being outcomes can be used to evaluate assumptions about indirect drivers (MA 2005a). In socio ecosystems in existing conditions provide negative feedbacks. Ecological resilience is linked to environmental sustainability proving that the booming cities around the world present greater risks for the future (Isenhour, 2011).

The resilience of the sustainability of any ecological system is influenced by sustainable social determinants. Ecological resilience can not be separated from social systems as both society and environment are coupled and interacting in non-linear manner. The concept of social resilience is relevant to understand other forms of resilience that can be individual, group, organizational and community resilience, including the psychological resilience at all levels. There are some developed tools to achieve psychological resilience as a form of vital and authentic life (Wagnild, 2010) and recovering from stress by increasing resistance (Bonanno et al, 2006; Fergus and Zimmerman, 2005).

Community resilience is a relevant aspect of social resilience, although it is usual to derive the notion from the pshychological resilience. Community resilience is defined as as the capacity to anticipate, minimize and absorb stresses and destructive forces through adaptation and resistance; as well as to manage and maintain basic functions and structures during disastrous events, and recover or 'bounce back' after an event (Twigg, 2009). The factors influencing community resilience are emerging cohesion, connectedness, community empowerment, collective efficacy, social justice, assessment, etc.

Institutional and organizational resilience is the recovery after disturbances and disruptions emerging from unespected harmful events, risks, stress and strain (Vogus and Sutcliffe, 2007). Organizational resilience is the capacity to unfold, respond and recover from unexpected events and external disturbances absorbing the extreme changes and impacts (Linnenluecke et al. 2012). Building organizational resilience aims to adjustments under challenging conditions from where the organization emerges strengthened and more resourceful (Vogus and Sutcliffe, 2007).

A resilient ecological system absorbs exogenous shocks and disturbances without changing its processes. Loosing resilience results in changes even from the small-scale disturbances (Witten et al, 2011).

Economic, social and ecological decline may be the result of disfunctional or lacking governance and leads downward to environmental conditions, welfare and human well-being (Folke et al., 2010). External disturbances resulting from storms, floods, etc., have structuring effects (Johnson & Miyanishi, 2007) and random internal effects pressure and drive change and instability. Food supply is experiencing ecological instability, reduction in biological diversity and agroecological unputs, climate change, biosecurity risks, transgenic drifts and crops genetically modified, risks of transboundary diseases, increasing of prices, etc.

The renewal capacity in dynamic environments generates an ecological protection against management failures of the system and allowing the managers to learn and change. These changes pose a major constraint to economic development, ecological needs, health and livelihoods. Healthy socio-ecological system increases the economic efficiency and reurns from supporting activities more focused on wealth creation. Restoring degradation of urban ecosystem resilience reduces vulnerability to natural disturbances and disasters, enhance the ecological and social networks and improves the quality of life.

Eco-innovative initiatives target at clean and renewable technologies to reduce impacts on the socio ecological system aimed to enhance resilience to environmental disturbances and to contribute to sustainable development. Eco-innovations on development of resilient cities, eco-innovative solutions are related to waste management, drinking fresh water, sustainable building materials and insulation (Eco-innovation action plan, 2011). Evolving policies fostering eco-innovations that boost efficiency, resource productivity and competitiveness safeguarding the environment (Eco-innovation action plan, 2011).

Resilience is an ecological concept to describe sustainability as an inherent property of the system. The fundamental properties for sustainable design of resilience systems in both engineered and larger systems are the diversity, adaptability, cohesion and ecological efficience. The resilient socio-ecological system functions on a broad spectrum of possible states created by disruptions, disasters and perturbations, gradually tending through evolution and adaptation to return to an equilibrium states. Under certain conditions, the resilient socio-ecosystem may shift to a different new equilibrium state with new structural and functional changes.

In sustainable urban planning, ecological policies for green interventions are implemented to improve urban ecosystems services quality, the environmental sustainability and to instrument the measure of the ecological value through the use of ecological indexes. Resilience can be measured at different scales: regional, city, infrastructure, neighboring, building. Environmental commitments recommended by international organizations such as the World Bank Group are being attended and to certain extent are being fulfilled. Adaptive ecological planning, management and governance assume that the ecological processes can be changed framed by the uncertainty (Peterson 2005).

Urban planning and design principles guide the land-use configurations to support the ecosystem processes and urban resilience. Protection and enhancement of urban ecosystems provide opportunities to increase urban resilience and promote public health and education by addressing climate change. The benefits of urban ecosystems can cantibute to build urban resilience by maximizing the qualities of robustnessresourcefulness and redundancy. Urban resilience enables to learn the best practices and opportunities to achieve synergies by mitigation cutting carbon pollution abd responde with adaptation to climate impacts.

The qualities of urban resilience are to be reflective learning and leveraging from past experiences; robustness of physical assets; redundancy, flexibility, resourcefulness, inclusiveness fir briad consultation, integrated processes aligning systems for consistency and enable them ri function and responde collectively. Urban spatial planning and design must have in the agenda the reliance building to avoid local extintion of species and eradication of native fauna and flora (McKinney, 2002). Manipulation of species composition to change the provisioning of ecosystem services may reduce the resilience of regulating services, increase the effects of extreme events and decrease future supply of ecosystem services.

The MA Scenarios is a step further the global ecological scenario analyses of ecosystem services and their ecological implications (Cumming et al. 2005, Raskin 2005). Maintenance of regulating services insures the adaptability against severe ecological changes and the human action that can decrease regulating services while increasing the ecosystems vulnerability. People tend to value resource provision and cultural values, but tend to undervalue regulating services, ecosystem processes and supporting ecosystem services. Ecosystem components that generate regulating services may undermine the supply of ecosystem services. The capacity of socio–ecological systems to cope with, and adapt to, varios types of disturbances and dissasters is related to regulating ecosystem services.

The possibilities for ecosystem services trajectories are dependent on access to data assimilated into the models to be assessed and projected. One example is the climate changes, one of the drivers of the ecosystem services which models only can be predict only the incremental but not the crucial changes that may transform socio-ecosystems (Smith 2002). This is one of the reasons why ecosystem services are uncertain, ambiguous, complex and difficult to predict (Walley 1991) and the range of possibilities are unknown.

The Millennium Ecosystem Assessment (MA) (MA 2003) has developed four MA ecological scenarios to address future ecological changes in ecosystems management, in the flows of ecosystem services, the trade-offs faced when making decisions on the different options and in their consequent changes in human well-being. Changes in ecosystem services and their trade-offs with poverty reduction and alleviation offer potential risks, costs and benefits. However, the capacity to substitute ecosystem services can be limited due to the costs or lacking availability of technology (Postel and Carpenter 1997). Nevertheless, technology may shape the trade-offs among ecosystem services (Rodriguez et al. 2006). For example, energy technologies have a potential for ecosystem services.

The MA scenarios incorporate detailed ecological dynamics (Raskin 2005, MA 2005a). The Millennium Ecosystem Assessment (MA) scenarios analyze the complex ecological dynamics implications of ecosystem services in human well-being and their trade-offs conections with poverty reduction and life quality of urban communities. The supply of ecosystem services eroded by poverty reduction makes more difficult the poverty alleviation.

Methodologies to support the MA scenarios had emerged from the interview, such as the analisys of green technology and TechnoGarden, locally based learning and Adapting Mosaic, security and Order from Strength (Cork et al. 2006). A quantitative analysis using models are based on a limited subset of ecosystem services have strong limitations because uncertainties and ambiguities are difficult to quantify if not impossible (MA 2005). An MA scenarios analysis involving both quantitative and qualitative results are cross-checked to find harmony with additional iterative analyses to improve the results. In fact, MA scenarios can provide cross-cutting insights to explain the ecosystem services dynamics.

Management strategies formulation and implementation to support the resilience of socio-ecosystems are crucial for the production of goods and environmental services for social systems. Policies, strategies and interventions should be designed to influence the evolution and behavior of socio-ecological systems and environmental resilience protection to enhance the heath and wellface of the overall socio ecosystem.

Ecological policies can be directed towards several ecological goals such as the reduction of energy consumption and use of renewable energies and towards green interventions such as water treatment, waste management, absorption of polluntants, etc. Material and energy use is resulting in growing emissions, pollution and waste materials. Public engagement of local authorities should develop the structures and develop the best management practices to connect urban population with nature awareness and ecological literacy and enable their implementation.

Design practices of systems resilience has developed many useful techniques. The Resilience Alliance (RA) is a research organization created and developed to join and collaborate with resources and efforts of a transdiciplinary group of practitioners and scientists aimed to explore the dynamics of socio-ecological systems. Members of the RA usually come from social, environmental and ecological sciences using theorethical and empirical frameworks of rigorous testing methodology in natural resource management, scenarios and model development, adaptive management techniques and partcipative approaches, self-organization, optimization and resilience systems, among other methods. The RA develops conceptual, theoretical and practical knowledge on biodiversity, resilience, adaptation and transformation of socio-ecosystems and sustainable development policy making and implementation.

The symbiotic relationships develop between large and small facilities in industrial ecology focusing on ecological efficiency. The methodology of BASF combines the ecological impacts of the life cycle with the economic analysis to find the eco-efficiency. The energy flows in industrial ecological systems develop sustainability indicators based on thermodynamic analysis where useful energy is a comparative assessment method of ecological efficiency in industrial processes.

The essence of resilience of any business company is to generate strategic options making sense of its environments in order to realign its resources taking advantage of its rivals (Hamel, Valikangas, 2003). Resilience has implications for the sustainability of business companies to develop beyond their own boundaries an intricated system for continuous renewal and innovation capable to ensure long term resilience. Within the supply chain of the manufacturer, renewable materials and energy have an impact on the economic and socio-ecology systems. Social responsibility should be an attribute of the company in continuos adaptation to achieve ecological integrity with improved quality of life through sustainable design principles.

Greening programs, community green areas and gardens managed by local authorities, as well as other local government initiatives are a source of civic engagement and community participation, a sense of an identity, transmiting and educating ecological knowledge. Resource users with local knowledge sharing the management and decision making may be more supportive for formulating and implementing strategies and policies. Urban green spaces have positive effects on the attractiveness of urban landscapes based on the interrelation between building features and other green elements (De Lotto, R. &M.L. Di Tolle 2013).

The different functions of land use are inserted in buildings shapes with high coverage ratio, while other functions are more specific for building typologies with lower coverage ratio. Ownership fragmentation of lots or buildings is related to the difficulty in finding agreed solutions.

City resolutions should facilitate urban farming and agriculture indicating to public owned property the suitability of land or the local restrictions for rooftop gardens and greenhouses (Brannen 2011). Any practice modification of urban agriculture and farming must begin with the socio economic boundary conditions (Lansing, 1991). Private and public gardening benefits the urban microclimate and change the gray contexts into green environments and can be incentivated by deducting accommodation and maintenance costs. A collective system intervention is the purpose to reduce the greenhouse emissions. Historic centers in urban contexts require applications to integrate natural and artificial solutions to increase the ecological quality and ecosystem resilience.

Local populations affected by any type of air, water, soil pollutions may be enable to mobilize in defense of the right to better environment and ecological resilience Environmental disputes adjudication synergistically links up democratic and ecological processes. The transformative power of ecological activists, environmental lawyers and the civil society organizations adviced by experts must fight back to deliver ecological justice on environmental outcomes and catalyze resilience though legal actions of regulating and monitoring institutions. The contributions to the resilience system should be assessed in terms of performance outcomes and intrinsic characteristics. The resilience of related systems is accomplished through design protocols.

Socio-ecological feedbacks intensify modifications of socio-ecosystems and changes the flow of ecosystem services received from nature, which in turn, increase the vulnerability of people to further cahnges (Cumming et al. 2005).

BIOLOGICAL DIVERSITY OR BIODIVERSITY

The concepts of biodiversity, redundancy or ecological variability, multiple equilibrium adaptation, hierarchy of interactions between spatial scales and temporal responses are elements that give support to the ecosystem resilience. Biodiversity increases the suitability between species and particular conditions (Chapin et al., 1995; Isbell et al., 2011). Biological diversity retains structural controls and provides functional redundancy when the system is facing disturbance.

Biodiversity conservation and natural resources principles must be included into sustainable governance and resilience policy initiatives. Conservation and enhancement of biodiversity can maintain the flows of ecosystem services and reduce socio ecological vulnerability. Diversification in socio-economic systems is aimed to reduce risk and is accomplished through the biodiversity maintenance and economic diversification.

Resilience is determined by preparation, response and recovery, representing the overall risk management efficiency and effectiveness. Resilience, vulnerability and criticality are all related with the notion of risk integrating causes and consequences and explanations of pre and post events (Philley, 2006; Shahriar et al, 2012). The risk index relates the components of critical infrastructure with the environment in preparation to react in case of disaster or crisis. The resilience index combines instruments of innovation and risk assessment simplifying the large scales areas of components and comparing the resilient capabilities of each component accourding to the different stakeholders and users. Risk monitoring and assessment based on the resilience index serve for warning, information processing and risk mitigation planning.

The resilience capacity to adapt to the ecosystem is based on the biodiversity, redundancy is ecological variability, cycles of adaptation to multiple equilibrium states, hierarchical interaction between spatial scales and temporal responses. Resilience is affected by biodiversity depending on the organization of species, response diversity, time and place scaling of ecosystem processes and spatial patterns. Response diversity is linked to resilience of the ecosystem services through biodiversity, in such a way that any change in species affects.

Other kind of response diversity is the spatial pattern (Elmqvist et al. 2003). A resilient city reacts to environmental dysturbances by protecting natural biodiversity and ecosystems. A resilient city for resilient people is related to a shared decision on developing framework. Building socio–ecological resilience through regulating biodiversity and ecosystem services moderate extreme events.

It is usual that local planning authorities have limited knowledge on biological conservation and biodiversity maintenance and management of natural resources in urban development settings in congruence with other biodiversity approaches (Sandstrom et al., 2006; von Haaren and Reich, 2006). Coordination among state-level agencies and city-level agencies may likely be the reason why planning urban farming and agriculture is related to the quality of the local food system, although it is required to design a strategy for harnessing the supporting resources.

The ecological land-use complementation benefits the biodiversity building from the land uses in urban green areas (Colding, 2007). Knowledge on the interactions between land-use changes and biodiversity is more abundant than the influence of landscape design in landscape functions in some specific contexts (Hobbs, 1993, 1997). Land uses have synergistic effects on supporting processes that are essential for the development of biodiversity. Land uses analysis based on ecological premises support biodiversity and ecosystem services. The ecological premises of ecological land use complementation spatially arranged as they are determinants of biodiversity protection. The ecosystem resilience can be promoted by adopting ecological land-use complementation in new urban areas supported by ecosystem services and respone diversity.

Decisions on supply of different ecosystem services involve trade-offs (Rodriguez et al. 2006) such as clearing the forest land to be used for agriculture improves the supply of food but leads to declines in biodiversity, climate regulation and water purification. Provision of some ecosystem services such as the production of food may reduce the supply of other ecosystem services such as water while removing income sources.

Trade-offs among ecosystem services have a substantial impact on the future, yet they are difficult to anticipate. Proper accounting for the ecosystem uncertainties drives the ecosystem services to low levels of discount rates and increasing the value of future benefits. Overexploitation of ecosystem services undervalue future benefits (Ludwig et al. 2006). An overexploitation of ecosystem services may degrade the regulating of other ecosystem services and decrease the yield of provisioning services, and increases environmental variability and vulnerability. Agrobiodiversity management and conservation of crop genetic resources, wild relatives and traditional seed varieties are a relevant factor for disaster risk reduction and recovery. There are some studies on post-disturbance ecosystem recovery (Connell and Slater, 1977; Odum, 1969).

ADAPTIVE CYCLE

The general theory of adaptive cycles has been developed to apply the concept of resilience to socio management ecosystems under the argument that exhibit similar patterns of increasing connectedness, decreasing resilience, slow accumulation of resources and crisis, transformation and renewal periods. The behiovioral quadrants differ in the degree of connectivity and the amount of capital.

Community resilience encompases disaster management process cycle and does not excludes psychological resilience dimension. adversities and disturbances. Community resilience manifests in the population mental and behavioural health, functioning dimensions and quality of life (Norris et al, 2008). Disaster management integrates the network if adaptive capacities considered as resources with dynamic attributes to achieve adaptation after. Regime in any dynamic system follows a given trajectory unles it is disturbed. Regime shifts can be found in different types of socio-ecosystems (Beisner et al., 2003).

Any scenarios address a set of assumptions concerned with ecosystem services. An analysis using the MA scenarios evaluates the changes of the regime shifts, its consequences and risks on ecosystem services and on human health and well-being. Ecosystem services must be considered in the MA scenarios in the context of global socio-ecosystems changes and their reciprocal feedbacks. The feedbacks producing ecosystem degradation and poverty are difficult to explain, but the poverty alleviation is dependent upon the access to supply of ecosystem services (Martinez-Alier 2002). One of the cross-cutting challenges emerging from the scenarios is the connection between ecosystems services and poverty, including the trade offs.

Interventions on the patterns of adaptive cycles enable to manage change of the system dynamics. Rigid structures prevent reorganization, while perturbations compromise the exploitation of the adaptive cycle (Biggs et al., 2010; Pelling & Manuel-Navarrete, 2011). Urban land uses are in continues flux of change governed with decisions at local government level (Theobald et al., 2000) but driven by no anticipated non-local drivers (Altieri et al., 1999).

The human action and government policies may enable people to decrease the risks of regime shifts and build resilience and create better conditions to develop ecosystem services. Urban sustainable planning policies implemented through adaptation strategies are able to mitigate climate change effects and manage the environmental impacts on urban socio ecosystems resilience of human communities and settlements.

GREEN ECONOMIC GROWTH

Green economy is aimed to improving social equity and human wellbeing while reducing environmental and ecological risks and scarcities (UNEP, 2010). The green economy concept is related to the principle of think globally, act locally. Economies must develop and strengthen the capacities to reduce the use of natural and environmental resources used by a growing green economy. The results sought are to reduce the levels of natural resources use, as well the levels of emissions and pollution in times when population, economic growth and consumption growth.

Green economic growth reduces environmental stress ensuring that the basic needs are met while increases competitiveness and profits. For example, ocean natural resources management can be a source of green and inclusive economic growth. However, thos reduction of stress may have other impacts.

The removal of predators in coastal areas has an impact on the degradation of resilience of these coastal ecosystems increasing vulnerability to storms and sunamis (Jackson et al. 2001; Adger et al. 2005).

Traditional production systems are moving to most modern production systems changing the life-styles of humankind, supported by new technologies and mineral materials towards more industrial and urban oriented systems and away from the use of biomass, natural and agricultural resources. Growth of financial, manufactured, human and social capital has been degradating natural capital and ecosystem services. Ecosystem services and human wellbeing may suffer irreversible severe declines if natural capital is not built and enhanced at the same time that financial, manufactured, social and human capital. However, many countries are experiencing dramatic changes in agricultural and biomass systems and away from urban/industrial systems and mineral materials. Adapting the mosaic approach emphasizes the multi-scale and cross-sectoral efforts to sustain ecosystem services.

The ability to transform the economic growth in economic green growth by taking the opportunities of an evolving policy landscape. The transition from traditional economy activities to green economy must be supported by effective, fair and inclusive governance. Greening of economic growth should be supported by integrated targets, strategies and indicators, aimed to give relevant policy data on systemic change for growing green. Greening economic growth strategies support integrated changes of the socio ecosystem. Green growth strategies strengthened through specific policies play a significant role in addressing the roots of persistent poverty and achieve reduction.

Global green economic liberalization requires investments in public goods and policies to reduce poverty and inequality reduction. Emerging green economies are experiencing industrialization, urbanizations and restructuration processes aimed to achieve better standards of welfare and living. These processes have changed the production, distribution and consumption systems, reduced the agricultural labor force and motivating the use of fossil-fuel-based energy systems which have enable and reinforced large-scale urbanization.

Low-income economies are experiencing activities and practices that usually have an intensive water resource usage and high degradation of the quality of water despite the technical improvements in water practices and systems and technologies fostering an effective use of water and higher economic added value. Technological innovation determines the ecosystem management and governance to increase the availability and to improve efficiency of supply and allocation of ecosystem services. Technology plays a relevant role for the provision of ecosystem services and have an impact on other connected ecosystems. However, these activities and practices are far away from the green economic growth perspective. Technological development and property rights for ecosystem services are co-evolving and augmenting the abilities and capacities to manage and monitor the ecosystems.

Green economic planning and design of urban greening elements create the opportunities to develop a smarter urban resilience and provide the additional benefits of a more sustainable city. Community economic greening programs allow residents to participate and take ownership and management developing a stewardship culture of urban natural resources and spaces. Poor urban and rural communities are vulnerable to negative impacts. Green economic growth policies may ensure a just transition for all the stakeholders in a community, ensuring mitigation of regressive impacts. Some specific measures on green growth can strenghtent the synergy poverty reduction strategies, although green economic growth policies cannot be a substitute of social policies addressing the causes of poverty.

To enable green growth is essential to create productive partnerships between the efforts of the different stakeholders for adaptive community co-management of natural resources. Relationships of cooperation between the different stakeholders dealing with interlinked challenges and aimed to close the gaps of green economic and social development as well as sustainable environment may contribute to achieve better quality of welfare.

Green economic growth initiatives require to be supported by an economic incentives framework. The green economic rationale is provided by investments in environmental improvements and protection. Green economic growth investments developing a time gap between short-term costs and long term benefits require collaborative action between public and private sectors.

A relationship between technical components and urban policies is critical to encourage investments in green interventions for improving urban environmental quality and efficiency. This collaborative action overcomes the financial risks and barriers allowing capital flows into investing in urban green sectors. Initiatives, investments and stimulus of green growth requires financing and economic mechanisms implemented through systemic reforms. Green growth achievements by initiatives, investments, and stimulus must be underlyed by financing mechanisms and economic forces.

Reduction of the price gap between the economic value and market prices of ecosystem goods, enhances and improves the investments in green economic sectors and reduces large scale environmental pressures. Green economic growth investments deliver large long term benefits after significant costs. Market failures should be corrected through the costs internalization of social and environmental externalities with negative impact by improving green investments and financing to achieve better natural resources efficiency and environmental protection.

New green jobs are demanded by skilled labor sectors that demand updated and scaled up skills training.

An economic green growth and urbanization resource-intensive processes are in transition from and agrarian to industrial-based resources increasing the demand for materials and energy. The transformation of urban land use is dynamic and therefore the benefits that the population receive from the ecosystem services, the regulated services, cultural services and the supporting services. The quality of all these services lead to the ecosystem resilience and environmental options (Folke et al., 2004).

The study of the ecosystem services feedbacks and regime shifts as part of a global system is relevant to change development paths (MA 2005) despite the difficulties to reduce uncertainties and the unexpected ecosystems phenomena in the projection of ecosystem services, a pattern that is likely to continue by the adoption of unexpected events (Bennett et al. 2003). For example, the vulnerability of drylands has increased with overcultivation and changes in rainfall, decreasing in the capacity of ecosystems to regulate and store water flows and maintain vegetation, leading to loss of resilience to external driven changes and invasive species (MA 2005c). Urban hydrological cycles represent a fundamental factor in providing and accumulating water, prevent soil loss and erosion, nutrient sequestration and recycling, remove toxins and sediments from water.

The new policies scenario is part of the commitment made by local governments related to renewable energy demand and subsidie reductions of fossil fuel and gass emission.

The reserved urban land also known as the urban greenbelts are the result of zoning planning for economic green growth used to protect areas that are environmental sensitive from urban development and control agricultural activities. Access to green spaces are linked to improving physical and mental health and reduce mortality.

CONCLUSIONS

Resilience is the most important point of view when planning the transition to new green economic growth. This must be based on the consideration and total understanding of its complexity, its distinct dimensions or social, infra-structural and environmental subsystems. On the basis of this methodological division it is possible to construct a resilience index based on the subindexes of the different dimensions of resilience. Nature is resilient but receive impacts from complex socio-ecosystems design.

Different mechanisms can be built to increase resilience. The redundancy together with the defense of the ecological balances present in urban ecosystems, constitues fundamental elements for building resilient cities. In general, this means integrating aspects of planning policies that take into account the advantages of urban growth based on the needs of native species, with the aim of keeping intact the ecosystem services that benefit the population, through the conservation of the local biodiversity. This can be done through some techniques, in particular the pursuable model of the BEF, together with the presence of ecological corridors to allow the species to migrate. However, resilience is a modern concept in spatial and economic planning. There are still a number of cultural and socio-economic obstacles that do not facilitate rapid adaptation.

Resilience in urban ecosystems is something that can be increased through planning. However, this is not just a physical planning of land use. First, it is about changing the paradigms of relationship with the territory, use of resources, and between human beings. To obtain long-term benefits, sustainable green growth must be able to count on resilient cities that grow in harmony with the rest of the territory.

REFERENCES

Adger, W. N., Hughes, T. P., Folke, C., Carpenter, S. R., & Rockstrom, J. (2005). Social– ecological resilience to coastal disasters. *Science, 309*(5737), 1036–1039. doi:10.1126cience.1112122 PMID:16099974

Alberti, M. (2006). *Advances in Urban Ecology. Integrating Humans and Ecological Processes in Urban Ecosystems*. Springer.

Altieri, M., & Nelso Companioni, A. (1999). The greening of the "barrios": Urbanagriculture for food security in Cuba. *Agriculture and Human Values, 16*(2), 131–140. doi:10.1023/A:1007545304561

Beisner, B. E., Dent, C. L., & Carpenter, S. R. (2003). Variability of lakes on the landscape: Roles of phosphorus, food webs, and dissolved organic carbon. *Ecology, 84*(6), 1563–1575. doi:10.1890/0012-9658(2003)084[1563:VOLOTL]2.0.CO;2

Bennett, E. M., Carpenter, S. R., Peterson, G. D., Cumming, G. S., Zurek, M., & Pingali, P. (2003). Why global scenarios need ecology. *Frontiers in Ecology and the Environment, 1*(6), 322–329. doi:10.1890/1540-9295(2003)001[0322:WGSNE]2.0.CO;2

Berkes, F., Colding, J., & Folke, C. (2003). *Navigating Social–Ecological Systems: Building Resilience for Complexity and Change*. Cambridge UniversityPress.

Biggs, R., Westley, F. R., & Carpenter, S. R. (2010). Navigating the back loop: Fostering social innovation and transformation in ecosystem management. *Ecology and Society, 15*(2), 9. doi:10.5751/ES-03411-150209

Bonanno, G. A., Galea, S., Bucciarelli, A., & Vlahov, D. (2006). Psychological Resilience after Disaster: New York City in the Aftermath of the September 11th Terrorist Attack. *Psychological Science*, *17*(3), 181–186. doi:10.1111/j.1467-9280.2006.01682.x PMID:16507055

Brannen, S. (2011). *Food works: A vision to improve NYC's food system*. The New York City Council.

Calkins, M. (2005). Strategy use and challenges of ecological design in landscapearchitecture. *Landscape and Urban Planning*, *73*(1), 29–48. doi:10.1016/j.landurbplan.2004.06.003

Cardinale, B. J., Ives, A. R., & Inchausti, P. (2004). Effects of speices diversity on the primary productivity of ecosystems: Extending our spatial and temporal scales of inference. *Oikos*, *104*(3), 437–450. doi:10.1111/j.0030-1299.2004.13254.x

Carpenter, S. R., & Folke, C. (2006). Ecology for transformation. *Trends in Ecology & Evolution*, *21*(6), 309–315. doi:10.1016/j.tree.2006.02.007 PMID:16769430

Center for Clean Air Policy. (2011). *The value of green infrastructure for urban climate adaptation*. Available at www.ccap.org

Chapin, F. S. III, Shaver, G. R., Giblin, A. E., Nadelhoffer, K. J., & Laundre, J. A. (1995). Response of arctic tundra to experimentaland observed changes in climate. *Ecology*, *76*(3), 694–711. doi:10.2307/1939337

Colding, J. (2007). Ecological land-use complementation for building resilience in urban ecosystems. *Landscape and Urban Planning*, *81*(1-2), 46–55. doi:10.1016/j.landurbplan.2006.10.016

Collins, E. M., & Kearins, K. (2010). Delivering on Sustainability's Global and Local orientation. *Academy of Management Learning & Education*, *9*(3), 499–506.

Connell & Slatyer. (1977). Mechanisms of succession in natural communities and their role in community stability and organization. *American Naturalist*, *111*, 1119–1144.

Conway-Cranos, L. L. (2012). Geographic variation in resilience: An experimental evaluation of four rocky intertidal assemblages. *Marine Ecology Progress Series*, *457*, 67–83. doi:10.3354/meps09715

Cooper, W. S. (1926). The fundamentals of vegetational change. *Ecology*, *7*, 391–413.

Cork, S. J., Peterson, G. D., Bennett, E. M., Petschel-Held, G., & Zurek, M. (2006). Synthesis of the storylines. *Ecology and Society*, *11*(2), 11. http://www.ecologyandsociety.org/vol11/ iss2/art11/

Cumming, G. S., Alcamo, J., Sala, O., Swart, R., Bennett, E. M., & Zurek, M. (2005). Are existing global scenarios consistent with ecological feedbacks? *Ecosystems (New York, N.Y.)*, *8*(2), 143–152. doi:10.100710021-004-0075-1

Cumming, G. S., & Child, M. F. (2009). Contrasting spatial patterns of taxonomic andfunctional richness offer insights into potential loss of ecosystem services. *Philosophical Transactions of the Royal Society of London. Series B, Biological Sciences*, *364*(1524), 1683–1692. doi:10.1098/rstb.2008.0317 PMID:19451119

Daniel, D., & Ortmann, J. (2011). Disambiguating Resilience. In Geoinformatik 2011 – Geochange (vol. 41, pp. 117–125). Heidelberg, Germany. AKA Verlag.

Daniel, D. C. (2011). *A Formal Theory of Resilience*. Dissertation. Master of Science in Geoinformatics, University of M˙unster, Germany.

De Lotto, R., & Di Tolle, M. L. (2013). Elementi di progettazione urbanistica-rigenerazione urbana nella città contemporanea. Maggioli Publisher.

De Lotto, R., Esopi, G., & Sturla, S. (2017). *Sustainable policies to improve urban ecosystem resilience* (Vol. 96). WIT Transactions on State of the Art in Science and Engineering.

Douglas, I., Goode, D., Houke, M., & Wang, R. (2011). *The Routledge Handbook of Urban Ecology*. Routledge.

Dunning, J. B., Danielson, B. J., & Pulliam, H. R. (1992). Ecological processes thataffect populations in complex landscapes. *Oikos*, *65*(1), 169–175. doi:10.2307/3544901

Eco-Innovation Action Plan. (2011). *Innovation for a Sustainable Future*. Communication from the Commission to the European Parliament, the Council, the European Economic and Social Committee and the Committee of the Regions. https://eur-lex.europa.eu/legal-content/EN/ALL/;ELX_SESSIONID=JBJhJh7RB5n2T7nTdZLLVLGRlJPCFRnTGM7ZvF8WhH9TJryvnsvL!1821898924?uri=CELEX:52011DC0899

Elmqvist, T., Folke, C., Nyström, M., Peterson, G., Bengtsson, J., Walker, B., & Norberg, J. (2003). Response diversity, ecosystem change, and resilience. *Frontiers in Ecology and the Environment*, *1*(9), 488–494. doi:10.1890/1540-9295(2003)001[0488:RDECAR]2.0.CO;2

Fergus, S., & Zimmerman, M. A. (2005). Adolescent resilience: A framework for understanding healthy development in the face of risk. *Annual Review of Public Health*, *26*(1), 399–419. doi:10.1146/annurev.publhealth.26.021304.144357 PMID:15760295

Folke, C., Carpenter, S., Walker, B., Scheffer, M., Elmqvist, T., Gunder-son, L., & Holling, C. S. (2004). Regime shifts, resilience, and biodiversityin ecosystem management. *Annual Review of Ecology, Evolution, and Systematics*, *35*(1), 557–558. doi:10.1146/annurev.ecolsys.35.021103.105711

Folke, C., Carpenter, S., Walker, B., Scheffer, M., Elmqvist, T., Gunderson, L., & Holling, C. S. (2005). Regime shifts, resilience and biodiversity in ecosystem management. *Annual Review of Ecology, Evolution, and Systematics*, *35*(1), 557–581. doi:10.1146/annurev.ecolsys.35.021103.105711

Folke, C., Carpenter, S. R., Walker, B., Scheffer, M., Chapin, T., & Rockström, J. (2010). Resilience thinking: integrating resilience, adaptability and transformability. *Ecology and Society*, *15*(4), 20. http://www.ecologyandsociety.org/vol15/iss4/ art20/

Groffman, P. M., Baron, J. S., Blett, T., Gold, A. J., Goodman, I., Gunderson, L. H., Levinson, B. M., Palmer, M. A., Paerl, H. W., Peterson, G. D., Poff, N. L., Rejeski, D. W., Reynolds, J. F., Turner, M. G., Weathers, K. C., & Weins, J. A. (2006). Ecological thresholds: the key to successful environmental management or an important concept with no practical application? *Ecosystems*, *9*(1), 1–13. http://springerlink.com/(mw0hzsfu fbpawv2rf24i3szk)/app/home/contribution.asp?referrer=parent&backto=issue,1,12;journal,2,60; linkingpublicationresults,1:101552,1

Gunderson, L. H., & Holling, C. S. (Eds.). (2002). *Panarchy: Understanding transformations in human and natural systems*. Island Press.

Hamel, G., & Valikangas, L. (2003). The Quest for Resilience. *Harvard Business Review*, *2003*(September), 52. PMID:12964393

Hobbs, R. J. (1993). Can revegetation assist in the conservation of biodiversity inagricultural areas? *Pacific Conservation Biology*, *1*, 29–38. doi:10.1071/PC930029

Hobbs, R. J. (1997, June). Future landscapes and the future of landscape ecology. *Landscape and Urban Planning*, *37*(1–2), 1–9. doi:10.1016/S0169-2046(96)00364-7

Holling, C. S. (1973). Resilience and stability of ecological systems. *Annual Review of Ecology and Systematics*, *4*(1), 1–23. doi:10.1146/annurev.es.04.110173.000245

Holling, C. S. (1996). Engineering resilience versus ecological resilience. In P. C. Schulze (Ed.), *Engineering within ecological constraints* (pp. 31–44). National Academy Press.

IFRC (International Federation of Red Cross and Red Crescent Societies). (2004). *World Disasters Report: Focus on Community Resilience*. International Federation of Red Cross and Red Crescent Societies.

International Energy Agency. (2010). *Energy Performance Certification of Buildings: a Policy Tool to Improve Energy Efficiency*. Author.

Isbell, F., Calcagno, V., Hector, A., Connolly, J., Harpole, W. S., Reich, P. B., Scherer-Lorenzen, M., Schmid, B., Tilman, D., van Ruijven, J., Weigelt, A., Wilsey, B. J., Zavaleta, E. S., & Loreau, M. (2011). High plant diversity is needed to maintain ecosystem services. *Nature*, *477*(7363), 199–202. doi:10.1038/nature10282 PMID:21832994

Isenhour, C. (2011). How the Grass Became Greener in the City: On Urban Imaginings and Practices of Sustainable Living in Sweden. *City & Society*, *23*(2), 117–134. doi:10.1111/j.1548-744X.2011.01058.x

Jackson, J. B. C., Kirby, M. X., Berger, W. H., Bjorndal, K. A., Botsford, L. W., Bourque, B. J., Bradbury, R. H., Cooke, R., Erlandson, J., Estes, J. A., Hughes, T. P., Kidwell, S., Lange, C. B., Lenihan, H. S., Pandolfi, J. M., Peterson, C. H., Steneck, R. S., Tegner, M. J., & Warner, R. R. (2001). Historical overfishing and the recent collapse of coastal ecosystems. *Science*, *293*(5530), 629–638. doi:10.1126cience.1059199 PMID:11474098

Johnson, E. A., & Miyanishi, K. (Eds.). (2007). *Plant disturbance ecology: The process and the response*. Academic Press.

Kennedy, C., Pincelt, S., & Bunje, P. (2011). The study of urban metabolism and its applications to urban planning and design. *Environmental Pollution*, *159*(8-9), 1965–1973. doi:10.1016/j.envpol.2010.10.022 PMID:21084139

Kuhn, W. (2003). Semantic reference systems. *International Journal of Geographical Information Science*, *17*(5), 405–409. doi:10.1080/1365881031000114116

Lansing, J. S. (1991). *Priests and Programmers: Technologies of Power in the Engineered Landscape of Bali*. Princeton University Press.

Linnenluecke, M. K., Griffiths, A., & Winn, M. (2012). Extreme Weather Events and the Critical Importance of Anticipatory Adaptation and Organizational Resilience in Responding to Impacts. *Business Strategy and the Environment*, *21*(1), 17–32. doi:10.1002/bse.708

Loreau, M., Mouquet, N., & Gonzalez, A. (2003). Biodiversity as spatial insurance in heterogeneous landscapes. *Proceedings of the National Academy of Sciences of the United States of America*, *100*(22), 12765–12770. doi:10.1073/pnas.2235465100 PMID:14569008

Ludwig, D., Brock, W. A., & Carpenter, S. R. (2006). Uncertainty in discount models and environmental accounting. *Ecology and Society*, 10.

MA (Millennium Ecosystem Assessment). (2005). *Ecosystems and Human Well-Being: Synthesis*. Island Press.

MacArthur, R. H., & Wilson, E. O. (1967). *The Theory of Island Biogeography*. Princeton University Press.

Martinez-Alier, J. (2002). *Environmentalism of the poor: a study of ecological conflicts & valuation.* Edward Elgar. doi:10.4337/9781843765486

Masolo, C., & Borgo, S. (2005). *Qualities in formal ontology. InFoundational Aspects ofOntologies (FOnt 2005)*. Workshop at KI.

Masolo, C., Borgo, S., Gangemi, A., Guarino, N., & Oltramari, A. (2003). *WonderWeb Deliverable D18 Ontology Library*. Final.

May, R. M. (1974). *Stability and Complexity in Model Ecosystems*. Princeton University Press.

McKinney, M. L. (2002). Urbanization, biodiversity, and conservation. *Biol. Sci.*, *52*, 883–890.

Millennium Ecosystem Assessment, M. A. (2003). *Ecosystems and their services. Chapter 2 in Ecosystems and human well-being*. Island Press.

Millennium Ecosystem Assessment, M. A. (2005c). *Ecosystems and human well-being: policy responses*. Island Press.

Norris, F. H., Stevens, S. P., Pfefferbaum, B., Wyche, K. F., & Pfefferbaum, R. L. (2008). Community Resilience as a Metaphor, Theory, Set of Capacities, and Strategy for Disaster Readiness. *American Journal of Community Psychology*, *41*(1-2), 127–150. doi:10.100710464-007-9156-6 PMID:18157631

Nyström, M., & Folke, C. (2001). Spatial resilience of coral reefs. *Ecosystems (New York, N.Y.)*, *4*(5), 406–417. doi:10.100710021-001-0019-y

O'Rourke, T. (2007). Critical Infrastructure, Interdependencies, and Resilience. *The Bridge*, *37*(1), 22–29.

Odum, E. P. (1969). The strategy of ecosystem development. –. *Science*, *164*(3877), 262–270. doi:10.1126cience.164.3877.262 PMID:5776636

Pelling, M., & Manuel-Navarrete, D. (2011). From resilience to transformation: The adaptive cycle in two Mexican urban centers. *Ecology and Society*, *16*(2), 11. doi:10.5751/ES-04038-160211

Peraboni, C. (2010). *Reti ecologiche e infrastrutture verdi*. Maggioli Publisher.

Peterson, G., Allen, C. R., & Holling, C. S. (1998). Ecological resilience, biodiversity, and scale. *Ecosystems (New York, N.Y.)*, *1*(1), 6–18. doi:10.1007100219900002

Peterson, G. D. (2005). Ecological management: control, uncertainty and understanding. In *Ecological paradigms lost: routes of theory change*. Elsevier. https://www.elsevier.com/wps/find/bookdescription.cws_home/705214/description#description

Philley, J. (2006). *Collar hazards with a bowtie: Chemical Processing*. Putman Media. http://www.nxtbook.com/nxtbooks/putman/cp0106/index.php?startid=27

Pope, S. E., Fahrig, L., & Merriam, H. G. (2000). Landscape complementation andmetapopulation effects on leopard frog populations. *Ecology*, *81*(9), 2498–2508. doi:10.1890/0012-9658(2000)081[2498:LCAMEO]2.0.CO;2

Postel, S., & Carpenter, S. R. (1997). Freshwater ecosystem services. In G. Daily (Ed.), *Nature's services* (pp. 195–214). Island Press.

Raskin, P. D. (2005). Global scenarios: Background review for the Millennium. *Ecosystem Assessment. Ecosystems (New York, N.Y.)*, *8*(2), 133–142. doi:10.100710021-004-0074-2

Reich, P. B., Tilman, D., Isbell, F., Mueller, K., Hobbie, S. E., Flynn, D. F., & Eisenhauer, N. (2012, May 4). Impacts of biodiversity loss escalate through time as redundancy fades. *Science*, *336*(6081), 589–592. doi:10.1126cience.1217909 PMID:22556253

Sandstr¨om, U.G., Angelstam, P., & Khakee, A., (2006). Urban comprehensiveplanning—identifying barriers for the maintenance of functional habitatnetworks. *Landsc. Urban Plann.*, *75*, 43–57.

Shahriar, A., Sadiq, R., & Tesfamariam, S. (2012). Risk analysis for oil & gas pipelines: A sustainability assessment approach using fuzzy based bowtie analysis. *Journal of Loss Prevention in the Process Industries*, *25*(3), 505–523. doi:10.1016/j.jlp.2011.12.007

Smith, L. (2002). What might we learn from climate forecasts? *Proceedings of the National Academy of Sciences of the United States of America*, *99*(Supplement 1), 2487–2492. doi:10.1073/pnas.012580599 PMID:11875200

Theobald, D. M., Hobbs, N. T., Bearly, T., Zack, J. A., Shenk, T., & Riebsame, W. E. (2000). Incorporating biological information in local land-use decisionmaking: Designing a system for conservation planning. *Landscape Ecology*, *15*(1), 35–45. doi:10.1023/A:1008165311026

Twigg, J. (2009). *Characteristics of disaster resilient community–a guidance note*. https://discovery.ucl.ac.uk/1346086/1/1346086.pdf

UNEP. (2010). *United Nations Environment Programme Green Economy Report*. http://www.unep.org/greeneconomy/GreenEconomyReport

Vogus, T. J., & Sutcliffe, K. M. (2007). The impact of safety organizing, trusted leadership, and care pathways on reported medication errors in hospital nursing units. *Medical Care*, *41*(10), 992–1002. doi:10.1097/MLR.0b013e318053674f PMID:17890998 von Haaren, C., & Reich, M. (2006). The German way to greenways and habitatnetworks. *Landscape and Urban Planning*, *76*(1-4), 7–22. doi:10.1016/j.landurbplan.2004.09.041

Wagnild, G. M. (2010): *Discovering Your Resilience* Core. https://www.pobal.ie/Publications/Documents/Discovering_Your_Resilience_Core.pdf

Walker, B., Holling, C. S., Carpenter, S. R., & Kinzig, A. (2004). Resilience, adaptability and transformability in social–ecological systems. *Ecology and Society*, *9*(2), 5. doi:10.5751/ES-00650-090205

Walker, B., & Salt, D. (2006). *Resilience thinking: Sustaining ecosystems and people in a changing world*. Island Press.

Walker, B., & Salt, D. (2012). *Resilience Practice: Engaging the Sources of Our Sustainability*. Island Press. doi:10.5822/978-1-61091-231-0

Walker, B. H., Kinzig, A. P., & Langridge, J. (1999). Plant attribute diversity, resilience, and ecosystem function: The nature and significance of dominant and minor species. *Ecosystems (New York, N.Y.)*, *2*(2), 95–113. doi:10.1007100219900062

Walley, M. P. (1991). *Statistical reasoning with imprecise probabilities*. Chapman and Hall.

Witten, I. H., Frank, E., & Hall, M. A. (2011). *Data Mining: Practical Machine Learning Tools and Techniques*. Morgan Kaufmann. doi:10.1007/978-1-4899-3472-7

KEY TERMS AND DEFINITIONS

Adaptive Cycle: The adaptive cycle, initially raised by Holling in 1986, is a metaphor for describing four phases that commonly occur in the processes of changing complex systems as a result of their internal dynamics and external influence: growth, conservation, liberation, or creative destruction. and reorganization

Biodiversity: Diversity of plant and animal species that live in a given space.

Green Economic: That which leads to the improvement of human well-being and social equality, while significantly reducing environmental risks and ecological scarcity.

Growth: It is the increase in the income or value of the goods and services that are generated in the economy of a country or region.

Resilience: Is the ability of human beings to adapt positively to adverse situations

Urban Ecology: It is a discipline whose object of study is the interrelationships between the inhabitants of an urban agglomeration and their multiple interactions with the environment.

Chapter 4
Sustainability of Biodegradable Polymers for the Environment:
An Alternative Approach for the Future

Pinki Saini
https://orcid.org/0000-0001-9083-6765
University of Allahabad, India

Anchal Singh
University of Allahabad, India

Mazia Ahmed
https://orcid.org/0000-0002-7006-3192
University of Allahabad, India

Unaiza Iqbal
University of Allahabad, India

Urvashi Srivastava
https://orcid.org/0000-0002-0779-155X
University of Allahabad, India

ABSTRACT

With the ever-increasing population, there is exploitation of our environment to meet the growing needs of the inhabitants. Along with pollution, climate change, global warming, etc., one of the major problems associated with overpopulation is the generation of a large amount of non-biodegradable solid wastes along with various other industrial effluents. Plastic, being inexpensive and durable, have found their way to be produced in large amounts for almost every purpose. This results in the accumulation of plastic products around the world causing plastic pollution. During the past few centuries, the idea of "green chemistry" has gained maximum attention, and therefore, efforts have been made by the researchers to reduce pollution with the help of some environmentally friendly solution. In this concern, bioplastics and biofilms have gained remarkable attention over the last two decades. The major advantages of bioplastics are their biodegradability and the reduced emission of carbon dioxide in their synthesis.

DOI: 10.4018/978-1-7998-4915-5.ch004

INTRODUCTION

Nowadays, there is an increasing awareness related to the degrading environment, the negative impact of fossil fuels and waste management problems all over the world. These problems are the reason why biopolymers and their use are promoted for developing a sustainable environment. In fact, without environmental sustainability, there will be no sustainable development in the economy. Various researches have proven that non-biodegradable resources pose a grave risk to the existence of humans, plants, and animals as they generate greenhouse gases and chlorofluorocarbons (CFCs). While observing this scenario critically, it can be say that it is not necessary to exaggerate the significance of a safe and healthy environment concerning global sustainable development. It is a prerequisite on which the existence of humans and other living being lies. Biodegradable polymers are not a threat to the environment (along with its inhabitants) compared to petroleum-based synthetic polymers. Therefore, it can be said that there is an immediate need for the development of biodegradable polymers globally.

Earlier, the durability of plastics was considered as its key characteristic which has now become the major reason for the ambivalence of the environment. During early 1980 the problem of solid waste generation in form of Municipal Solid Waste (MSW), shrinking of the landfill capacity and the rising cost of solid waste dumping (Philp et al., 2013; Regan, 1990) came out as a major problem in various regions of the United States. In the early 1960s, plastic goods made up to less than half a percent of MSW generation in the US. A report released by Central Pollution Control Board of India (2013), about 15342 tons of plastic waste is generated every day all over the country. In this only 5% plastics are recyclable and the rest pollute the all beings of environment. In several countries, plastics are found to be the fastest-growing element of solid waste. In cities, right after the food and paper wastes, plastic waste attains the third spot as a major component of municipal and industrial litter. For instance, Japan is also dealing with a similar problem of handling new waste because of the growing pressure on land and increasing public concern over environmental sustainability and health protection (Philp et al., 2013; Ishizaka and Tanaka, 2003).

The generation of greenhouse gases is another environmental factor in manufacturing plastic goods. The Intergovernmental Panel on Climate Change (IPCC) postulates a reduction in the emission of CO_2 by 80% (compared to 1990 level) to stabilize the GHG concentration at 450 ppm in the atmosphere by the year 2050 (Bashmakov et al., 2007). At the beginning of this century, the bio-based economy emerged first as a policy conception under the Organization for Economic Co-operation and Development (OECD), that linked renewable resources and bioprocesses by applying industrial biotechnologies for producing sustainable products, as well as job opportunities and income (Philp et al., 2013; OECD, 2001). It has been recently realized that 'microplastics' are getting accumulated in the environment, particularly oceanic environments (GESAMP, 2010). Microplastics potentially facilitate the exchange of various contaminants and have been shown to get consumed by various sea creatures. It is the greatest uncertainty that whether this will lead to the bioaccumulation of pollutant load (absorbed and plastic additives), or whether the microplastics corresponds as an additional and major vector for transferring contaminants.

Hence, for the elimination of such environmental issues and for improving the competence of biomaterials, there is an increasing need for the development of new bio-based substances and other groundbreaking technologies to minimize the extensive dependency on fossil fuels. The biodegradable polymers and plastics were first introduced in 1980s. biodegradable plastics can be obtained from synthetic as well as natural polymers. Natural polymers are obtained in large quantities from different renewable sources but synthetic polymers are generated from non renewable petrolium resources. Environment friendly bio-based products include industrially synthesized materials, manufactured from renewable agricultural products and forestry supply, such as wood including its wastes and residues, grasses, crops, and byproducts of the crops are novel and innovative substances of the twenty-first century. Whether it is renewable, recyclable, sustainable, or triggered biodegradable, these all have the potential to create a positive change in the natural environment today as well as tomorrow. A material obtained from renewable resources, capable of getting recycled and triggered biodegradable i.e., potentially stable in its proposed lifespan but would easily biodegrade after getting disposed off in a compost environment, and have profitable feasibility and ecological acceptability is defined as a “sustainable” bio-based product.

METHODOLOGY

The present chapter has been prepared after reviewing various data sources like PubMed, BioOne, Scopus, Science Open, Web of Science etc, using search terms and search strategies like biopolymers, sustainability, ecofriendly, green plastics. The selection criteria taken into account were year of publication, impact factor, relevance of the content discussed in the papers; about 180 studies were screened and 82 number of studies finally included in the article.

RATIONALE OF THE STUDY

The application of biomass, such as starch, cellulose, wood, and sugar, are used as an alternate of fossil resources for the manufacuturing of plastics, is a broadly accepted typology towards sustainability. There has been considerable interest in the progress and production of biodegradable polymer to solve the problem of pollution triggered by the regular use of synthetic polymer of petroleum origin. Consequently, a significant reduction of non renewable energy consumption and carbon dioxide emission will be accomplished. The use of biodegradable polymer was favoured by a second environmental motive, related to the issues of waste disposal i.e. waste management can be achieved to a greater extent. Presently, the market suggests several strategies of biopolymers and the most crucial are based on cellulosic esters, starch derivatives, polyhydroxybutyrate, polylactic acid, and polycaprolactone.

BIOPOLYMERS

Biopolymers/biodegradable polymers are obtained from renewable resources, synthesized with the help of microorganisms, or manufactured from petro-based substances. A novel biopolymer can be designed by blending two or more biopolymers which can be used to serve any specific purpose. Hence, the biodegradable nature of polymers is not just a function of origin but also of chemical configuration and degrading ecosystem. The biodegradation process is based on the immediate environmental condition such as location or temperature, on the substance and also, on its application. The biodegradable nature is based on its chemical structure. Based on the need of packaging different products, biodegradable packaging is produced in various forms viz., gels, film, bag and box, Gel as hydrogel applied as a coating on fruits and vegetables, helps prevent microbial contamination(Ambrose 2020).

A biopolymer is made partially or fully from polymers derived from biological sources like potato, sugarcane, or cellulose coming from trees, straw, and cotton. Biopolymers encompass an entire family of substances having different properties and applications, and are not merely a single substance. Hence, the word, "bio-based polymers", seems to be more appropriate to define traditional polymers that are made up of vegetable sources. Alternatively, the term "biopolymer" appears to be more fitting for describing those innovative products that are biobased and biodegradable. The meaning of the term "biobased" is that the material is moderately derived from plant biomass, while "biodegradation" is defined as a chemical process in which microbes present in the environment alter products into natural materials like water, CO_2, methane, inorganic compounds and last compost.

GLOBAL PRODUCTION OF BIOPOLYMERS

The global drift in the production of biopolymers has been predicted which demonstrates a major shift dominated by the bio-based, long-lasting thermopolymers, with a minor input by biodegradable polymers (Figure 1). For a prospective bio-economy, a considerable percentage of petro-based polymers can be replaced with bio-based polymers, though the tangible percent is doubtful. Conversely, a relatively recent revision by European Bioplastics has put the past prognosis as obsolete. The market of roughly 1.2 million tons in the year 2011 may witness a five times increment in the production amount by 2016, to reach near about 6 million tons and in 2018 it was about 6.73 million tones. The Leader of the market is partly bio-based PET (Polyethylene terephthalate), which has already made up approximately 40% of the universal biopolymer production capability. Bio-based PET (partially) would maintain to expand this lead for about 4.6 million tons by the year 2016 (Philp et al., 2013).

The market is clearly dominated by biobased and non-biodegradable polymers. Drop in biobased polymers such as polyethylene terephthalate (PET) and polyethylene (PE) lead this category. Figure 2. shows the production capacities increased from 37% in 2013 to 74% in 2018. As a consequence, the biobased nonbiodegradable polymers market is expected to grow strongly as well since biobased PET is part of this category. However, in 2018, PLA became second, biodegradable polymer. Figure 3. show the composition of the global biomass supply the cellulose (49%) is the dominant type of biomass used, followed by sugar & starch (24%), proteins (11%), oils and fats (4%) and other biomass (12%), such as rubber. figure 3(A,B) shows the Global production capacity of biopolymers.

CLASSIFICATION OF BIOPOLYMERS

The concern about biopolymers has outgrown far beyond simple applications in packaging to more sophisticated applications such as engineering materials in intense environmental conditions. This needs the discovery of new-fangled molecules, the amalgamation of molecules, exploring novel plasticizers, and lately, the efforts to manufacture indistinguishable thermo-polymers to the petro-polymers with the help of using renewable raw substances obtained from biological sources. figure 4 shows classification of biopolymers results in a progressively more diverse series of biopolymers demonstrating various characteristics.

Figure 1. Global production capacity of biopolymers (a,b) (Source: Aeschelmann et al., 2015)

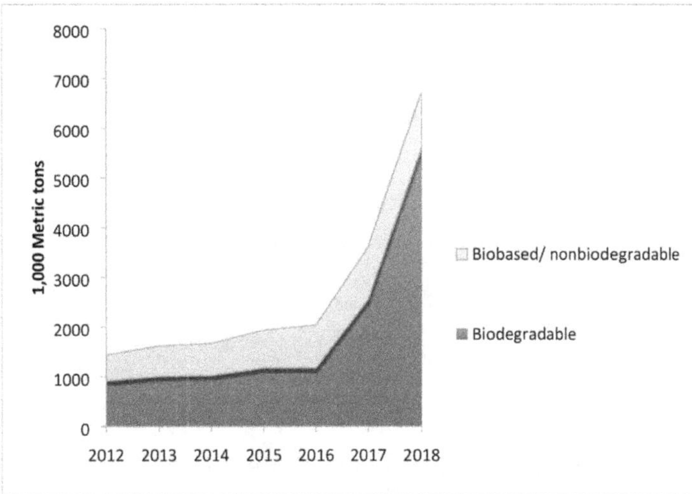

(a)

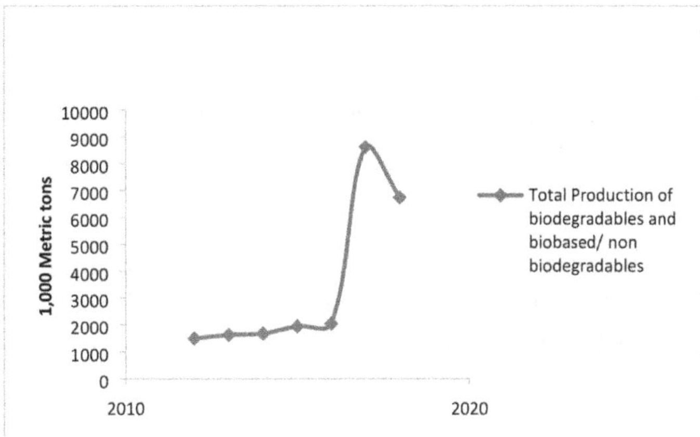

(b)

Polymers From Natural Resources

Cellulose

Cellulose is one of the most abundant biopolymers on earth. It is the major constituent of plant cell walls, and more than half of the organic carbon on earth is found in cellulose. It is composed of unbranched, linear chains of D-glucose molecules, linked to one another by 1,4-β-D glycosidic bonds.The hydrophilic nature, poor solubility characteristics, and a highly crystalline structure make it difficult to use in packaging. So after esterification of cellulose produced thermoplastic materials like cellulose acetate

Figure 2.Global production capacities of different bioplastics, Year 2013 to Year 2018 (Source: Aeschelmann et al., 2015)

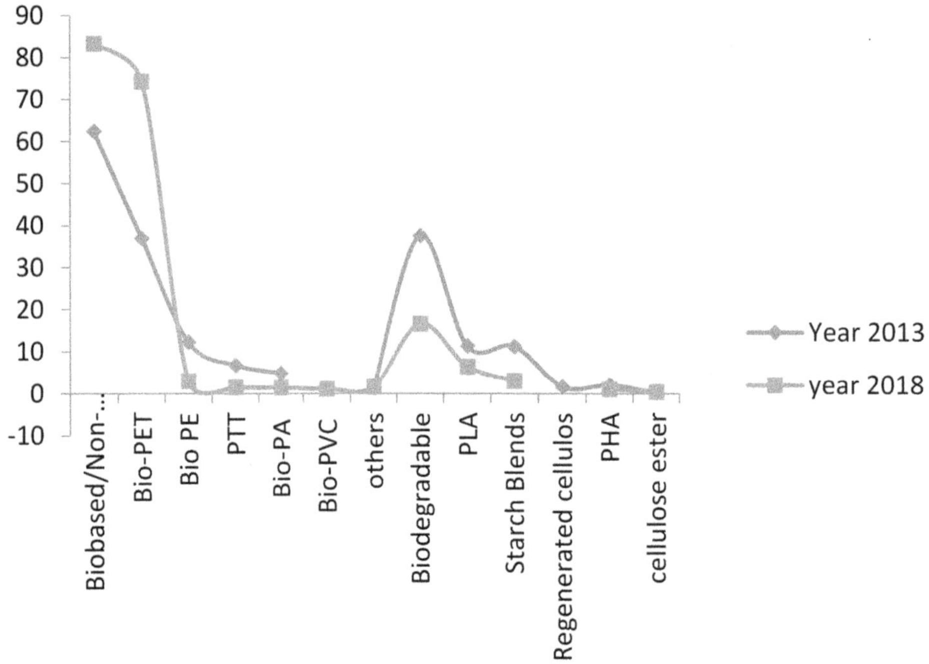

(CA) cellulose acetate propionate (CAP), and cellulose acetate butyrate (CAB) are widely used as food packaging materials for fresh products and baked goods (Pawar and Purwar, 2013).

Figure 3. Global biomass production in 2018 by biomass constituents: Total 12.3 billion tone dry matter (Source: Nova Institute, Bioplastic Magazine.Com)

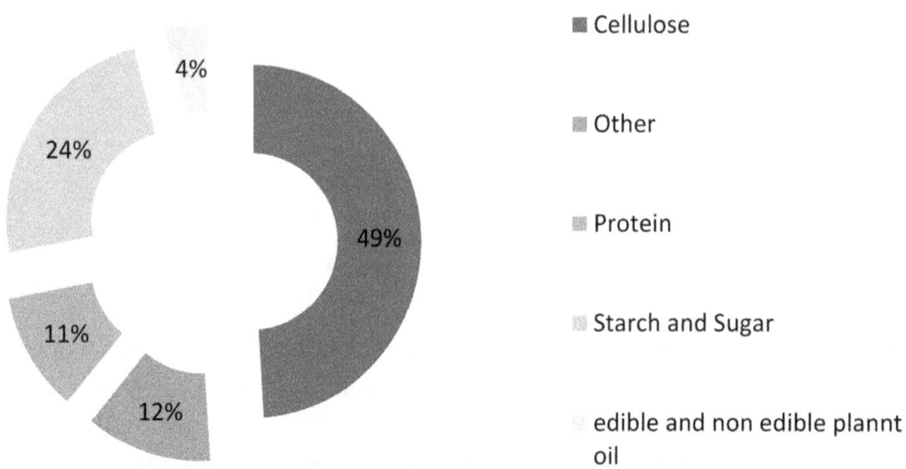

Figure 4. Concept of sustainable bio-based product

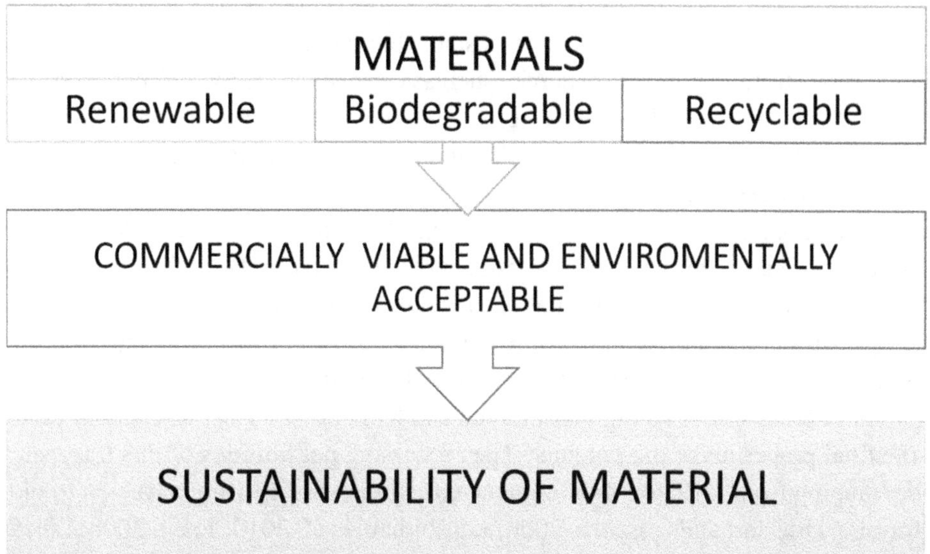

Formation of Cellulose Based Plastic: From Native Cellulose

Formation of cellulose based plastic has been reported by Wang et al., (2013). Cellulose (cotton linter pulp) was dissolved in 7 wt% NaOH pre-cooled to 12°C to form a 5 wt% transparent cellulose solution within 3 min and get, a cellulose bioplastic sheet with a thickness of about 0.3 mm was obtained, coded as Cellulose Based Plastic. Zhang et al., (2016) developed new cellulose films via low temperature solvents. The cellulose solution in NaOH (or LiOH)/urea was cast on a glass plate to give a gel sheet with thickness of 200–300 micro metre, excellent optical transmittance (90% at 800 nm), good tensile strength (100 MPa) and could be completely biodegraded in soil at about 30°C within one month, showing a promising application in packaging field.

Formation of Cellulose Based Plastic: From Modified Cellulose

To improve the quality and functionality of bioplastics, cellulose is acetylated to form cellulose acetate. Cellulose acetate (CA) is typically made from wood pulp through reaction with acetic acid and acetic anhydride in the presence of sulfuric acid to form cellulose triacetate (Bikales et al., 1971; Gedon et al., 1993). CA is 50% biobased. Thus, the successful preparation of such bioplastic would provide a promising opportunity to select the agricultural residues as a potent and competitive candidate for industrial utilization. Currently CA is generally recognized as a biodegradable polymer (Buchanan et al., 1993; Komarek et al., 1993; Rivard et al., 1992; Gu et al., 1993). The main drawback of cellulose acetate plastic is that its melt processing temperature is close to its decomposition temperature. This means that cellulose acetates should be plasticized if they are to be used in thermoplastic processing applications.

Poly Lactic Acid

Polylactic acid (PLA) is 100% biobased and compostable belonging to family of biodegradable thermoplastic polyester made from renewable resources such as corn, sugar beets, and potato starch (Garlotta, 2001; Kale et al., 2007). It is nowadays seen as one of the most promising polymers for commercial use as a substitute for low-density polyethylene (LDPE) and high-density polyethylene (HDPE), polystyrene (PS) and poly ethylene terephthalate (PET).

Since lactic acid exist as two optical isomers, L- and D-lactic acid, three different stereochemical compositions of lactide can be found, i.e. L,L-lactide, D,Dlactide and L,D-lactide. In packaging applications, poly(D,L-lactide) with 90% L-lactide has been widely used. Increasing the D-lactide concentration produces PLA polymers with a more crystalline structure and PLA films with better thermal stability, mechanical strength, and barrier properties. However, PLA with a high D-lactide concentration is not commercially viable today due to its high cost.(Byun and Kim, 2014). This stereochemical composition determines the final properties of the polymer. The processing possibilities of this transparent material are very wide, ranging from injection molding and extrusion over cast film extrusion to blow molding and thermoforming (Bogaert and Coszach, 2000; Jamshidian, et al, 2010; Joshi, 2008; Liu, 2006; Rasal et al., 2010; Siracusa et al., 2008; Södergård and Stolt, 2002). Koide and Shi (2007) evaluated the microbiological and physicochemical quality of whole green peppers packed in a PLA based film (Tohcello, 25 mm, Japan) compared to LDPE and perforated LDPE. Results showed no significant difference in hardness, ascorbic acid concentration and color after 1 week of storage at 10°C, but for the PLA film coliform bacteria counts were lower. PLA is a very brittle material, however, with less than 10% elongation at break. Its low deformation at break, high modulus, and hydrophilic properties have limited its application primarily to rigid thermoformed packaging (Okamoto et al., 2009; Pillin et al., 2006).

Soy Based Biopolymer

A number of proteins of plant origin have received attention for the production of biodegradable polymers. These proteins are corn zein, wheat gluten, soy protein, and peanut protein. Protein-based materials have been proved to be completely degrading in 50 days when buried in farmland soils (Domenek et al., 2004). Soy protein is the major coproduct of soybean oil and is one of the cheapest proteins in nature (Tian et al., 2012). Soy protein polymers are macromolecules that contain a number of amino acids and side chains that can be used in the manufacture of plastic (Sun et al., 1999). The mechanical properties of soy protein based plastics can be controlled and optimized by adjusting the initial moisture content as well as some processing parameters such as the molding temperature and/or pressure (Wang et al., 2007; Liang et al., 1999; Mo et al., 1999; Jane and Wang, 1996; Pateau et al., 1994). A great deal of research of soybean plastics was conducted in the 1930s and 1940s. Park et al., (2000) reported an extensive study on soy protein films during composting where they found that the samples degraded with 50% mass loss in about 10 days and with up to 95% mass loss in 30 days. Soy protein isolates films do not show satisfactory mechanical properties or water vapor barrier properties in practical applications due to the inherent hydrophilicity as well as the strong molecular interaction of natural protein, and these properties become poor under highly humid conditions (Gontard et al., 1998).

Starch Polymer

Starch is made up of repeating units of glucose and comprised of amylose and amylopectin (Gallant et al., 1997). Amylose is a mostly linear α-D-(1-4) glucan and has a molecular mass of about 10^5 -10^6 g mol$^{-1.}$ Amylopectin is a very large, highly branched α-D-(1-4) glucan with α-D-(1-6) linkages at the branch points and has a molecular mass in the range of 10^7 to 10^9 g mol^{-1} (Galliard, 1987; Imam et al., 2008; Verbeek and Bier, 2011).

Composites of Starch

Starch composites are complex blends of starch with compostable plastics such as Polylactic Acid (PLA), Polybutylene adipate terephthalate (PBAT), Polybutylene succinate (PBS), Poly-epsilon(ε)- caprolactone (PCL) and Polyhydroxyalkanoates (PHAs). Blending of starch with plastics improves water resistance, processing properties and mechanical properties. Liao and Wu (2009) produced a ternary blend composed of Polylactic acid PLA, Polycaprolactone PCL, and starch, and the PLA-PCL-Starch composites demonstrated better biodegradation in a soil environment than pure PLA.

Thermoplastic starch (TPS) can be produced after destructuration of starch in the presence of heat and plasticizers (Gandini, 2008; Wang et al., 2003). Destructuration transforms the starch from semi-crystalline granules into a homogeneous amorphous polymeric material. TPS has properties similar to those of common synthetic thermoplastics and can be processed as a traditional plastic; however, its sensitivity to moisture and poor mechanical properties limit its packaging application (Averous et al., 2000). The drawback of starch is its high water vapor permeability and poor mechanical properties when compared to synthetic plastics. To modulate these properties, other ingredients or property enhancers such as carboxymethylcellulose (CMC), microcrystalline cellulose (MC), nanoclays, fibers, carbon nanotubes, etc., have been added to the starchpolymer matrix (Gironès et al., 2012; Ma et al., 2008a,b; Müller et al., 2011).

To summarize, in order to prepare plastic from various biopolymers, plasticizers are added. Plasticizers improve or modify the mechanical properties, reduces the deformation tension, rigidness, density and viscosity and enhances the flexibility of polymer chain as well as provide resistance to fracture. Thereby, enabling the polymers from biomass to be used effectively on a larger scale.

Microbial Biopolymers

Biopolymers in this category produced with the help of microorganism, mainly bacteria. The bacteria produce the polymers when a carbon source is in excess and at least one other nutrient essential for growth have been depleted (Chen, 2009). The polymers accumulate in the form of intercellular granules as a carbon reserve for energy (Rincones et al., 2009). The bacteria use sugar from plant feed stocks to fuel this cellular process. This growth and accumulation of polymers within the bacteria occur in a bioreactor during the fermentation process. The polymer from the microorganism is extracted by a sequence of isolating the bacteria by means of centrifugation followed by press filtration and subsequent drying of the final biopolymer. Bacteria used for this purpose are sometimes genetically engineered to metabolize different feed stocks as well as to increase conversion efficiency of feedstock into polymer. Important variables in choosing bacteria strains include high density growth, high polymer content, rapid growth rate as well as the ability to produce controllable structures (Chen, 2009). The most widely studied bioplastics produced from this process are polyhydroxylalkanoates (PHAs).

Polyhydroxyalkanoate (PHA)

Polyhydroxyalkanoates are linear polymers that can be produced via bacterial fermentation of plant-derived feedstocks such as sugars or lipids (Byun and Kim, 2014; Galliard, 1987). They are biopolyesters of hydroxyalkanoates (HAs). PHAs have a broad spectrum of monomers that provide PHAs with variable and distinct properties. The most common type of PHA is poly (hydroxybutyrate) (PHB) and poly (hydroxyvalerate) (PHV). PHAs are classified in terms of short chain length (PHAscl) possessing thermoplastic properties and medium chain length (PHAmcl) with elastomeric properties. PHAs have the potential to be a competitive biopolymer due to the range of possible monomers that can be polymerized by the microorganisms. Types of PHAs that are produced commercially include poly-(R)-3-hydroxybutyrate (PHB), copolymer of (R)-3-hydroxybutyrate (3HB) and (R)-hydroxyvalerate (3HV) termed PHBV, copolymer of (R)-3-hydroxybutyrate (3HB) and (R)-hydroxyhex- anoate (3HHx) termed PHBHHx (Chen, 2009).

PHAs can be degraded at a high rate (3-9 months) by many microorganisms into CO_2 and water using their own secreted PHA depolymerases (Saharan and Ankita, 2012; Jendrossek and Handrick, 2002). PHAs extracted from bacterial cells have properties similar to conventional plastics, such as polypropylene. Therefore, they are very good substitute of petrochemical thermoplastics (Saharan and Ankita, 2012; Poirier, 1999). PHAs are produced by many different bacterial cultures. *Cupriavidus necator* (formerly known as *Ralstonia eutropha* or *Alcaligenes eutrophus*) is the one that has been most extensively studied (Vaneechoutte et al., 2004). A few important other strains that were recently studied include: Bacillus sp., Alcaligenes sp., Pseudomonas spp, Aeromonas hydrophila, Rhodopseudomonas palustris, *Escherichia coli*, *Burkholderia sacchari* and *Halomonas boliviensis*. The use of open mixed cultures, such as activated sludge (Lemos et al., 2006) assist in decrease of PHAs cost, thus enhancing their market potential (Patnaik, 2005).

Bio-Based Polyethylene, Polypropylene and Polyethylene Terephthalate

Ethylene is a base monomer for polyethylene (PE) which can be produced from bio-based ethanol via fermentation of sugarcane or corn. PE is a 100% biobased drop-in polymer. bio-based Polypropylene (PP) is more complicated than the production of biobased PE. Bio-based PE and PP are mechanically and physically identical to their traditional counterparts currently being used by plastic packaging manufacturers, which mean that manufacturers can continue to utilize their existing tools and knowledge for the production of bio-based PE and PP packaging. Bio-based PE and PP are not biodegradable; however, they can be recycled like conventional PE and PP are. When compared to petroleum-based PE and PP, bio-based PE and PP can significantly lower the levels of carbon dioxide produced.

In general, Polyethylene Terephthalate (PET) is comprised of 15 to 30% ethylene glycol (EG) and 70% terephthalic acid (TPA). Coca-Cola has produced a bio-based PET bottle, named the PlantBottles, made up of bio-based EG and petroleum-based TPA. PlantBottles can reduce the usage of petroleum-based materials up to 30% due to the composition of monomers. Bio-based TPA also can be produced by the bioroute, although commercial TPA is still being produced using petroleum-based resources. Utilization of isobutanol is the most viable process to produce biobased TPA on a commercial scale.

Bio-based TPA also can be produced from isobutanol, n-butanol, isobutylene, muconic acid, limonene, terpenes, and carbohydrates such as glucose and fructose (Berti et al., 2010; Timothy, 2010). Bio-based PET has exactly the same properties as petroleum-based PET and it can be recycled by conventional recycling systems. Also, the manufacture of biobased PET produces significantly less carbon dioxide when compared to conventional petroleum-based PET.

APPLICATIONS OF BIOPOLYMERS

Cellulose

Blackistone and Sand (2007) developed sustainable packaging i.e. bio based trays wrapped with cellulose film, reduce the emissions of green house gases. Makino and Hirata (1997) showed that a laminate of chitosan-cellulose and Polycaprolactone laminates are possible MAP packaging for fresh produce (shredded lettuce, cabbage, tomatoes, sweet corn and broccoli). Popa and Belc (2007) found that coating of cellophane with nitrocellulose or PVdC (polyvinylidine chloride) improved barrier properties and this film could be used for packaging of candies, processed meat, cheese and baked goods.

Poly Lactic Acid (PLA)

Rhim and Kim (2009) stated that paperboard coated with PLA could be used as a substitute for PE-coated paperboard in manufacturing 1-way paper cups or containers for high moisture foods (beverage cartons, ice cream containers).Fresh cut romaine lettuce can be packed in thermoformed oriented PLA (OPLA) during storage at 10°C (Benyathiar et al., 2009). Almenar et al., (2010) stated that PLA containers (Versapack, 8 oz., Wilkinson Industries Inc., USA) are applicable for packaging of blueberries. An equilibrium modified atmosphere could be developed inside the PLA containers. This increased the shelf life of the blueberries. Sensory evaluations showed that consumers preferred blueberries packed in PLA containers for one or two weeks over the blueberries packed in conventional containers. Haugaard et al., (2002) reported that PLA packaging (thermoformed cups, Autobar, France) is suitable for storage of fresh, unpasteurized orange juice at 4°C for 14 days. Color changes, ascorbic acid (AA) degradation and limonene scalping were most effectively prevented by PLA when compared to PS and HDPE.

Starch, as an abundant raw material with low cost, has been applied in the field of degradable plastics, and blend films containing. starch are potential materials in the agriculture, medicine, and packaging industries (Funke et al., 1998; Hulleman et al.,1998; Lu et al., 2005). Starch blended with synthetic polymer polyvinyl alcohol (PVA) has been studied as a potential biodegradable polymer (Follain et al., 2005; Xiao and Yang, 2006; Zhai et al., 2002; Zhai et al., 2003). Depending on the degree of biodegradability, it seems that PVA might provide a stable support medium for starch films (Jayasekara et al., 2004).

Starch and poly (vinyl alcohol) (PVOH)-based packaging is of great importance, availability, biodegradability and compatibility. Heat sealing capacity of native and acetylated corn starch based films was evaluated to develop biodegradable packages, such as bags. Acetylated starch addition reinforced 80% sealing resistance of starch films (Gadhave et al., 2018). Starch-based films are suitable alternatives to conventional plastics for different food products. Cannarsi et al., (2005) demonstrated that two biodegradable films based on starch (1 blend of starch and polyester and 1 blend of 3 biodegradable/biobased polyesters, Novamont, Italy) could be used to replace PVC films to pack fresh cut beef steaks.

Furthermore, Ifezue (2009) found Mater-Bi (blend of starch with biodegradable synthetic polymers like PCL or PVOH) to be superior to perforated LDPE, PLA and Ecoflex to pack whole fresh celery, especially regarding mechanical performance. Also Kantola and Helen (2001) stated that the quality of tomatoes packed in PLA-coated cardboard covered with a perforated Mater-Bi bag remained as good as when packed in LDPE (low-density polyethylene) bags during 3 weeks.

SUSTAINABILITY OF BIO-BASED POLYMERS

Biocomposites, or more specifically the "green composites," consist of biofiber and biopolymer from renewable resources and thus are expected to be biodegradable. However, polymer derived from renewable resources may be nonbiodegradable, depending on the structure and curing nature of polymer during fabrication of biocomposites. Thermoplastic is having more environmental impact than thermosets because of its recyclibility. Again bio-based products obtained from renewable resources also maintain carbon dioxide neutrality (Figure 5).

Figure 5. Carbon dioxide sequestration

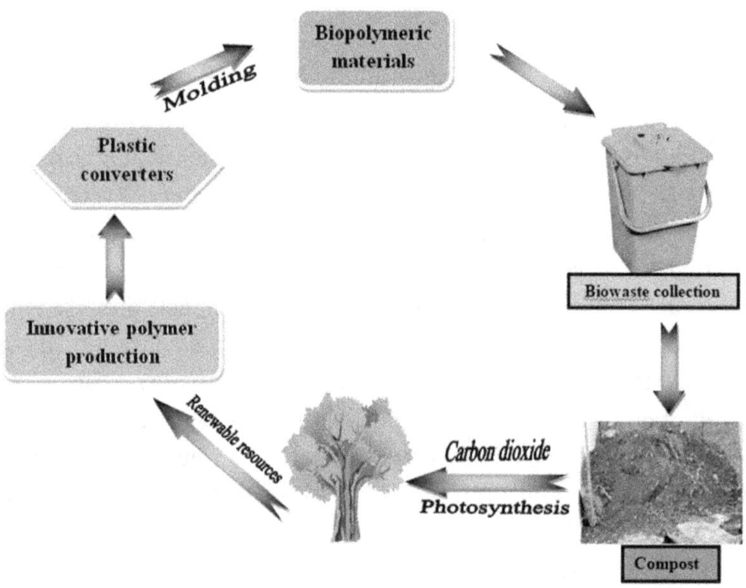

The "sustainability" issues of biopolymers, e.g., PLA, cellulosic polymers (cellulose esters), and PHA (bacterial polyesters) are undergoing considerable debate in scientific literature, providing divergent views (Mohanty et al., 2003). The sustainability issue of each specific biopolymer is a complex problem, and several parameters must be considered, including the raw materials from which the biopolymer is generated, the energy consumed during biopolymer conversion, and its life cycle assessment analysis from production to ultimate disposal or recycle, with due recognition to the design and engineering of the biopolymer. It is encouraging to develop cost-effective bio-based products or bio-composites through inexpensive natural/biofiber reinforcements. Most of the biopolymers cannot compete economically at their present state of technological development with the currently dominating petroleum-based polymers. The effective bio-composite formulations of such biopolymers with natural fiber reinforcements, through careful design and engineering, can result in new commercial attractions applications. The emergence of new applications of bio-composites would necessitate a large-scale demand of biopolymers, which would promote sustainability.

COUNTRY EXPERIENCES REGARDING SUSTAINABILITY

It is believed that the current growth rate of population is around 83 million people and it will be continuously increasing each year and the population will be reach about 8.5 billion by 2030(Trend G 2017). So the demands and requirement for food, energy and economic development will be continuously increasing. The production of "environmental friendly" materials has attracted extensive interest because of the increasing environmental concerns or plastic disposal majorly occupying the landfills. Countries around the world have already begun to incorporate these materials into their technologies. In America, McDonald is making biodegradable containers for their fast food. Other companies such as Bayer, DuPont, Dow Cargill, Nike, Danone, etc. are also manufacturing biodegradable packaging (Krzan et al., 2006). The global biopolymer market is thought to be growing about 20%~25 annually. By 2020, approx. 10%~15% biopolymer of the total will increase its market share to 25%~30% (Arikan and Ozsoy 2015). More and more companies are entering and investing in this market.

In terms of consumption of biodegradable polymers, Europe is the biggest market. As compared to the other markets, biopolymers find more place in the food industry. According to global market watchers, global Poly Lactic Acid market is estimated to reach US$2.6 billion by 2016 at a Compounded Annual Growth Rate (CAGR) of 28%. Volume based studies reveal that the major share of growth rate is expected from Asia Pacific region. Large petro-chemical, chemical, agriculture and polysaccharide companies are investing in establishing plants for bioplastics generation, parallel to establishing businessess and forming R&D partnerships to produce the next generation of bioplastics (Queiroz and Queiroz 2009). This shows the acceptance of the biopolymeric material throughout the country and it can be concluded that the domestic and the industrial sector are focusing majorly on the expansion of these biodegradable materials for a sustainable environment.

SOCIAL, ECONOMIC AND ENVIRONMENTAL BENEFITS OF BIOPOLYMERS

Social Benefits

It is significant for the global community to have a substitute for the product derived from petro-based product including plastics. Polymers from renewable sources will be a solution for various industries and society, which mostly depend on materials made from plastic. No new inventions can rescue from the limitations and biodegradable polymers too have some drawbacks. The most important disadvantage for biopolymer utility is its production cost, but the benefit is that the price of biopolymer is decreasing, whereas, petroleum oil price is increasing constantly. Biopolymer cover various range of products such as coating and packaging, bottles, cosmetic containers. PHAs have been processed into fibers, for a non woven fabrics material (Son et al., 1996). Biomass such as starch, cellulose, PLA, and PHAs can be utilized for all types of biodegradable packaging materials, including composting bags, food packaging, and also posses several applications in pharmacy and medical science.

Economic Benefits

According to European Bio-plastics Association, the global production capacity for bioplastics is estimated to grow four times by 2020. The prices of any biopolymer tend to be high when it is produced on a small scale. Biopolymer has more growth and stability in the food industry, nearly around 52% as compared to the other markets such as 18% beverage, 5% pharma and 5% cosmetics (Reddy et al., 2013). Comparing the growth and application of biopolymer, the industries, textiles and electronics are going to be the major contributors of this market. As more and more companies are entering and investing in this market, the economy will continue to grow slowly and steadily.

Environment Benefits

Fossil-Fuel Consumption

NatureWorks, a producer of polylactic acid (PLA), claims a 68% reduction of fossil fuel use over traditional plastics. HDPE requires 49 GJ/metric ton of feedstock in a cradle-to-factory analysis. PET requires 39 GJ/metric ton in feedstocks and Polystyrene requires 48 GJ/metric ton (Narayan and Patel, 2003). These are all significant amounts of petroleum and a switch to biopolymer would lessen these feedstock demands. If all plastics in the world were substituted by biodegradable and the energy used in the process came from biomass, the fossil fuel savings would be approximately 3.49 million barrels a day. That is 4% of the world's fossil fuel usage (Narayan and Patel, 2003).

Carbon-Emissions

The manufacture of high and low-density polyethylene emits 2.5 and 3.0 kilograms of carbon dioxide per kilogram of plastic respectively as shown in the table1. below (Vink et al., 2003; Narayan and Patel, 2003). PLA production emits less carbon dioxide than most other polymers. Manufacture of polypropylene, LDPE, and nylon all result in more carbon emissions. Thermoplastic Starch (TPS) requires little processing and thus has the lowest Carbon emissions of the three bioplastics given here as examples. To summarize, starch based plastics are the most favorable in terms of carbon emissions. Hence concluding that switching to biopolymer will be a judicious step towards the sustainable environment.

THE FUTURE PROSPECTS OF BIODEGRADABLE POLYMERS

Due to the adverse effect on environment, health and economy, it has become essential to look into the impact of petro-based polymers. The ill-effect of fossil-based polymers turn out to be a source of motivation for researchers and various industries to dig into the area of biopolymers. As the outcomes of petro-based polymers are unexpected, universities and research institutions are vigorously working on developing new biocomposites (Ibrahim et al., 2018; Hogg, 2001). In the present time, the biopolymer industries are well-invested and expanding, issues related with production, cost and quality has to be figured out as research on developing better substitutes are in progress. Also, nanopolymers along with biocomposites are still a subject of research. Researchers are fully aware and accordingly focusing on various areas of concern. For example, multilayered films made up of starch and/or any other natural fibre are likely to have adhesion-related problems and therefore, fibers coming from natural sources are also regarded as an alternative for reinforcing medium (Martin et al., 2001). Standard setting organizations like ASTM (American Society for Testing and Materials) and ISO (International Organization for Standardization) along with many others have issued methodologies for testing materials so as to have environment-friendly and pocket friendly biopolymers for sustainable development. Considering the nature, scope and characteristics of natural materials, there is a great need to give, consideration and due observation to them as given to synthetic materials. Thus, biopolymer industry has a great positive future, which is driven mainly by the environmental and economic benefits of using natural resources for production of biomaterials.

Nowadays, bio-based packaging materials are mostly used to pack short shelf life products, like fresh fruits and vegetables, and long shelf life products, like pasta and chips, which do not need very high oxygen and/or water barrier properties. However, the inventory of films shows a wide variety in properties, which could make them also applicable as a packaging material for other food products with stricter conditions, like MAP packaging. Storage tests and tests on the industrial packaging machines should be performed to make sure that these films can be used commercially. Bio-based materials offer great potential for the packaging industry. However, a thorough evaluation of the functional properties of a bio-based material is essential before it can be used as an alternative for traditional film materials.

RESEARCH GAP

Biopolymers are novel materials of the 21st century and would be highly significant to the material world. High cost, recycling methods, reduction of raw materials, lack of legislation, knowledge of energy consumption during conversion of biopolymers for its application needs to be highly investigated. Clearly, the bio-based packaging materials offers various potential in case of packaging industry however, there is need of certain storage tests to be performed in order to certify the use of these packaging films on a broader scale. Hence, critical evaluation is needed to access the functionality of bio-based packaging materials before they are introduced into the market as a primary substitute for conventional packaging materials. Therefore a thorough research for the feasibility of the biopolymer to be utilized effectively throughout the globe is required.

CONCLUSION

As we all know that ecological concerns are becoming more prominent these days, it encourages greater debates and bigger researches in the segment of biopolymer which in turn increase the range of lucrative biodegradable materials and products. Biopolymers are taken seriously now as a consequence of several recognized advantages; for example, conserving fossil-based resources, environment security and generating new jobs in future-oriented divisions. As the origin of biopolymers is renewable resource, they are believed to be much more environment friendly than conventional 'non-renewable' petro-based plastics. Bio-composites have the potential to supplement and even replace petroleum-based materials in various applications, which eventually offers new environmental, agricultural, manufacturing, and costomer benefits. As biopolymers are a derivation of renewable resource, the cost of the raw materials can be clearly decreased with their extensive use. The recent advancement in genetic engineering, composite science, and development of natural fibre offers considerable prospective for manufacturing improved profitable substances from renewable sources with enhanced support for universal sustainability. Fibres coming from natural sources are biodegradable, but polymers based on renewable resource can be drafted as biodegradable or not, based on the precise requirement of a specified application. The unique properties of biopolymers promises to open-up new opportunities for developing market of biocomposites in the current world of green materials.

REFERENCES

Aeschelmann, F., & Carus, M. (2015). Bio-based building blocks and polymers in the world. *Capacities, production and applications: Status quo and trends towards 2020.*

Almenar, E., Samsudin, H., Auras, R., & Harte, J. (2010). Consumer acceptance of fresh blueberries in bio-based packages. *Journal of the Science of Food and Agriculture, 90*(7), 1121–1128. doi:10.1002/jsfa.3922 PMID:20393992

Ambrose, D. C. (2020). Biodegradable Packaging – An Eco-Friendly Approach. *Current Agriculture Research Journa, 8*(1), 4-6.

Averous, L., Fauconnier, N., Moro, L., & Fringant, C. (2000). Blends of thermoplastic starch and polyesteramide: Processing and properties. *Journal of Applied Polymer Science, 76*(7), 1117–1128. doi:10.1002/(SICI)1097-4628(20000516)76:7<1117::AID-APP16>3.0.CO;2-W

Bashmakov, I., Bernstein, L., Bogner, J. E., Bosch, P., Dave, R., Davidson, O., ... Ribeiro, S. K. (2007). *Technical summary. Climate Change 2007: Mitigation.* Contribution of Working Group III to the Fourth Assessment Report of the Intergovernmental Panel on Climate Change, 70.

Benyathiar, P., Almenar, E., Auras, R., Ryser, E., Harte, J., & Siddiq, M. (2009). A sustainable package for fresh cut green salad. *United Fresh 2009 Convention.*

Berti, C., Binassi, E., Colonna, M., Fiorini, M., Kannan, G., Karanam, S., . . . Vannini, M. (2010). *U.S. Patent Application No. 12/347,423.* Washington, DC: US Patent Office.

Bikales, N. M., Segal, L., Ott, E., Spurlin, H. M., & Grafflin, M. W. (1954). *Cellulose and cellulose derivatives.* Wiley-Interscience.

Blakistone, B., & Sand, C. (2007, March). Using sustainable packaging technologies to respond to consumer, retailer, and seafood industry needs. In *International Smoked Seafood Conference Proceedings* (pp. 75-79). Academic Press.

Bogaert, J. C., & Coszach, P. (2000, March). Poly (lactic acids): A potential solution to plastic waste dilemma. *Macromolecular Symposia, 153*(1), 287–303. doi:10.1002/1521-3900(200003)153:1<287::AID-MASY287>3.0.CO;2-E

Buchanan, C. M., Gardner, R. M., & Komarek, R. J. (1993). Aerobic biodegradation of cellulose acetate. *Journal of Applied Polymer Science*, *47*(10), 1709–1719. doi:10.1002/app.1993.070471001

Byun, Y., & Kim, Y. T. (2014). Bioplastics for food packaging: chemistry and physics. In *Innovations in food packaging* (pp. 353–368). Academic Press. doi:10.1016/B978-0-12-394601-0.00014-X

Cannarsi, M., Baiano, A., Marino, R., Sinigaglia, M., & Del Nobile, M. A. (2005). Use of biodegradable films for fresh cut beef steaks packaging. *Meat Science*, *70*(2), 259–265. doi:10.1016/j.meatsci.2005.01.006 PMID:22063482

Chen, G. Q. (2009). A microbial polyhydroxyalkanoates (PHA) based bio-and materials industry. *Chemical Society Reviews*, *38*(8), 2434–2446. doi:10.1039/b812677c PMID:19623359

Domenek, S., Feuilloley, P., Gratraud, J., Morel, M. H., & Guilbert, S. (2004). Biodegradability of wheat gluten based bioplastics. *Chemosphere*, *54*(4), 551–559. doi:10.1016/S0045-6535(03)00760-4 PMID:14581057

Follain, N., Joly, C., Dole, P., & Bliard, C. (2005). Properties of starch based blends. Part 2. Influence of poly vinyl alcohol addition and photocrosslinking on starch based materials mechanical properties. *Carbohydrate Polymers*, *60*(2), 185–192. doi:10.1016/j.carbpol.2004.12.003

Funke, U., Bergthaller, W., & Lindhauer, M. G. (1998). Processing and characterization of biodegradable products based on starch. *Polymer Degradation & Stability*, *59*(1-3), 293–296. doi:10.1016/S0141-3910(97)00163-8

Gadhave, R. V., Das, A., Mahanwar, P. A., & Gadekar, P. T. (2018). *Starch based bio-plastics: the future of sustainable packaging*. Academic Press.

Gallant, D. J., Bouchet, B., & Baldwin, P. M. (1997). Microscopy of starch: Evidence of a new level of granule organization. *Carbohydrate Polymers*, *32*(3-4), 177–191. doi:10.1016/S0144-8617(97)00008-8

Galliard, T. (Ed.). (1987). *Starch: properties and potential*. John Wiley & Sons Incorporated.

Gandini, A. (2008). Polymers from renewable resources: A challenge for the future of macromolecular materials. *Macromolecules*, *41*(24), 9491–9504. doi:10.1021/ma801735u

Garlotta, D. (2001). A literature review of poly (lactic acid). *Journal of Polymers and the Environment*, *9*(2), 63–84.

Gedon, S., & Fengl, T. (1993). Encyclopedia of Chemical Technology. New York: Wiley.

GESAMP. (2010). IMO/FAO/UNESCOIOC/UNIDO/WMO/IAEA/UN/UNEP Joint Group of Experts on the Scientific Aspects of Marine Environmental Protection. In *Proceedings of the GESAMP International Workshop on plastic particles as a vector in transporting persistent, bio-accumulating and toxic substances in the oceans* (*Vol. 82*, p. 68). Academic Press.

Gironès, J., López, J. P., Mutjé, P., Carvalho, A. J. F. D., Curvelo, A. A. D. S., & Vilaseca, F. (2012). Natural fiber-reinforced thermoplastic starch composites obtained by melt processing. *Composites Science and Technology*, *72*(7), 858–863. doi:10.1016/j.compscitech.2012.02.019

Gontard, N., Guilbert, S., & Cuq, J.-L. (1993). Water and glycerol as plasticizers affect mechanical and water vapor barrier properties of an edible wheat gluten film. *Journal of Food Science*, *58*(1), 206–211. doi:10.1111/j.1365-2621.1993.tb03246.x

Gu, J. D., Eberiel, D. T., McCarthy, S. P., & Gross, R. A. (1993). Cellulose acetate biodegradability upon exposure to simulated aerobic composting and anaerobic bioreactor environments. *Journal of Environmental Polymer Degradation*, *1*(2), 143–153. doi:10.1007/BF01418207

Haugaard, V., Weber, C., Danielsen, B., & Bertelsen, G. (2002). Quality changes in orange juice packed in materials based on polylactate. *European Food Research and Technology*, *214*(5), 423–428. doi:10.100700217-001-0474-x

Hogg, P. (2001). *Plastics, rubber and composites at Queen Mary*. Academic Press.

Hulleman, S. H., Janssen, F. H., & Feil, H. (1998). The role of water during plasticization of native starches. *Polymer*, *39*(10), 2043–2048. doi:10.1016/S0032-3861(97)00301-7

Ibrahim, M. S., Sani, N., Adamu, M., & Abubakar, M. K. (2018). Biodegradable polymers for sustainable environmental and economic development. *MOJ Biorg. Org. Chem*, *2*, 192–194.

Ifezue, C. (2009). *The effect of bio-based films on quality and shelf life of fresh celery*. Clemson University.

Imam, S., Glenn, G., Chiou, B. S., Shey, J., Narayan, R., & Orts, W. (2008). Types, production and assessment of biobased food packaging materials. In *Environmentally compatible food packaging* (pp. 29–62). Woodhead Publishing. doi:10.1533/9781845694784.1.29

Ishizaka, K., & Tanaka, M. (2003). Resolving public conflict in site selection process—A risk communication approach. *Waste Management (New York, N.Y.)*, *23*(5), 385–396. doi:10.1016/S0956-053X(03)00094-1 PMID:12893010

Jamshidian, M., Tehrany, E. A., Imran, M., Jacquot, M., & Desobry, S. (2010). Poly-Lactic Acid: Production, applications, nanocomposites, and release studies. *Comprehensive Reviews in Food Science and Food Safety*, *9*(5), 552–571. doi:10.1111/j.1541-4337.2010.00126.x

Jane, J. L., & Wang, S. (1996). *U.S. Patent No. 5,523,293*. Washington, DC: U.S. Patent and Trademark Office.

Jayasekara, R., Harding, I., Bowater, I., Christie, G. B. Y., & Lonergan, G. T. (2004). Preparation, surface modification and characterisation of solution cast starch PVA blended films. *Polymer Testing*, *23*(1), 17–27. doi:10.1016/S0142-9418(03)00049-7

Jendrossek, D., & Handrick, R. (2002). Microbial degradation of polyhydroxyalkanoates. *Annual Review of Microbiology*, *56*(1), 403–432. doi:10.1146/annurev.micro.56.012302.160838 PMID:12213937

Joshi, S. (2008). Can nanotechnology improve the sustainability of biobased products? The case of layered silicate biopolymer nanocomposites. *Journal of Industrial Ecology*, *12*(3), 474–489. doi:10.1111/j.1530-9290.2008.00039.x

Kale, G., Auras, R., & Singh, S. P. (2007). Comparison of the degradability of poly (lactide) packages in composting and ambient exposure conditions. *Packaging Technology and Science: An International Journal*, *20*(1), 49–70. doi:10.1002/pts.742

Kantola, M., & Helén, H. (2001). Quality changes in organic tomatoes packaged in biodegradable plastic films. *Journal of Food Quality*, *24*(2), 167–176. doi:10.1111/j.1745-4557.2001.tb00599.x

Koide, S., & Shi, J. (2007). Microbial and quality evaluation of green peppers stored in biodegradable film packaging. *Food Control*, *18*(9), 1121–1125. doi:10.1016/j.foodcont.2006.07.013

Komarek, R. J., Gardner, R. M., Buchanan, C. M., & Gedon, S. (1993). Biodegradation of radiolabeled cellulose acetate and cellulose propionate. *Journal of Applied Polymer Science*, *50*(10), 1739–1746. doi:10.1002/app.1993.070501009

Krzan, A., Hemjinda, S., Miertus, S., Corti, A., & Chiellini, E. (2006). Standardization and certification in the area of environmentally degradable plastics. *Polymer Degradation & Stability*, *91*(12), 2819–2833. doi:10.1016/j.polymdegradstab.2006.04.034

Lemos, P. C., Serafim, L. S., & Reis, M. A. (2006). Synthesis of polyhydroxyalkanoates from different short-chain fatty acids by mixed cultures submitted to aerobic dynamic feeding. *Journal of Biotechnology*, *122*(2), 226–238. doi:10.1016/j.jbiotec.2005.09.006 PMID:16253370

Liang, F., Wang, Y., & Sun, X. S. (1999). Curing process and mechanical properties of protein-based polymers. *Journal of Polymer Engineering, 19*(6), 383-394.

Liao, H. T., & Wu, C. S. (2009). Preparation and characterization of ternary blends composed of polylactide, poly (ε-caprolactone) and starch. *Materials Science and Engineering A*, *515*(1-2), 207–214. doi:10.1016/j.msea.2009.03.003

Liu, L. (2006). Bioplastics in food packaging: Innovative technologies for biodegradable packaging. *San Jose State University Packaging Engineering*, *13*, 1348–1368.

Lu, Y., Tighzert, L., Dole, P., & Erre, D. (2005). Preparation and properties of starch thermoplastics modified with waterborne polyurethane from renewable resources. *Polymer*, *46*(23), 9863–9870. doi:10.1016/j.polymer.2005.08.026

Ma, X., Chang, P. R., & Yu, J. (2008). Properties of biodegradable thermoplastic pea starch/carboxymethyl cellulose and pea starch/microcrystalline cellulose composites. *Carbohydrate Polymers*, *72*(3), 369–375. doi:10.1016/j.carbpol.2007.09.002

Ma, X., Yu, J., & Wang, N. (2008). Glycerol plasticized-starch/multiwall carbon nanotube composites for electroactive polymers. *Composites Science and Technology*, *68*(1), 268–273. doi:10.1016/j.compscitech.2007.03.016

Makino, Y., & Hirata, T. (1997). Modified atmosphere packaging of fresh produce with a biodegradable laminate of chitosan-cellulose and polycaprolactone. *Postharvest Biology and Technology*, *10*(3), 247–254. doi:10.1016/S0925-5214(96)01402-0

Martin, O., Schwach, E., Averous, L., & Couturier, Y. (2001). Properties of biodegradable multilayer films based on plasticized wheat starch. *Stärke*, *53*(8), 372–380. doi:10.1002/1521-379X(200108)53:8<372::AID-STAR372>3.0.CO;2-F

Mo, X., Sun, X. S., & Wang, Y. (1999). Effects of molding temperature and pressure on properties of soy protein polymers. *Journal of Applied Polymer Science*, *73*(13), 2595–2602. doi:10.1002/(SICI)1097-4628(19990923)73:13<2595::AID-APP6>3.0.CO;2-I

Mohanty, A. K., Wibowo, A., Misra, M., & Drzal, L. T. (2003). Development of renewable resource–based cellulose acetate bioplastic: Effect of process engineering on the performance of cellulosic plastics. *Polymer Engineering and Science*, *43*(5), 1151–1161. doi:10.1002/pen.10097

Müller, C. M., Laurindo, J. B., & Yamashita, F. (2011). Effect of nanoclay incorporation method on mechanical and water vapor barrier properties of starch-based films. *Industrial Crops and Products*, *33*(3), 605–610. doi:10.1016/j.indcrop.2010.12.021

Narayan, R., & Patel, M. (2003). Review and analysis of bio-based product LCA's. In *Proceedings of the International Workshop Assessing the Sustainability of Bio-based Products*. Institute for Science & Public Policy.

OECD. (2001). The Application of Biotechnology to Industrial Sustainability. OECD Publishing.

Okamoto, K., Ichikawa, T., Yokohara, T., & Yamaguchi, M. (2009). Miscibility, mechanical and thermal properties of poly (lactic acid)/polyester-diol blends. *European Polymer Journal*, *45*(8), 2304–2312. doi:10.1016/j.eurpolymj.2009.05.011

Paetau, I., Chen, C. Z., & Jane, J. L. (1994). Biodegradable plastic made from soybean products. 1. Effect of preparation and processing on mechanical properties and water absorption. *Industrial & Engineering Chemistry Research*, *33*(7), 1821–1827. doi:10.1021/ie00031a023

Park, S. K., Hettiarachchy, N. S., & Were, L. (2000). Degradation Behavior of Soy Protein– Wheat Gluten Films in Simulated Soil Conditions. *Journal of Agricultural and Food Chemistry*, *48*(7), 3027–3031. doi:10.1021/jf0000272 PMID:10898660

Patnaik, P. R. (2005). Perspectives in the modeling and optimization of PHB production by pure and mixed cultures. *Critical Reviews in Biotechnology*, *25*(3), 153–171. doi:10.1080/07388550500301438 PMID:16294831

Pawar, P. A., & Purwar, A. H. (2013). Biodegradable polymers in food packaging. *American Journal of Engineering Research*, *2*(5), 151–164.

Philp, J. C., Bartsev, A., Ritchie, R. J., Baucher, M. A., & Guy, K. (2013). Bioplastics science from a policy vantage point. *New Biotechnology*, *30*(6), 635–646. doi:10.1016/j.nbt.2012.11.021 PMID:23220474

Pillin, I., Montrelay, N., & Grohens, Y. (2006). Thermo-mechanical characterization of plasticized PLA: Is the miscibility the only significant factor? *Polymer*, *47*(13), 4676–4682. doi:10.1016/j.polymer.2006.04.013

Poirier, Y. (1999). Green chemistry yields a better plastic. *Nature Biotechnology*, *17*(10), 960–961. doi:10.1038/13652 PMID:10504694

Popa, M., & Belc, N. (2007). Packaging. In *Food Safety* (pp. 68–87). Springer. doi:10.1007/978-0-387-33957-3_4

Queiroz, A. U., & Collares-Queiroz, F. P. (2009). Innovation and industrial trends in bioplastics. Journal of Macromolecular Science®, Part C. *Polymer Reviews (Philadelphia, Pa.)*, *49*(2), 65–78. doi:10.1080/15583720902834759

Rasal, R. M., Janorkar, A. V., & Hirt, D. E. (2010). Poly (lactic acid) modifications. *Progress in Polymer Science*, *35*(3), 338–356. doi:10.1016/j.progpolymsci.2009.12.003

Reddy, R. L., Reddy, V. S., & Gupta, G. A. (2013). Study of bio-plastics as green and sustainable alternative to plastics. *International Journal of Emerging Technology and Advanced Engineering*, *3*(5), 76–81.

Regan, M. J. (1990). *Sites for our solid waste: a guidebook for effective public involvement.* Office of Solid Waste.

Rhim, J. W., & Kim, J. H. (2009). Properties of poly (lactide)-coated paperboard for the use of 1-way paper cup. *Journal of Food Science*, *74*(2), E105–E111. doi:10.1111/j.1750-3841.2009.01073.x PMID:19323738

Rincones, J., Zeidler, A. F., Grassi, M. C. B., Carazzolle, M. F., & Pereira, G. A. (2009). The golden bridge for nature: The new biology applied to bioplastics. Journal of Macromolecular Science®, Part C. *Polymer Reviews (Philadelphia, Pa.)*, *49*(2), 85–106. doi:10.1080/15583720902834817

Rivard, C. J., Adney, W. S., Himmel, M. E., Mitchell, D. J., Vinzant, T. B., Grohmann, K., Moens, L., & Chum, H. (1992). Effects of natural polymer acetylation on the anaerobic bioconversion to methane and carbon dioxide. *Applied Biochemistry and Biotechnology*, *34*(1), 725–736. doi:10.1007/BF02920592

Saharan, B. S., & Ankita, S. D. (2012). Bioplastics-for sustainable development: A review. *Int J Microbial Res Technol*, *1*, 11–23.

Siracusa, V., Rocculi, P., Romani, S., & Dalla Rosa, M. (2008). Biodegradable polymers for food packaging: A review. *Trends in Food Science & Technology*, *19*(12), 634–643. doi:10.1016/j.tifs.2008.07.003

Södergård, A., & Stolt, M. (2002). Properties of lactic acid based polymers and their correlation with composition. *Progress in Polymer Science*, *27*(6), 1123–1163. doi:10.1016/S0079-6700(02)00012-6

Son, H., Park, G., & Lee, S. (1996). Growth-associated production of poly-β-hydroxybutyrate from glucose or alcoholic distillery wastewater by Actinobacillus sp. EL-9. *Biotechnology Letters*, *18*(11), 1229–1234. doi:10.1007/BF00129945

Sun, X. S., Kim, H. R., & Mo, X. (1999). Plastic performance of soybean protein components. *Journal of the American Oil Chemists' Society*, *76*(1), 117–123. doi:10.100711746-999-0057-8

Tian, H., Wu, W., Guo, G., Gaolun, B., Jia, Q., & Xiang, A. (2012). Microstructure and properties of glycerol plasticized soy protein plastics containing castor oil. *Journal of Food Engineering*, *109*(3), 496–500. doi:10.1016/j.jfoodeng.2011.10.033

Timothy, A. (2010). WO Patent Application No. 2010/151346.

Trends, G. (2017). *Challenges and Opportunities in the Implementation of the Sustainable Development Goals.* Academic Press.

Vaneechoutte, M., Kämpfer, P., De Baere, T., Falsen, E., & Verschraegen, G. (2004). Wautersia gen. nov., a novel genus accommodating the phylogenetic lineage including Ralstonia eutropha and related species, and proposal of Ralstonia [Pseudomonas] syzygii (Roberts et al. 1990) comb. nov. *International Journal of Systematic and Evolutionary Microbiology*, *54*(2), 317–327. doi:10.1099/ijs.0.02754-0 PMID:15023939

Verbeek, C. J. R., & Bier, J. M. (2011). Synthesis and characterization of thermoplastic agro-polymers. In *Handbook of Applied Biopolymer Technology* (pp. 197–242). RSC Publishing. doi:10.1039/9781849733458-00197

Vink, E. T., Rabago, K. R., Glassner, D. A., & Gruber, P. R. (2003). Applications of life cycle assessment to NatureWorks™ polylactide (PLA) production. *Polymer Degradation & Stability*, *80*(3), 403–419. doi:10.1016/S0141-3910(02)00372-5

Wang, H., Jiang, L., & Fu, L. (2007). Properties of molded soy protein isolate plastics. *Journal of Applied Polymer Science*, *106*(6), 3716–3720. doi:10.1002/app.27064

Wang, Q., Cai, J., Zhang, L., Xu, M., Cheng, H., Han, C. C., Kuga, S., Xiao, J., & Xiao, R. (2013). A bioplastic with high strength constructed from a cellulose hydrogel by changing the aggregated structure. *Journal of Materials Chemistry. A, Materials for Energy and Sustainability*, *1*(22), 6678–6686. doi:10.1039/c3ta11130j

Wang, X. L., Yang, K. K., & Wang, Y. Z. (2003). Properties of starch blends with biodegradable polymers. *Journal of Macromolecular Science, Part C: Polymer Reviews*, *43*(3), 385–409. doi:10.1081/MC-120023911

Xiao, C., & Yang, M. (2006). Controlled preparation of physical cross-linked starch-g-PVA hydrogel. *Carbohydrate Polymers*, *64*(1), 37–40. doi:10.1016/j.carbpol.2005.10.020

Zhai, M., Yoshii, F., & Kume, T. (2003). Radiation modification of starch-based plastic sheets. *Carbohydrate Polymers*, *52*(3), 311–317. doi:10.1016/S0144-8617(02)00292-8

Zhai, M., Yoshii, F., Kume, T., & Hashim, K. (2002). Syntheses of PVA/starch grafted hydrogels by irradiation. *Carbohydrate Polymers*, *50*(3), 295–303. doi:10.1016/S0144-8617(02)00031-0

Zhang, J., & Reineccius, G. A. (2016). Factors controlling the turbidity of submicron emulsions stabilized by food biopolymers and natural surfactant. *Lebensmittel-Wissenschaft + Technologie*, *71*, 162–168. doi:10.1016/j.lwt.2016.03.035

APPENDIX

Table 1. Carbon di oxide emission during polythene manufacturing

Types of Polymer	Petro-Based Polymer			Biopolymer		
Polymer	LDPE	PP	HDPE	PHB	PLA	TPS
CO_2 emissions (kg CO_2 / kg polymer)	3.0	3.4	2.5	2.6	1.8	1.14

Chapter 5
3R Principle in Household Energy Consumption Patterns:
A Case of Turkish Households

Seda Yildirim
https://orcid.org/0000-0003-4367-6652
Tekirdag Namık Kemal University, Turkey

ABSTRACT

Energy is an important issue for achieving sustainable development. Sustainable Development Goals emphasize renewable energy sources, green energy technologies, and energy efficiency for sustainable world. In addition, energy consumption patterns can influence energy efficiency as much as green energy policies. In this point, sustainable consumption patterns should be adapted into energy consumption. 3R principle can be an alternative model for sustainable energy consumption. This study aims to explore how household can adapt 3R principle in energy consumption patterns in Turkey. Turkey aims to achieve sustainable development goals and especially works to achieve energy security and efficiency in the long term. In this point, the contribution of households will play a critical role for energy efficiency and security in a country. The study analyzed open access data from TURKSTAT (Turkish Statistical Institute) to determine the framework of household energy consumption patterns.

INTRODUCTION

Energy plays a key role in achieving sustainable development goals in the long term. Affordable, accessible and sustainable energy provides higher life-standards for all people (citizens) in the world. Energy supports well-being of people and improves usual life-standards through meeting basic needs such as illumination, heating, hot meal etc. Today's living conditions makes people almost energy-addicted because people don't want leave modern conveniences such as electronics, small home appliances and smart phones etc. On the other hand, there are still many countries or people who suffer from lack of energy access (Yıldırım and Yıldırım, 2019). Sustainable development aims to get balance between economic, social and environmental issues (Munasinghe, 2002) globally that energy sources should be accessible,

DOI: 10.4018/978-1-7998-4915-5.ch005

affordable and sustainable for the future. Recently, the critical value of energy issues is accepted more important to achieve sustainable development (Johnston, 2001; Ahuja and Tatsutani, 2009; Minelli, 2017; Owusu and Asumadu-Sarkodie, 2016; Yoshida and Zusman, 2015; Yıldırım et.al., 2019). In 2015, the United Nations set 2030 Sustainable Development Goals which included 17 main targets to be achieved by 2030 (LEAD, 2016). Energy issues were included in 2030 SDGs as a separate theme. In Millennium Development Goals, energy issues or themes weren't included directly. MDGs had 8 main goals to be achieved by 2015 (Easterly, 2008). These 8 goals were categorized as "reducing poverty and hunger, achieving equal primary education, gender equality, reducing child mortality, improving health, fighting against diseases (HIV/AIDS, malaria etc.), achieving environmental sustainability and developing global partnerships" (WHO, 2015). 2030 SDGs have more expanded and detailed goals. Table 1 summarizes the general framework of 2030 SDGs as below.

As seen in table 1, 2030 SDGs added new themes as sustainable consumption and production, green energy, peace-justice, sustainable industry and sustainable urban life. Every goal contributes sustainable development dimensions directly or indirectly. Some of them are related with environmental dimension directly such as Goal-13, Goal-14 and Goal-15. Some of them are related with social dimension directly such as Goal-3 and Goal-4. It is seen that 2030 Sustainable Development Goals points out that energy sources should be sustainable-renewable and also affordable and accessible (UN, 2015). It is a fact that energy consumption patterns have a significant impact on three dimensions of sustainable development (see table 2).

Achieving Goal-7 (Affordable and Clean Energy) contributes economic, social and environmental issues such as (IAEA, 2005; UN, 2015):

- **Economic Issues and Energy:** The share of energy sources and consumption rates are both important indicators for economic development. The price of energy sources determines whether the energy is affordable or not in a country.
- **Social Issues and Energy:** Accessible energy is related with social dimension of sustainable development. When all citizens (people) can access energy sources such as electricity or fuel in a country and when citizens (people) meet their basic energy needs such as illumination, heating or cook, it can be said that Goal-7 is achieved socially.
- **Environmental Issues and Energy:** Clean, renewable and sustainable energy sources and technologies are related with environmental dimension of sustainable development. Promoting renewable energy systems in a country and strengthen sustainable energy sources will support environmental issues in the context of 2030 Sustainable Development Goals.

Investigating households' energy consumption patterns will contribute to achieve Goal-7 and other 2030 SDGs. Researchers can use different indicators or factors to determine energy consumption behavior. This study prefers to investigate 3R principle and energy consumption patterns. 3R principle is almost new concept using in sustainable consumption patterns. Essentially, 3R principle is belonged to circular economy. Circular economy combines economy and environment through 3R principle (Heshmati, 2015). To reduce, recycle and reuse are all main concept of 3R principle. Slawinska et.al. (2014) presented a report for adults that guiding them to adapt 3R principle. According to Slawinska et.al. (2014), the below list shows how consumers can adapt 3R principle in their consumption patterns (Slawinska et.al., 2014:19-20):

Table 1. The framework of 2030 sustainable development goals

Themes	2030 SDGs Goals	Aims[1]	Included by Millennium Development Goals (MDGs)[2]
Poverty	Goal 1: End poverty	Reducing the rate of poor people in the world	Included by Goal-1 in MDGs
Hunger	Goal 2:End hunger	Achieving sustainable food and nutrition for everybody in the world	Included by Goal-1 in MDGs
Health	Goal 3:Ensure healthy lives for everybody	Reducing the rate of mortality from many diseases, preventing deaths of newborns and children (under 5 years). Ensuring standard well-being for people by 2030.	Included by Goal-4, Goal-5 and Goal-6 in MDGs
Education	Goal 4: Ensure equal education	Ensuring free and equitable education for all girls and boys by 2030.	Included by Goal-2 in MDGs
Gender equality	Goal 5: Achieve gender equality	Ending discrimination against all women and girls in the world.	Included by Goal-3 in MDGs
Clean and sustainable water sources	Goal 6: Ensure sustainable water and sanitation sources	Achieving safe and clean water sources for all people in the world by 2030.	Partially included by Goal-7 in MDGs
Energy	Goal 7: Ensure sustainable and affordable energy	Ensuring clean, affordable and accessible energy for all people	Not included
Work and employment	Goal 8: Ensure employment and decent work	Achieving higher economic productivity and full-productive employment by 2030	Partially included by Goal-3 in MDGs
Industry-sustainable industry	Goal 9: Sustainable industrialization	Achieving sustainable production systems and technologies in industry by 2030	Not included
Inequality	Goal 10: Reducing inequalities	By 2030, reducing unequal conditions between countries	Not included
City life and sustainable cities	Goal 11: Achieving Sustainable cities and communities	Ensuring affordable and safe housing services for all people and achieving sustainable cities by 2030	Not included
Consumption and production patterns	Goal 12: Ensure sustainable consumption and production	Promoting sustainable consumption and production patterns.	Not included
Climate changes	Goal 13: Prevent climate changes	Fighting against negative impacts of climate changes	Partially included by Goal-7 in MDGs
Seas and oceans	Goal 14: Sustainable seas and oceans	Reducing marine pollution and endings over-fishing and keeping biodiversity of seas/oceans	Partially included by Goal-7 in MDGs
Forestry	Goal 15: Sustainable forest areas and landing	Increasing protection of forest areas and its biodiversity.	Partially included by Goal-7 in MDGs
Society-peace-justice	Goal 16: Promoting peaceful society through justice	Reducing violence and abuse, promoting justice and legal implications.	Not included
Global partnerships	Goal 17: Promoting global partnerships and collaboration	Strengthen global partnerships and supporting less developed and poor countries.	Included by Goal-8 in MDGs

[1]Adapted from UN, 2015
[2]Adapted from Millennium Development Goals Report 2015

Table 2. The link between energy and sustainable development's dimensions

Social dimension		Economic dimension		Environmental dimension	
The Link with	**Energy issue**	**The Link with**	**Energy issue**	**The Link with**	**Energy issue**
Accessibility	Share of households with/ without energy	Usage	Energy use- total consumption Energy use per capita	Climate change	CO_2 emissions from energy, Population and GDP
Affordability	Share of household income spent on energy	Productivity	Energy use per unit of GDP	Air quality	Ambient concentrations of air pollutants in urban areas, Air pollutant emissions from energy systems
Disparities	The usage of various energy sources by households	Supply efficiency	Losses in distribution process	Water quality	Contaminant discharges in liquid effluents from energy systems including oil discharges
		Production	Reserves-production Total energy production	Soil quality	Affected soil area Critical load
		End use	Industrial demand, agricultural energy demand, Household demand, Service/commercial demand, Transport demand	Forest	Rate of deforestation
		Diversification	Fuel shares in energy and electricity, Non-carbon energy share, Renewable energy share	Waste management	Rate of waste from energy consumption and production
		Prices	End-use prices (with tax and without tax)		
		Imports	Energy imports		
		Strategic Fuel Stocks	Stocks of critical fuels		

Source: adapted from IAEA (International Atomic Energy Agency), 2015, 11-15.

- Reduce: It means reducing every kind of consumption activities. It is the first step in 3R principle. Examples for reducing consumption activities:
    - Buying products without any packaging or less packaged.
    - Reducing or end using one-time used plastic bags and preferring long-time used bags for shopping.
    - Reducing buying unnecessary products and preferring to borrow.
    - Reducing personal drive and preferring to use public transportation or walk, ride a bike etc.
    - Reducing the rate of your waste through using digital products such as online magazine, news, e-book etc.

    - Saving energy and reducing the usage of excessive electricity at home.
    - Saving water at home through not using excessive water and be careful to use efficiently.
- Reuse: It is the second step for sustainability principle. It means finding new ways to use same products for a long time. Examples:
    - Using imagination to create new usage area of used products such as coffee cans, show boxes etc.
    - Don't throw used things such as clothes, toys or furniture. Donate them or give them to other people.
    - Using every product fully.
- Recycle: The third step in 3R principle is to recycle. Recycled items help us to waste less and support environmental quality. Examples:
    - Preferring to use recycling products or recycling packaged products.

3R principle is the core of circular economy. Reduce means in economic production as: reducing materials, the usage of water and energy at first side. Then, reducing production and reducing demand are important activities. Reuse means producing multi-usable products. With reuse activity, waste can be transformed into raw material for industry and renewable resources are prior for production. Recycle principle keeps using waste and materials in a production process (Xuan et.al., 2011:1399-1400).

This study aims to explain energy consumption behavior through 3R principle. As a sample case, this study investigates Turkish households' energy consumption behavior. In addition, this study explains 3R principle usage in energy consumption and suggests an alternative model that explaining 3R principle usage in energy consumption. In this chapter, the below issues are explained:

- What is the link between 2030 Sustainable Development Goals and Energy Issue?
- What is the usage of 3R principle in consumption patterns?
- What is the current profile of Turkish households' energy consumption?
- How Turkish households can adapt 3R principle in energy consumption?

HOUSEHOLD ENERGY CONSUMPTION PATTERNS

Increasing awareness of environmental problems and the concern about global climate change have forced countries to transform into sustainable economies (Yıldırım and Yıldırım, 2020). Accordingly, increasing focus on sustainability has lead citizens (people) to change their consumption patterns such as using or preferring energy-saving products or using energy more efficiently (Grønhøj and Thøgersen, 2011; Rasanen et.al., 2008). Energy consumption patterns can influence sustainable development positively or negatively. The literature provides many researches that investigating households energy consumption patterns as below:

- Wei et.al. (2007) investigated the link between consumers' energy consumption behavior and lifestyles and they found that there was a significant link between personal lifestyle and energy consumption.
- Hu et.al. (2017) studied on the relationship between energy consumption and energy usage behavior of households in China. They used survey method to collect data from households. The study

showed that Chinese urban households were smaller-sized families in general. Compared with other countries, the rate of Chinese urban households' energy consumption was lower. This study proved that the lifestyles, traditional behaviors and culture were important factors for achieving energy efficiency policies in a country.

- Reiss and White (2008) determined that household energy consumption was related with energy prices. When the electricity price increased, households reduced the rate of electricity consumption.
- Vassileva (2012) investigated household energy consumption behavior in Sweden. The study (thesis) found that seasonal and hourly changes were effected on household energy consumption. For example, long and dark days/nights made households spent more electricity. The study suggested that household's habits and consumption patterns should be transformed into sustainable consumption patterns to reduce energy consumption in general.
- Aldabas et.al. (2015) studied on changing energy consumption behavior. They suggested that habits should be changed immediately. Communication channels were helpful for sending message to consumers when creating the awareness of environmental change.
- Bin and Dowlatabad (2009) found that implication of energy-saving policies and reducing CO2 emission were related with understanding of consumer's energy consumption patterns. They investigated the relationship between consumption patterns and environmental impacts through Consumer Lifestyle Approach in the US. It was seen that direct influences from household activities such as home energy use and personal travel were so low in the US GDP (%4) but the usage of energy (%28) and CO2 emission (%41) were higher in a total share. Indirect influences from household activities such as housing operations, transportation, food etc.) had higher share of energy use and CO2 emissions.
- Singh et.al. (2017) studied on household energy consumption behavior through their appliance preferences. They suggested that understanding of household energy consumption behavior and their preference of appliance contributed for setting successful energy-saving policies in a country. According to theoretical view, the usage of appliance can influence energy consumption and can support energy-saving.
- He and Greenberg (2008) studied on the importance of giving information about sustainable energy usage at home for consumers. They investigated household sustainable energy consumption behavior through some psychological factors.
- Sapci and Considine (2014) investigated the relationship between household energy consumption behavior and their environmental attitudes in USA. They found that households who were environmentally concerned, also were more careful to save energy.
- Zhou and Yang (2016) explained household energy consumption behavior through big data analyze. The study suggested that big data could be used to determine household energy consumption behavior. Household energy consumption behavior can be change based on individuals.
- Jain et.al. (2013) investigated the impact of eco-feedback systems and household energy consumption behavior. The study determined that information representation had a significant effect on household energy reduction and energy-saving activities.
- Al-Mumin et.al. (2003) investigated why household highly consumed electricity and the general attitudes of energy consumption in Kuwaiti residences. The study showed that Kuwaiti households didn't turn off all lights when the rooms were vacant and they used thermostat with hot

santigrat. According to simulation studies, Kuwaiti households could reduce the annual electricity consumption through energy-saving behaviors.

- Vassileva et.al. (2012) investigated energy consumption patterns of Swedish households. In the study, households were informed about their daily electricity consumption by previous months and years and also they were informed how they can reduce electricity consumption. The study showed that it was important to informed households about their energy consumption.
- Malla (2013) investigated the relationship between household energy consumption patterns and environmental quality in Nepal. The study suggested increase of biomass energy consumption can reduce energy poverty and improve energy access in Nepal.
- Sahakian (2011) explored household electricity consumption patterns in Metro Manila, Philippines. The results showed that household electricity consumption influenced environmental quality. In addition, global trends in energy consumption and structural factors such as building type had a significant impact on local household energy consumption patterns.
- Druckman and Jackson (2008) showed the relationship between geographical factors, socio-economic factors, household energy consumption patterns and carbon emissions in the UK.

RESEARCH DESIGN AND AIM

This study aims to explain how household adapts 3R principle in energy consumption behavior. In this context, this study investigates Turkish households' energy consumption patterns through 3R principle. It is thought to be the first study that explaining the usage of 3R principle in energy consumption behavior in the literature.

The study employs quantitative research design and presents current profile for energy consumption behavior of Turkish Households. The secondary data (such as articles, books, chapters, proceedings etc.) was used for literature survey. Then, the data from Turkish Statistical Institute was analyzed to determine Turkish household energy consumption habits. As a sample, it is thought to prefer Turkish households will be a great case. The Turkish government accepted 2030 Sustainable Development Goals and began to achieve sustainable development goals for a long time ago. Table 3 presents what Turkey did for sustainability according to climate changes and global agreements.

Turkey also pays more attention to energy issues according to 2030 SDGs. As seen in table 3 and table 4, Turkey keeps 2030 Sustainable Development Goals and achieving sustainable energy policies. Turkey also developed alternative energy sources and energy policies to achieve energy security such as TANAP (Yildirim et.al., 2017). It can be said that Turkey has focused on energy policies, recently. Accordingly, it is thought that Turkish household will be reliable sample case to investigate 3R principle in household energy consumption pattern.

Achieving sustainable energy will support 2030 SGDs directly or indirectly in the long term (OECD, 2007). To determine the significant effects of energy consumption patterns on sustainable development, the indicators of household energy consumption should be explained. Household energy consumption is generally based on the usage of electricity at home (Rasanen et.al., 2008). So, the reduction of electricity usage will contribute positively to energy efficiency in the long term. Hu et.al. (2017) determined that urban residential buildings influenced energy consumption patterns. For example, the rate of energy consumption in China's urban residential buildings was lower than the United States and Japan. Accord-

Table 3. Turkey's position in the global agreement on climate change

Years	Agreements and Positions
1992	Turkey joined into the list for Annex-1 and Annex-2 of The United Nations Framework Convention on Climate Change (UNFCCC).
1995-2000	As being a developing country, Turkey attempted to exit from Annexes of the UNFCC in the periods between COP1 and COP6. Then, Turkey wanted to drop out Annex-2 and to join into Annex-1 under special conditions.
29 October-6 November 2001	Turkey dropped out the Annex-2 by the 7th Conference of the Parties (COP7) and proceed with Annex-1 through special conditions.
2004	Turkey was recorded as 189th country that officially participating in the UNFCCC. Today, Turkey is still on the list in Annex-1 with the special conditions.
5 February 2009	Turkey began to work for the Kyoto Protocol and Turkish Parliament passed a bill to join the Kyoto Protocol in February 2009.
26 August 2009	Turkey became an official part in the Kyoto Protocol.
2010	It was confirmed that Turkey had a different position in the Annex-1when comparing other countries and there was no obligation to provide technology transfer by COP 16.
2011	It was determined that Turkey should continue to ensure negotiations on emission reduction, adaptation to climate change, providing support for the necessary technology and financing areas by COP 17.
2012	It was decided to start the Kyoto Protocol by 2013 and to end by 2020 in COP 18. According to the special conditions of Annex-2, Turkey should take technological and financial supports to achieve the low carbon development strategies.
2013	The negotiations about technological financial supports for Turkey was accepted by COP 19 when fighting against Climate Change
2014	It was decided that Turkey should be supported with technology and financial sources until 2020 by COP 20.
2015	Paris Agreement emerged by COP 21.
22 April 2016	Turkey signed Paris Agreement on April 22, 2016. There were negotiations about missing point of Paris Agreement in COP 22.
2017	The study on the implementation of the Paris Agreement was continued at COP 23.
2018	The results from negotiations on the implementation of the Paris Agreement were waited at COP 24.
* COP: Parties Agreement Information: 1: Compiled from the information of Turkish Ministry of Energy and Natural Resources.	

Source: adapted from Yıldırım and Yıldırım, 2019.

ing to International Atomic Energy Agency (2015)'s study, final energy consumption of households can be classified as below (IAEA, 2015:157):

- Space Heating
- Water Heating
- Cooking
- Lighting
- Refrigerators
- Freezers
- Clothes Washers
- Clothes Dryers
- Dishwashers
- Other Appliances

Table 4.The historical development of Turkish master plan for energy efficiency

Related years	Policies
2007	It was determined that the main purpose of the Energy Efficiency Law No. 5627 is to increase efficiency in the use of energy resources through using energy effectively, preventing waste, reducing the burden of energy costs on the economy and protecting the environment.
2010-2023	Within the scope of the National Climate Change Strategy Document, it was aimed to increase energy efficiency and reduce greenhouse gas emissions in the building, industry, transportation and energy sectors.
2012-2023	The main policy set and related actions were determined to achieve Energy Efficiency Strategy Programs.
2014-2018	The precautions about energy efficiency were determined in the context of "Energy Efficiency Improvement Program" which was numbered (1.14) of the Tenth Development Plan.
2015-2019	The Ministry of Energy and Natural Resources defined main goals through "Goal 4: Energy Efficient Using a Turkey", and "Goal 5: Advanced Capacity Building for the Energy Efficiency and Savings" which were sub-goals of "Theme 2: Energy Efficiency and Savings" in 2015-2019 Strategic Plan
2017	The goals of the National Energy Efficiency Action Plan were also included in the National Energy and Mining Policy which was prepared by the Ministry of Energy and Natural Resources by 2017.

Source: adapted from The Ministry of Energy and Natural Resources, 2018.

This study figures household energy consumption behavior through 2 main categories. Firstly, this study considers usage of energy such as electricity, natural gas and fuel. Secondly, the usage of electronics and batteries are considered for 3R principle. The study suggests the below model to explain how 3R principle can be adapted into energy consumption behavior:

3R PRINCIPLE AND ENERGY CONSUMPTION PATTERNS: A CASE OF TURKISH HOUSEHOLD

Turkey is a crowded developing country so it is expected that the future energy demand will be higher. Compared to the previous year (2019), it is seen that Turkish population increased and it reached 83.154.997 people. While the male population was 41.721.136 people, the female population was 41.433.861 people. Population growth rate was 13.9 per thousand. While the annual population growth rate was 14.7 per thousand in 2018, it was recorded as 13.9 per thousand in 2019 (TURKSTAT, 2020a). As showed in table 6, Turkish population has increasing trend recently. Also, the population density was increased by 2019.

Table 5. The research model: 3R principle in energy consumption pattern at home

<table>
<tr><th>Reduce</th><th>Recycle</th><th>Reuse</th></tr>
<tr><td>Reducing electricity consumption</td><td>Electronics recycling</td><td>Maintenance of electronics</td></tr>
<tr><td>Reducing natural gas consumption</td><td>Battery recycling</td><td>Usage of rechargeable electronics</td></tr>
<tr><td>Reducing electronics usage and waste</td><td>Having Energy recycling systems in houses</td><td rowspan="3">Usage of rechargeable batteries</td></tr>
<tr><td>The usage of Energy-saving products</td><td rowspan="2">Collecting bio-waste</td></tr>
<tr><td>Reducing battery usage and waste</td></tr>
</table>

Table 6.Annual growth rate and population density of provinces by years, 2007-2019

Years	Annual growth rate of population (%)	Years	Population density
2007-2008	13,1	2007	92
2008-2009	14,5	2008	93
2009-2010	15,9	2009	94
2010-2011	13,5	2010	96
2011-2012	12,0	2011	97
2012-2013	13,7	2012	98
2013-2014	13,3	2013	100
2014-2015	13,4	2014	101
2015-2016	13,5	2015	102
2016-2017	12,4	2016	104
2017-2018	14,7	2017	105
2018-2019	13,9	2018	107
		2019	108
Info: The results of Address Based Population Registration System (ABPRS), 2007-2019			

Source: adapted from TURKSTAT, (a).

The last recorded data for energy consumption of Turkish market is presented in table 7. The percentage distribution of energy consumption includes parts as household, commercial, government, industrial, illumination and others. Two parts that have the highest share in energy consumption are seen as industrial and household. So, it can be said that industrial and household have the first responsible for energy efficiency.

During the period between January and March, 341.038 houses were sold (see table 8). Compared to the same period of the previous year, there was a 33.0% increase in house sales.

The fact that household has a big share in energy consumption. The average energy consumption per person was recorded as 3.149 by 2018 and the number of household was recorded as 23.221.218. By 2017, the average energy consumption per person was recorded as (see table 9). It can be said that the higher numbers of household increase energy consumption per capita in a country.

According to Environmental Indicators (2016), population and energy indicators and developments can be shown in the table 9. It is seen that there was a negative increase in the consumption parts. The population increased negatively in terms of sustainability. The energy efficiency index was evaluated as a response indicator in the study. Response indicators included the reactions of the society and individuals against changes in the state of the environment. In other words, indicators related to solutions against environmental pollution. The purpose of the “Energy Efficiency Index” study was to determine how much savings was made according to primary and final energy consumption by calculating the amount of energy savings provided in the manufacturing, housing and transportation sectors, and how much contribution was made to the national economy by improving energy efficiency. As a result, 1.8% (annually) improvements were recorded as the energy consumption of manufacturing industry. In addition, 1.9% improvements were recorded as the energy consumption of household and 2.7% improvements were recorded as the energy consumption of transportation in Turkey. The energy efficiency and saving

Table 7. Distribution of net electricity consumption by sectors (GWh and percentage)

Years	Total	Household	Commercial	Government	Industrial	Illumination	Others
2000	98.296	24,3	9,5	4,2	49,7	4,6	7,7
2001	97.070	24,3	10,2	4,5	48,4	5,0	7,6
2002	102.948	22,9	10,6	4,4	49,0	5,0	8,1
2003	111.766	22,5	11,5	4,1	49,3	4,5	8,1
2004	121.142	22,8	12,9	3,7	49,2	3,7	7,7
2005	130.263	23,7	14,2	3,6	47,8	3,2	7,5
2006	143.071	24,1	14,2	4,2	47,5	2,8	7,2
2007	155.135	23,5	14,9	4,5	47,6	2,6	6,9
2008	161.948	24,4	14,8	4,5	46,2	2,5	7,6
2009	156.894	25,0	15,9	4,5	44,9	2,5	7,2
2010	172.051	24,1	16,1	4,1	46,1	2,2	7,4
2011	186.100	23,8	16,4	3,9	47,3	2,1	6,5
2012	194.923	23,3	16,3	4,5	47,4	2,0	6,5
2013	198.045	22,7	18,9	4,1	47,1	1,9	5,3
2014	207.375	22,3	19,2	3,9	47,2	1,9	5,5
2015	217.312	22,0	19,1	3,7	47,6	1,9	5,7
2016	231.204	22,2	18,8	3,9	46,9	1,8	6,4
2017	249.023	21,8	19,8	4,1	46,8	1,8	5,7

Source: TURKSTAT,(2018).

potential of the building sector reached 50% which was compared to current consumption (Environmental Indicators, 2016:103-104).

The electricity demand has increased in parallel with the rising of living standards and the population in Turkey, recently. The gross electrical energy consumption, which increased by 3.3% compared to the

Table 8.Total house sales (2013-2020)

Years	Total Sales	Total number of January-February-March
2013	1 157 190	273 919
2014	1 165 381	257 853
2015	1 289 320	297 218
2016	1 341 453	303 464
2017	1 409 314	325 780
2018	1 375 398	303 877
2019	1 348 729	256 433
2020	341.038*	341.038
*Total sales of 2020 includes January, February, March		

Source: adapted from TURKSTAT, 2020b.

Table 9. Electricity consumption per capita and number of households

Years	Consumption per capita (kWh)	Number of households
2007	2.198	-
2008	2.264	-
2009	2.162	-
2010	2.334	-
2011	2.490	-
2012	2.577	19.842.850
2013	2.583	20.489.721
2014	2.669	21.091.075
2015	2.760	21.662.260
2016	2.897	22.206.776
2017	3.082	22.676.186
2018	3.149	23.221.218
2019	-	24.001.940

Source: TURKSTAT

previous year (2015), reached 278.3 billion kWh in 2016. The future demand scenario suggests that the demand will be 407.9 billion kWh by 2023. However, Turkey has still lower electricity consumption in the OECD countries (Energy and Natural Resources Ministry, 2017: 22). To meet future energy demand, Turkey keeps 2023 National Strategy and Goals and energy-saving policies and adaptation of green energy systems are all promoted by Turkish governments (Energy and Natural Resources Ministry.a). Energy efficiency programs aim to reduce energy consumption per unit or product amount, without causing a

Table 10.Indicators of Population and Energy 2016

Energy Indicators	Result	Population Indicators	Result
Total energy consumption	Negative increase	Population	Negative increase
Share of coal and its derivatives in primary energy consumption	Negative increase	Population growth rate	Negative increase
Total final energy consumption	Negative increase	Urban-rural population ratio	Negative increase
Energy consumption per capita	Negative increase	Migrating population	Positive decrease
Primary energy production	Negative increase		
Share of renewable energy sources in consumption	Positive increase		
Electricity generated from renewable sources			
Primary energy density	Positive decrease		
Final energy density	indifferent		
Energy efficiency	indifferent		

Source: Environmental Indicators, 2016, 8-9.

decrease in living standards and service quality in buildings and manufacturing sector. Energy saving and efficiency are one of the most important components of "Turkish 2023 National Strategy Goals and Energy Policies", such as ensuring energy supply security, reducing the risks of external dependence, protecting the environment and increasing the effectiveness of the fight against climate change. In this point, the main goal is to reduce at least 20% of energy consumption by 2023 (Energy and Natural Resources Ministry, b). Turkey Waste Report (2018) noted that Turkish household could contribute positively to energy efficiency programs in the long term. In particular, any sustainable consumption patterns in household heating, enlightenment, cleaning, personal care can improved energy-saving. For example, approximately 20 percent of the energy used in the buildings was consumed by electrical appliances. Accordingly, the use of energy efficient appliances can minimize energy consumption. Not only reducing energy consumption, also household should reduce usage of electronics and batteries which are big threats for the environment. Household should be careful when using electronic and batteries to extend the exposure time. According to the Report of Turkish Environment Condition by Ministry of Environment and Urbanization (2016), the collected waste of battery was recorded as 13.488 tonnes by 2013 and it was recorded as 11.982 tonnes by 2014. Waste of electronics was recorded as 4.911 tonnes by 2013 and it was recorded as 6.817 tonnes by 2014. Accordingly, reduction of battery usage and the quantity of waste will support environmental quality in the long term.

Table 11.The most common appliances used by Turkish household, 2016

Appliances	Total share
Desktop computer	22,9
Portable computer (Laptop, netbook)	36,4
Tablet computer	29,6
Mobile phone (incl. smart phone)	96,9
Game console	5,6
Fixed line telephone	25,6
Digital camera/photography	20,6
DVD / VCD / DivX player	20,4
Smart TV	24,6
None of above	3,5
Info: TurkStat, Survey on Information and Communication Technology (ICT) Usage Survey in Households and by Individuals, 2016	

Source: TURKSTAT, 2016.

Table 11 shows the most common appliances used by Turkish household by 2016. Turkish household mostly (96.9%) had mobile phone. The 36.4 percentage of household had portable computer, the 29.6 percentage of them had tablet computer and the 25.6 percentage of them had fixed line telephone by 2016 (TURKSTAT, 2016). In addition, white goods have a big share in housing electricity consumption. White goods are one the most purchased products at home. In particular, the housing sales increase the demand for white goods in Turkey. For example, Turkish housing sales increased by 5.1% and the demand of white goods increased in 2017 (TSKB, 2018).

Table 12. Turkish domestic market sales for white goods

White Goods	2015	2016	2017	2018	2019
Refrigerator	1.976.199	2.039.575	2.170.836	1.810.238	1.719.567
Deep Freeze	571.160	648.289	936.489	818.218	766.617
Washing Machine	2.026.292	2.121.619	2.354.768	1.948.671	1.853.693
Dishwasher	1.483.435	1.572.950	1.803.554	1.460.700	1.332.131
Oven	951.231	994.833	1.122.249	946.768	816.737
Dryer	81.734	92.530	145.117	125.598	166.410
Total	7.090.051	7.469.796	8.533.013	7.110.193	6.655.155

Source: TURKBESD.

In the last five years, domestic market of white goods had peak and fall in sales of Refrigerator, Deep Freeze, Washing Machine, Dishwasher, Oven and Dryer. On the other side, TURKBESD (2020) reported that Turkish domestic sales of white goods (including 6 main goods as Refrigerator, Deep Freeze, Washing Machine, Dishwasher, Oven and Dryer) increased by 11% in the first quarter of the year. During COVID-19 crises, the domestic sales of white goods increased by 14% in March 2020 (TURKBESD, 2020). In addition, Turkish household purchased small home appliances much more than previous years during the quarantine period. Accordingly, electricity consumption of household increased through increased usage of small home appliances (DHA, 2020).

SOLUTIONS AND RECOMMENDATIONS

This study explains a brief framework for household energy consumption patterns and also it suggests using 3R principle in energy consumption to support 2030 Sustainable Development Goals. This study aims to provide an alternative approach to guide household to get sustainable energy consumption patterns through 3R principle. In the study, 3R principle includes sample activities of reduce, reuse and recycle in energy consumption such as:

- Household can adapt 3R principle into energy consumption behavior to support 2030 Sustainable development Goals and Energy-saving policies in a country.
- Household can reduce buying of electronics such as mobile telephone, home appliances, TV etc.
- Household can maintain every kind of home appliance and mobile telephone or laptop/Desktop.
- Household can reduce the usage of battery and be careful about the waste.
- Household can reduce electricity consumption and save energy through turning off lights and home devices when there is no need.
- Household can adapt new technologies to reuse or recycle energy at home.
- Household should know that they have a big responsibility in energy-saving policies.

FUTURE RESEARCH DIRECTIONS

This study contributes the related literature through showing how households can use 3R principle in energy consumption patterns. In addition, this study suggests an alternative model that explaining the usage of 3R principle in energy consumption patterns. According to theoretical approach, this study concludes some findings that consumers are responsible for sustainable development goals. This study focuses on how households can use 3R principle in energy consumption patterns and contribute 2030 Sustainable Development Goals (SDGs). In the context of theoretical approach, I can suggest the below list for future research directions:

- Future studies can use survey method to explain how household can use 3R principle in energy consumption.
- Future studies can use qualitative research methods and statistical analyzes to determine significant relationships between household energy consumption and 2030 Sustainable Development Goals.
- Future studies can develop alternative models to determine household energy consumption patterns.
- Understanding of household energy consumption behavior will help setting successful energy-saving policies in a country.
- Future studies can investigate household energy consumption patterns during COVID-19 and post-COVID-19 term.
- Future studies can analyze the link between life styles, living standards, culture, conditional factors and household energy consumption patterns.

CONCLUSION

2030 Sustainable Development Goals guides developed and developing countries to transform into green economy. In other words, transforming into green economy is a main aim in the long term (Yıldırım and Yıldırım, 2020). Energy access is a critical problem for many countries and achieving sustainable energy policies is a still big challenge for the world. Energy efficiency or energy-saving policies should be supported by household as same as industry and local governments. In general, the energy consumption share of household has higher proportion in total energy consumption. This study explores Turkish household energy consumption behavior and aims to explain how 3R principle can be adapted into energy consumption patterns. Turkish household energy consumption has a big share in total energy consumption and energy-saving policies should be supported by Turkish household to achieve energy efficiency in the long term. The open access data from TURKSTAT (Turkish Statistical Institute and the Ministry of Energy and Natural Resources was used to determine Turkish household energy consumption behavior. The rising population and house sale are all increase the rate of household energy consumption. In addition, living standards, life style, culture and conditional factors such as COVID-19 crises can influence buying behavior of electronics, small home appliances and white goods that consuming electricity at home. According to Turkish energy consumption patterns, it can be said that household's support for saving energy is so critical. The study can summarize how Turkish household can adapt 3R principle in energy consumption behavior as below table:

Table 13. 3R principle adaptation in household energy consumption

3R principles	Activities by household
Reduce	Reducing electricity consumption. For example, household can get habits of turning off lights and home devices when there is no need.
	Reducing natural gas consumption through cooking efficiently or using heat control by optimum temperature at home.
	Reducing the usage of batteries. For example, household can get habits of removing the battery when electronic equipment is not in use
	Lowering the air conditioner temperature. For example, household can get habits of living with lower and usual heating.
	Buying and using energy-saving electronics
	Buying and using energy-saving lamb for lighting at home
	Reducing the use of fast electricity consuming products such as irons, heaters, toasters, etc.
Reuse	Extending the exposure time of electronic devices
	Extending the exposure time of batteries
	Maintaining electronics such as computes, mobile phone, laptop etc.
	Maintaining home appliances such as iron, heaters, food processor etc.
Recycle	Supporting governments and local governments for electronics recycling. Household can collect their electronic waste and deliver it to formal recycling organizations.
	Collecting battery waste and deliver it to formal recycling organizations
	Preferring to use energy recycling systems at home

REFERENCES

Ahuja, D., & Tatsutani, M. (2009). Sustainable energy for developing countries. *SAPIENS, 2*(1). Retrieved from https://journals.openedition.org/sapiens/823

Al-Mumin, A., Khattab, O., & Sridhar, G. (2003). Occupants' behavior and activity patterns influencing the energy consumption in the Kuwaiti residences. *Energy and Building*, *35*(6), 549–559. doi:10.1016/S0378-7788(02)00167-6

Aldabas, M., Gstrein, M., & Teufel, S. (2015). Changing Energy Consumption Behaviour: Individuals' Responsibility and Government Role. *Journal of Electronic Science and Technology*, *13*(4), 343–348.

Bin, S. & Dowlatabad, H. (2009). Consumer lifestyle approach to US energy use and the related CO2 emissions. *Energy Policy*, *33*, 197–208.

DHA (Dogan News Agency). (2020). *Electricity consumption increased in the quarantine period*. Retrieved from https://www.cnnturk.com/ekonomi/karantinada-elektrik-tuketimi-artti

Druckman, A., & Jackson, T. (2008). Household energy consumption in the UK: A highly geographically and socio-economically disaggregated model. *Energy Policy*, *36*(8), 3177–3192. doi:10.1016/j.enpol.2008.03.021

Grønhøj, A., & Thøgersen, J. (2011). Feedback on household electricity consumption: Learning and social influence processes. *International Journal of Consumer Studies, 35*(2), 138–145. doi:10.1111/j.1470-6431.2010.00967.x

He, H. A., & Greenberg, S. (2008). *Motivating Sustainable Energy Consumption in the Home. Motivating Sustainable Energy Consumption in the Home*. Technical Report 2008-914-27, Department of Computer Science, University of Calgary, Calgary, AB, Canada.

Heshmati, A. (2015). *A Review of the Circular Economy and its Implementation*. Discussion Paper Series, December 2015, IZA DP No. 9611.

Hu, S., Da Yan, D., Guo, S., Cui, Y., & Dong, B. (2017). A survey on energy consumption and energy usage behavior of households and residential building in urban China. *Energy and Building, 148*, 366–378. doi:10.1016/j.enbuild.2017.03.064

IAEA. (2015). *Energy Indicators for Sustainable Development: Guidelines and Methodologies*. International Atomic Energy Agency, United Nations Department Of Economic And Social Affairs, International Energy Agency, Eurostat And European Environment Agency, Austria April 2005, STI/PUB/1222. Retrieved from https://www.iaea.org/Publications/index.html

Jain, R. K., Taylor, J. E., & Culligan, P. J. (2013). Investigating the impact eco-feedback information representation has on building occupant energy consumption behavior and savings. *Energy and Building, 64*, 408–414. doi:10.1016/j.enbuild.2013.05.011

Johnston, D. (2001). *Sustainable energy for future generations*. NEA News 2001 – No. 19. Retrieved from https://www.oecd-nea.org/pub/newsletter/2001/sustainable-energy19-1.pdf

LEAD (Leadership for Environment and Development). (2016). *Sustainable Development Goals (SDGs) 2015-2030*. June 2016 Briefing Note.

Malla, S. (2013). Household energy consumption patterns and its environmental implications: Assessment of energy access and poverty in Nepal. *Energy Policy, 61*, 990–1002. doi:10.1016/j.enpol.2013.06.023

Minelli, F. (2017). *Achieving energy security And Environmental sustainable development*. ODUMUNC Issue Brief Economic and Social Council (ECOSOC). Retrieved from https://www.odu.edu/content/dam/odu/offices/mun/2017/ib-2017-ecosoc-energy-security.pdf

Munasinghe, M. (2002). The Sustainomics Trans-disciplinary Meta-framework for making Development more Sustainable: Applications to energy issues. *International Journal of Sustainable Development, 4*(2), 6–54. doi:10.1504/IJSD.2002.002563

OECD. (2007). OECD Contribution to the United Nations Commission on Sustainable Development 15 Energy for Sustainable Development. OECD Publishing.

Owusu, P. A., & Asumadu-Sarkodie, S. (2016). A review of renewable energy sources, sustainability issues and climate change mitigation. *Cogent Engineering, 3*(1), 1167990. doi:10.1080/23311916.2016.1167990

Rasanen, T., Juhani Ruuskanen, J., & Kolehmainen, M. (2008). Reducing energy consumption by using self-organizing maps to create more personalized electricity use information. *Applied Energy*, *85*(9), 830–840. doi:10.1016/j.apenergy.2007.10.012

Reiss, P. C., & White, M. W. (2008). What changes energy consumption? Prices and public pressures. *The RAND Journal of Economics*, *39*(3), 636–663. doi:10.1111/j.1756-2171.2008.00032.x

Sahakian, M. D. (2011). Understanding household energy consumption patterns: When "West Is Best" in Metro Manila. *Energy Policy*, *39*(2), 596–602. doi:10.1016/j.enpol.2010.10.032

Sapci, O., & Considine, T. (2014). The link between environmental attitudes and energy consumption behavior. *Journal of Behavioral and Experimental Economics*, *52*, 29–34. doi:10.1016/j.socec.2014.06.001

Singh, S., & Yassine, A. (2017). Mining Energy Consumption Behavior Patterns for Households in Smart Grid. *IEEE Transactions on Emerging Topics in Computing*, *7*(3), 404–419. doi:10.1109/TETC.2017.2692098

Slawinska, K., Soltysiak, M., Davies, S., Bénard, L., Finelli, T., & Tomassone, A. (2014). A toolkit On Promoting Sustainable Development Practices To Adults Learners 3R's – RECYCLE, REDUCE, REUSE. Library of Work Pedagogy Publish.

The Energy and Natural Resources Ministry. (n.d.a). *Electricity*. Retrieved from https://www.enerji.gov.tr/tr-TR/Sayfalar/Elektrik

The Energy and Natural Resources Ministry. (n.d.b). *Energy Efficiency*. Retrieved from https://www.enerji.gov.tr/tr-TR/Sayfalar/Enerji-Verimliligi

The Ministry of Commerce. (2018). *Turkey Waste Report*. General Directorate Of Consumer Protection And Market Surveillance.

The Ministry of Energy and Natural Resources. (2018). National Energy Efficiency Action Plan 2017-2023. Author.

The Ministry of Environment And Urban, Environmental Impact Assessment, General Directorate Of Authorization And Audit, Environmental Inventory And Information Management Department. (2018). *Environmental Indicators 2016*. Publication No: 38-1.

TSKB (Turkish Industrial Development Bank). (2018). *Sectoral Outlook*. White Goods February 2018. Retrieved from http://www.tskb.com.tr/i/assets/document/pdf/beyaz-esya-sektorel-gorunum-subat-2018.pdf

TURKBESD. (2020). *White goods sales increased by 11 percent in the first quarter.* Retrieved from http://www.turkbesd.org/Haberler.php?P=37

TURKBESD (White Goods Manufacturers' Association of Turkey). (n.d.). *TÜRKBESD Last 5 Years Figures, Domestic Sales*. Retrieved from http://www.turkbesd.org/bilgiler.php?P=25

TURKSTAT. (2018). *The Statistics of Energy Consumption by Sectors, 2017*. Retrieved from http://www.tuik.gov.tr/PreHaberBultenleri.do?id=27667

TURKSTAT. (2020a). *Address Based Population Registration System Results 2019*. Retrieved from http://www.tuik.gov.tr/HbGetirHTML.do?id=33705

TURKSTAT. (2020b). *Housing Sales Statistics, March 2020*. Retrieved from http://www.tuik.gov.tr/PreHaberBultenleri.do?id=33878

TURKSTAT. (n.d.). *Basic Statistics, Population Projections*. Retrieved from http://www.tuik.gov.tr/UstMenu.do?metod=temelist

TURKSTAT (Turkish Statistical Institute). (2016). *Household Information Technologies Usage Research, 2016*. Retrieved from http://tuik.gov.tr/PreHaberBultenleri.do?id=21779

UN (United Nations). (2015). *Transforming Our World: The 2030 Agenda for Sustainable Development, A/RES/70/1*. Retrieved from https://sustainabledevelopment.un.org/content/documents/21252030%20Agenda%20for%20Sustainable%20Development%20web.pdf

Vassileva, I. (2012). *Characterization Of Household Energy Consumption In Sweden*. School of Sustainable Development of Society and Technology Mälardalen University Press Dissertations No. 129.

Vassileva, I., Wallin, F. & Dahlquist, E. (2012). Understanding energy consumption behavior for future demand response strategy development. *Energy, 46*, 94-100.

Wei, Y. M., Liu, L. C., Fan, Y., & Wu, G. (2007). The impact of lifestyle on energy use and CO2 emission: An empirical analysis of China's residents. *Energy Policy*, *35*(1), 247–257. doi:10.1016/j.enpol.2005.11.020

WHO. (n.d.). *Millenium Development Goals (MDGs)*. Retrieved from https://www.who.int/topics/millennium_development_goals/about/en/

Xuan, L., Baotong, D., & Hua, Y. (2011). The Research Based on the 3-R Principle of Agro-circular Economy Model-The Erhai Lake Basin as an Example. *Energy*, *5*, 1399–1404.

Yildirim, D. C., Erdogan, S., Yildirim, S., & Can, H. (2017). The effect of the Trans-Anatolian Natural Gas Pipeline Project (TANAP) on industrial production in Turkey. *International Journal of Energy Sector Management*, *11*(3), 404–415. doi:10.1108/IJESM-10-2016-0005

Yıldırım, D. Ç., Yıldırım, S., & Demirtas, I. (2019). Investigating energy consumption and economic growth for BRICS-T countries, *World Journal of Science. Technology and Sustainable Development*, *16*(4), 184–195.

Yıldırım, S., & Yıldırım, D. Ç. (2019). Sürdürülebilir Kalkınma Kapsamında Türkiye'nin Sürdürülebilir Enerji Kaynakları Profili [Energy Resources Profile of Turkey in the context of Sustainable Development]. In İktisadi Teknik ve Strateji Boyutları ile Türkiye'de Enerji Sorunsalı [Economic, Technical And Strategy Dimensions in Turkey by The Energy Problems]. Ekin pub.

Yıldırım, S., & Yıldırım, D. Ç. (2020). Achieving Sustainable Development Through a Green Economy Approach. In Advanced Integrated Approaches to Environmental Economics and Policy: Emerging Research and Opportunities. IGI Global Pub.

Yoshida, T., & Zusman, E. (2015). Achieving the multiple benefits of a sustainable development goal for energy. In Achieving the Sustainable Development Goals: From Agenda to Action. Academic Press.

Zhou, K., & Yang, S. (2016). Understanding household energy consumption behavior: The contribution of energy big data analytics. *Renewable & Sustainable Energy Reviews*, *56*, 810–819. doi:10.1016/j.rser.2015.12.001

KEY TERMS AND DEFINITIONS

2030 Sustainable Development Goals: There are main 17 sustainable targets that are expected to be achieved by 2030.

3R Principle: It includes activities of reduce, reuse, and recycle based on sustainability principle.

Energy Consumption: Consumption or usage of energy sources.

Sustainable Consumption: Consumers prefer buying green products or reducing consumption to be less harmful for the environment.

Sustainable Development: A development model that aims to achieve economic development through protecting natural environment and keeping social welfare.

Chapter 6
Insight of Green Economy in Algeria

Zina Arabeche
https://orcid.org/0000-0003-3893-3378
Ahmed Zabana University of Relizane, Algeria

Mohammed El Amine Abdelli
https://orcid.org/0000-0002-5784-4306
University of Salamanca, Spain

ABSTRACT

Since the emergence of the industrial revolution, the use of energy resources has increased considerably, particularly non renewable (coal, oil and gas), so these resources are no longer sufficient to cover the different energy needs, and this has become a challenge to the energy independance of many gouvernment now and in the future. This has caused the world to scramble for other ways to satisfy these needs in which the results of scientific research and development envolved from alternative uses of old energy sources and named green economy, and many think about the future of energy despite the barriers that still hinder the general application of this type of economy. Energy consumption in Algeria is focused almost entirely on fossil fuels, hydrocarbons, and gas in particular. In recent years, Algeria is deciding to move intro the green economy, the unique solution for meeting future energy requirements and helping reduce environmental risks.

INTRODUCTION

It is clear that humanity is living on earth farther form sustainably. We are doing very few activities in a completely sustainble way.Therefore, changes need to be made around the board. Thankfully, there are plenty of suggestions, good practices and success stories that span almost all industries, demonstrating how we can embrace more green technology and begin living their lives.

The notion of Green Economy is not completely a new term. It was first proposed by the London Environmental Economics Centre (LEEC) in 1989 written by David Pearce, Anil Markandya, and Ed

DOI: 10.4018/978-1-7998-4915-5.ch006

Barbier[1]. Nevertheless, at that period the term did not receive universal acceptance. With the emergence of the economic crisis in 2007 and the inability of many countries to step towards a sustainable development direction, it has become clear that the recent development model does not produce the desired results on all sides economic, social, and environmental. One of the explanations for the lack of significant change was the failure to make a real business case environmental investment. In order to persuade strategy and decision makers to engage in the environment, they must be assured that such a transformation would also bring economic benefits. Such advantgaes involve creating additional jobs, increasing productivity, creating new business niches and developing trade, and having a positive impact on GDP.

It is therefore important to show that investment in the environment, socioeconomic and green development have a clear connection. As the ultimate objective of any development strategy should be human wellbeing, a well-designed green economy strategy will contribute to the eradiction of poverty. Green Economy could be seen as a strategy that emphasizes these links. It could therefore be regarded as a process or tool that drives the transition to sustainable development.

Embracing an ecosystem view instead of dealing with various environmental biomes like forests, deserts, and aquatic life as different and isolated environment elements. The adoption of an ecosystem approach allows for a good understanding and appreciation of the reality that the different biomes are part of an embedded whole i.e. part of an environment. Such an method guarantees a better understanding of the connections and interrelationship between the various biomes. As a result, it leads to better green economy.

The Problem of Paper and Questions Subcommittee: Through the above highlights we have the following problem: *What is the reality and the prospects of green Economy in Algeria ?*

And through this problem highlights the following sub-questions:

- *what are we talking about when we are talk about Green innovation ?*
- *What is the difference between green innovation and theses concepts: eco-innovation, Sustainable development and Environmental innovation* ?
- *what is the concept of green economy ?*
- *what is the requirements for achieving the green economy ?*

Objectives of The study: This study aims to:

- Identify concepts related to green innovation and green economy .
- Know the reality of Algeria in the field of Green economy and renewable energies .

The Importance of the Study: demonstrating the importance of the paper through the importance of the issues addressed by:

- The importance of the subject of Green innovation
- The importance of the subject of renewable energy as the energy of the future.
- The importance of the subject of Green Economy in Algeria.

The Methodology of the Work: the preparation of this paper was based on the descriptive and analytical method, in which, we find the descriptive method in the definition of variables of the research, the analytical method applied in the case study.

This study is structured as follows: Part 1 contains the conceptual framwork of this paper. A review of existing literature regarding innovation and green innovation. Part2 will describe the dimensions of green innovation and the similar words of this concept. A business model of green innovation will be presented in part 3, followed by a brief discussion and the conclusion of this paper in part 4.

GREEN ECONOMY CONCEPT

The United Nations Environmental Programme (UNEP) has led the development and propagation of the notion of the Green Economy. According to UNEP, Green Economy aims *"to improve human well-being and social equity while significantly reducing environmental risks and ecological scarcity"*. Throughout this form of economy, revenue growth and jobs come from public and private projects that reduce emissions and pollution, improve resource efficiency and avoid loss of biodiversity and technical services[2].

The notion of a green economy was developed as a response to the inability of neoclassical systems to effectively integrate the importance of natural resources and ecological degradation in pricing and other business mechanisms . Over the past 50 years, Global economic development has been followed by an unpercedented environmental decline. The global Gross Domestic Product (GDP) more than doubled from 1981 to 2005, but 60% of the world's habitats have been destroyed or unsustainably used. In essence, the world's economies have so far overexploited natural resources and underestimated the ecological products and services that form the basis of all business activity .

Economic development has put unprecedented demands on the Earth's natural resource base and energy constained environment already. Most resources, like water, fertile soil and fossil fuels, are rapidly puching the limits of their production capacity. Furthermore, the effects of climate change involving unpredictable weather motifs. The green economy has became a global priority in response to the many challenges that the world has faced in recent years: climate change, the nutrition and financial crises, and the increase of poverty and social injustice. The United Nations Environment Program (UNEP) has suggested a 'Global Green New Deal' to restore the global economy and increase employment while speending up the fight against climate change, ecological degradation and poverty. A number of intergovernmental green economic projets have consequently been developed, such as: The Green Economy Initiative of the UNEP's; The International Labour Organisation's (ILO) Green Jobs Initiative ; and the Organisation for Economic Cooperation and Development's (OECD) Green Growth Strategy .

The sources of the climate change, nutrition, social and economic issues that have occurred over the past decade differ, but all sharing a common feature- the massive misallocation of ressources. Much capital has been invested in property development, fossil fuels and corporate finance over the past two decades.

In contrast, comparatively little investment has been made in the production of renewable energy, increasing energy efficiency, encouraging and enhacing effective public transport, sustainable agriculture, the preservation of habitats and and biodiversity, land the conversationn of land and water.

Years and years of creating new wealth through this 'brown economy' model injected by fossil fuels, has not achieved in substantially directing social marginalization, ecological degradation and resource depletion.

The green economy needs a decoupling of economic development from resource utilization and environmental degradation. Convergence refers to the environmental impact correlated with any economic activity (both in terms of resource use and the emissions and wastes generation). The Word Green Economy Organisation (WGEO) describes a thriving Green Economy as one that is: *"low carbon, climate-resilient, resource-efficient and socially-inclusive"*[3]. This is the description that it uses in the report of the world green economy organization. With the understanding that it is an economy able of delivering the United Nations Sustainable Development Goals (SDGs, also known as the Global Goals). The strategy arranges the 17 SDGs into six goals that should be achieved by the economy to meet the objectives: basic needs, well-being, decent jobs, climate stability, resource security and healthy ecosystems (Figure 1). Therefore, a green economy acknowledges that our economies are beyond natural and social structures and are bounded by them. If these structures are placed under undue stress it can undermined the long-term viability / stability of the economies upon which we are all dependent. The three systems are closely interlinked and the scientific community is clear that a number of our environmental processes are now being affected by global economy (Rockström, et al, in World Green Economy Organisation(2018)

A difference can be made between relative divergence (a reduction in environmental effects per unit of economic production), and actual decoupling (a decrease in the overall environmental effects of an economy).

Figure 1. CISL six outcomes for the 17 SDGs
Source: World Green Economy Organisation(2018), "World green economy report" university of Cambridge and UNDP

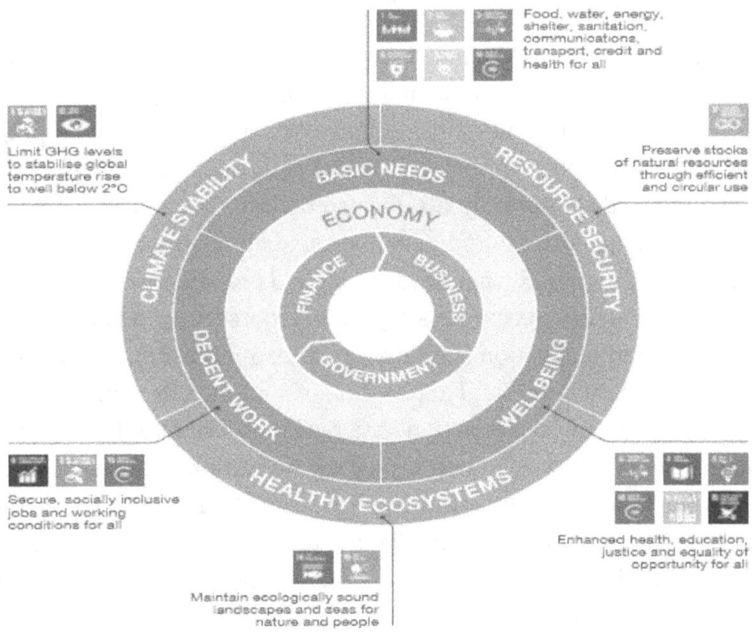

Green economy Principles

The most important green economy principles established from a research of eight published sets of principles:

1. The green economy is a way to achieve sustainability ;
2. The green economy should build proper work and green jobs ;
3. The green economy is efficient inn terms of resource and energy ;

4. The green economy respects geological or environmental limits or scarcity ;
5. The green economy is using collaborative decision making ;
6. The green economy tests development beyond GDP using appropriate indices/metrics ;
7. The green economy is equal, fair and responsable, between nations and generations.
8. The green economy preserves habitats and biodiversity ;
9. The green economy offers poverty alleviation, well-being, living conditions, social support and accessibilty to basic services ;
10. The green economy promotes governance and the rule of law. It is progressive; democratic; collaborative; responsible and clear ;
11. The green economy is internalizing externalities.

SOME WORDS SIMILAR TO GREEN ECONOMY

Green Innovation

In 1939 Shumpeter made one of the earliest concepts of innovation. Innovation is not an invention, but it is the application of creativity in technology or organization. Innovation is defined in the OECD Manual as the realization on organizational implemntations of new or highly improved goods or processes a new marketing startegy or a new organizational method (OECD,2005) .

It is now well understood that innovation is a source of economic and social progress on a national (macro) scale as well as a catalyst of company success and competitive advantage at the business (micro) scale. Michael Porter claimed that "innovation is the central issue in economic prosperity". However, if nations want to move towards a more environmentally sound and harmonious society, it is necessary to promote specific areas of innovation. Such innovation should provide for new ways to address potential future environmental issues and increasing energy and resource use, while encouraging sustainable economic activity. This type of innovation is referred to as eco-innovation or green innovation.

As per Larsen (2005), the first successful application of a product or method is known as innovation to create, develop, and adapt a new idea or action for a business. The OECD has developed a classification of innovation based on four criteria, known as product innovation, system innovation, organizational innovation and advertising innovation. According to Rave and al.,(2011) Green innovation is characterized as envionmental quality production or using naturel resources efficiently. Some authors have noted this definition by suggesting that specific definitive terms are mostly used in literature such as: organic, ecological, environmental and sustainable (Schiederig, Tietze & Herstatt, 2011).

Although, depending on applied innovation type, green innovation appears to emerge out when environmental pressures exist (Huber, 2008).In general, Green innovation aims at reducing pollution, energy productivity, reducing waste, replacing limited resources with sustainable resources and recycling (Kemp & Arundel, 1998). Indeed, this concept play an important role in the environmental performance of the firm and in the implementation of the environmental sustainability in business (Rave, Goetzke &Larch, 2011). So As per Ramus (2002) green innovation is categorized in three types according to its application method and potential effects

- Green innovation that reduce the firm's environmental impact (reuse and recycling)

- Green innovation that solves environmental problems of businesses (diminishing use of hazardous elements)
- Green innovation that envolves environmental friendly/effective products/processes (less resource / energy using) The environmental agenda has become the most important topic in the global manufacturing industry. Consequently, the development of technologies that minimize or prevent pollution in the production processes is becoming a necessary condition for survival. Green innovation has emerged as one of the solutions to this growing economic and environmental pressure.

Accordine to Driessen and Hillebrand (2002) green innovation is:"a rather pragmatic definition" stating that it "does not have to be developed with the goal of reducing the environmental burden. [...] It does however, yield significant environmental benefits". Chen, Lai and al. (2006: 534) said:"as hardware or software innovation that is related to green products or processes, including the innovation in technologies that are involved in energy-saving, pollution-prevention, waste recycling, green product designs, or corporate environmental management".

Many producers's issues have recently changed from "How do we produce products efficiently?" to "How do we develop more eco-friendly materials or reduce the generation of pollutants in the production process?". Manufacturing firms have started to view themselves as an agency that diligently performs the society's needs, and the societal expectations have risen as manufacturing firms are considered to contribute to improving the environment or the quality of life in society.

Green Innovation may occur either to resolve local or global environmental concerns or to develop the sector's environmental leadership, Also, Green innovation can have ecological or economical motivation, and as other types of innovation, it can be incremental or radical. Indeed, to order to create or maintain their competitive advantage, This practice helps businesses insert environmental issues into their strategies. Furthermore, there are many empirical studies that confirm a positive relationship between green innovation and firm performance. Green innovation is closely linked to corporate environmental management and eco-target achievement; thus, green innovation is generally believed to improve environmental performance.

In fact, Green product and process innovation not only minimize negative environmental impact, but they also growth the economic and social performance of a firm through waste and cost reduction. Businesses introduce green process innovation in the manufacturing process to reduce production time and reduce costs . Therefore, a good product innovation improves market position, confirms brand names, creates advances, and attracts new clients.

According to Carrillo-Hermosilla (2012), two different options to innovation in design can be distinguished from the perspective of the environment: one is to find most human activities incompatible with the natural environment and to concentrate on reducing these environmental impacts; and the other is to regard incompatible human activities as 'technology flaws ' and to focus on renovation human-made structures towards biocompatibility and positive environmental impacts. Biocompatibility relates to the performance of human-made products have no poisonous or otherwise adverse effects on biological systems, e.g. materials like lead and mercury are incompatible, hence harmful for species. When the above two views are integrated with the perspectives of the gradual or radical nature of the technological change generated and the degree of impacts on the process, three different approaches can be applied to identify the role and impacts of the eco-innovation.

Eco-Innovation: Eco-innovation is any form of innovation designed to achieve meaningful and demonstrable progress towards the objectif of sustainable development, that reducing effects on the environment or achieving a more efficient and responsible use of natural resources, particularly energy. [4] 'Eco-innovation is the creation of new and competitively priced products, technologies, sturctures, services, and methods designed to meet human wants and provide a better quality of life for all, with minimal use of natural resources (including materials like energy, and surface area) per unit production and a minimal release of toxic chemicals. [5] Eco-innovation is the process of developing new goods, processes or services that deliver value to consumers and industry but significantly reduce the impact of the environment. [6]

Sustainable Development: In 1987, the Bruntland Commission presented its report, « *Our Common Future, in an effort to link the issues of economic development and environmental stability* ». In doing so, this report established the frequently cited concept of sustainable development as "*development that meets the needs of the present without compromising the ability of future generations to meet their own needs*" (United Nations General Assembly, 1987, p. 43) this word of sustainable development aims to promote economic progress and development while maintaining the environment's long-term value; it provides a mechanism for environmental policy implementation and growth strategies. (United Nations General Assembly, 1987). The overall objectif of sustainable development is economic and environmental long-term stability; this can only be accomplished through the incorporation and consideration of politicla, environmental and social issues throughout the decision making process.[7]

Environmental Innovation: Environmental innovation can be defined as innovations that consist of new or changed processes, procedures, systems ans goods that benefit the environment and relate the conservation of the environment (Rennings, 2000). Clearly, the key element of the concept is the positive environmental effect of innovation. But this environmental impact may be deliberate or not, local or international, and more or less significant compared to current or conventional developments. Empirical research on environmental innovation generally concentrate on deliberate, or environmentally driven, innovations because they generally focus on a selected and well-defined sample of « environmental » environmental technologies. For example, the intentional nature of the environmental impact is clear when researching what is called the 'eco-industry', in which environmental innovation is the core business.

What is at risk here is the very term of environmental innovation and the frontiers of its meaning. In addition, while distinguishing environmental developments, it is difficult to assess the environmental impact. Other parameters can be used to evaluate an innovation's environmental impact: greenhouse gas emissions, air pollution, energy consumption, water pollution, noise, soil degradation and waste generation. Given the number of environmental parameters, the global environmental influence of an innovation is very hard to assess. In general terms, the use of an environmental innovation may or may not lead to an complete reduction in environmental damage. In particular, it depends on rebound impacts, the classic example being cost- effective inventions that have a rebound effect through increased investment. So it is not possible to define environmental innovations in terms of absolute environmental impact, but in terms of alternative technologies.

These criteria contibute to the following concept of an environmental innovation: "The production, assimilation or exploitation of a product, production process, service or management or business methods that is novel to the organization (developing or adopting it) and which results, throughout its life cycle, in a reduction of environmental risk, pollution and other negative impacts of resources use (including energy use) compared to relevant alternatives" (MEI Report, 2008). [8]

Green/environmental innovation can be the outcome of pro-active environmental management, that can establish innovation in products/services, processes or markets, but on the other level, can provide the advancement of corporate environmental management which promotes their environmental pro-activity. Therefore, environmental management and innovation have bilateral with causal relationships premises.

Corporate Social Responsability (CSR)

According to Ismail, (2011) CSR, generally referred to as Social Responsability (SR), but now often referred to as corporate Responsability (CR), has achieved immense momentum in industry and public discussion over the decades and become a strategic issue that crosses departmental boundaries and affects a company's way of doing business. CSR can be descriped as ethically or responsibly treating the company's stakeholders (Hopkins 2004). The definition given by Koestoer (2007) is: organizations are recognized as corporate social Responsibility (CSR) to solve various social problems in their operating places, invidually or collectively.

World Business Council for Sustainable Development, (WBCSD), which describe the CSR as "*The continuing commitment by business to behave ethically and contribute to sustainable economic development while improving the quality of life of the workforce and their families as well as of the local community and society at large*." Ismail (2011) stated that company has a wider obligation as it is not limited to investors alone but applies to various stakeholders. It is backed by the argument the government alone can certainly not afford to be solely responsible for improving their human lives as it exceeds their abilities.

To sum up, the definitions of the four concepts of sustainable, eco, environmental and green economy mentioned above show small differences in their descriptive precision. We see to be looking at the same subject in terms of content and can be used mostly interchangeably. However, in the different definitions we detected six important elements[9]:

1. Innovation object: Product, process, service, method
2. Market orientation: Satisfy needs/ be competitive on the market
3. Environmental aspect: Reduce negative impact (optimum = zero impact)
4. Phase: Full life cycle must be considered (for material flow reduction)
5. Impulse: Intention for reduction may be economical or ecological
6. Level: Setting a new innovation/ green standard to the firm

DIMENSIONS OF GREEN INNOVATION

Greater research on environmental innovation can be tested in an institutional contexte as an attempt to differentiate environmental organizational management (Chapple et al., 2011), thud, according to the term of environmental innovation we can group them into three aspects: goods/services; processes and market, like, respectively, recyclable products, new technologies that require less natural resources (water, natural inputs, electricity) to be generated; current market and brand niche for unique green products for example, green consumers (Chapple et al., 2011).

Table 1. Some papers on environmental innovation (three papers from each term to environmental/ green/eco-innovation

Research	Method	Term used	Context
Del Brio and junquera (2003)	Theoetical	Environmental innovation	The study was a review of the literature that identified the main environmental management system barriers for environmental innovation
Wagner (2007)	Empirical	Environmental innovation	This study reveals the relations between environmental management, environmental innovations and patents. The studyconcluded that environmental innovation can be identified using patent data and that the environmental innovation defined in this manner is less consolidated than just as environmental innovation
Brunnermeier and Cohen (2003)	Empirical	Environmental innovation	The study reveals that it is more probable for environmental innovation to occur in sectors that are internationally competitive
Ming-Ji and Ching-Hsun (2009)	Empirical	Green innovation	The study identifies that the green learning process has a positive effect on the company's environmental ethics and a positive effect on green innovation
Yu-Shan (2008)	Empirical	Green innovation	The study identified that the investment in key competence development at companies on behalf of environmental commitment are fundamental factors for a company's capacity for green innovation
Hillestad et al. (2010)	Empirical	Green innovation	The study encompasses the importance of the manager's role, which influences the company's image, as well as his capacity for environmental innovation
Yang and Chen (2011)	Empirical	Eco-innovation	Eco-innovation offers technical support for different fields of technology to increase creativity in product innovation and environmental commitment
Pujari (2006	Empirical	Eco-innovation	The product innovation process has been studied by researchers and the products have a good performance on the market
CarrilloHermosilla et al. (2010	Empirical	Eco-innovation	The capacity for eco-innovation provides new business opportunities and contributes toward the transformation to a sustainable society, depending in the interaction of these dimensions and the engagement of interested parties in the innovation process

Source: Angelo,FD and Al, (2012), « Environmental innovation: in search of a meaning » world journal of Entrepreneurship, Management and Sustainable Development, vol 8,iss 2,Emerald.

GREEN ECONOMY AND BUSINESS MODELS

According to Bryson and Lombardi (2009) developing markets for greener goods and services on the one side and the emergence of sustainability and green innovation policies in corporate management on the other are progressively leading companies to integrate non-financial indicators into their decision-making processes, to revisit the terms of value and productivity that guide their business models, and to rethink the balance between two goals of short-term profitability and long-term sustainability.

Searching at how business opportunities are to be established in the long-term, the World Business Council for Sustainable Development (WBCSD) formed the Vision 2050 with multinational firms (Figure 2). The anticipated economic changes reflect opportunities in a wide range of business areas as the challenges of expansion, industrialization, resource scarcity and global climate change become major strategic drivers for companies in the coming decades. Prospects range from developing and implementing low-carbon, zero-waste cities and infrastructure to boosting environment and lifestyles. Making these transformations is also considered to be creating opportunities for all sectors.

Main elements in assessing the success of eco-innovation, a special attract needs to be on the business model, which takes out eco-innovation to the market and boosts its dissemination.

According to Osterwalder et al. (2010), "*a business model describes the rationale of how an organization creates, delivers, and captures economic, social, and other forms of values*". A business model is also known as a comprehensive approach to understanding how businessess operate (Zott et al., 2010).

Figure 2. WBCSD's vision 2050

Source: WBCSD (in OECD/European Commission/Nordic innovation joint workshop « the future of eco innovation: the role of business models in green transformation », 2012, copenhagen, danmark)

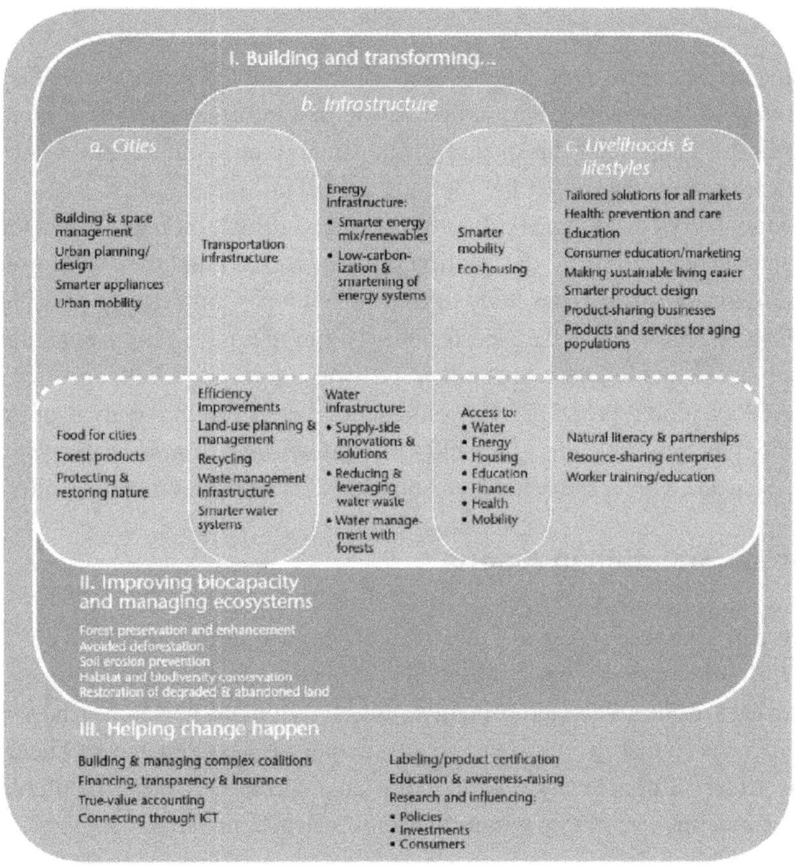

The business model approach provides a systematic way of explaining how value is made and transmitted. Green innovation is designed to promote economic and environmental benefit, and business models serve as a value engine and facilitator of green technologies and solutions. focusing on business models allows consumers to a better explain how environmental quality is measured, converted into successful products and services, and provides comfort and satisfaction. In concrete terms, the study of eco-innovation areas can shed light on whether, to what degree and how environmental principales are expressed in company quality intiatives, customer segmentation, resource usage, patterns of cooperation and revenue source management.[10]

By replacing old business methods, innovative business models also enable companies to reorganize their value chain and create new types of production company-consumer interactions, and change the consumer culture and discipline. Moreove, The business model perspective especially relevant for radical and structural green innovation, including how business models and approaches can encourage and help spread radical green innovation and allow systemic change and transition. it is also necessary to better explain how policy can affect and promote the development of new business models that are successful in promoting eco-innovation.[11]

METHODOLOGY

Because this paper is descriptive, qualitative study is necessary. The review of the literature already includes a great deal of information, but in this section that information will be reviewed the aid of an illustrative case study . because of this, a case of the strategy of RS in Algeria has taken place. In this particular case, it is the best way of getting the most useful information.

Given the abundance of Algerian wind resources, geothermal, biomass, solar represent a growing market for sustainable development . this section presents a study and the use of green economy position in Algeria . often addressed in the study of the current renewable energy position and the potential target. Considering Algeria's rising demand for energy and increasing environmental issues due to the use of fossil fuels, sustainable energy resources provide interesting challenges for Algeria. In 2011, Algeria established a national program to encourage concrete steps in the areas of green economy and sustainable development for the period 2011-2030. ***What strategy should be developed for sustainable development in Algeria?*** This is the central question for which we are attempt to provide some answers.

Green Economy Situation in Algeria

Encouraging a green economy is part of the current investment program (2015-2019) which promotes investment growth, entrepreneurship and enhanced industrial production (excluding hydrocarbons). However, establishing high-value green sectors involves the adoption of a new industrial strategy that benefits the economy, enhances competitiveness and boosts industry innovation and technology acquisition. [12]

By 2035, the Algerian people enjoy a reduction in social inequalities, the elimination of regional disparities, and competitive economy within the sense of sustainable development and environmental preservation.

Solar Energy in Algeria

Algeria implemented solar energy to the Southern project for the first time in 1988. Algeria began to prepare larger cities, such as Skikda and Oran, with appropriate equipment to enhance solar energy potential. Solar energy can either be producted by using the CSP (Concentrated Solar Power Plan) or PhotoVoltaic method[13].

Wind Energy in Algeria

Typically, Wind energy varies from one topographic region to another,which also differs on the atmosphere. The climate in Algeria varies widly between Algeria's northern and southern halves. The Northern part, is special because it has an ideal Mediterranean location, the Atlas Mountains and other high plains. But the winds of the north are not as strong as those of the south. The speeds of the southern winds range from 4m/s to 6m/s, however most southern lands are higher in latitude than the northern area, whereas the desert account more the 70% of the total surface area of Algeria. Adrar is known to be the most suitable location as it's popular for operating, and having strong winds. providing strong winds across a high hill can provide a good power plant.

Geothermal Energy in Algeria

Algeria is renowned for its hot springs cluster, as there are over 200 hot springs throughout the region. They were often used for recreational and therapeutic goals, widely viewed by local residents. With these amount of hot springs in a region, taking advantage of the pressurized heat and making useful work out of it can be taken into account.

Biomass Fuel in Algeria

Biological processes can involve some restrictions to produce fuel. To be effective in producing biofuel, we need vast parts of greenland, and not just that, but we need also need to be prepared to use them in the generation process. Fortunately, Algeria has plenty of agricultural land and excellent quality unpolluted soil that is completely rich in minerals, making it a good call for soybeans,corn and wheat to be planted, for the use of fuel. Originally, The Deglet Nour date arrives from Algeria, which is still the world's biggest producer of Deglet Nour dates. It is primarily grown in the Biskra region of Algeria, in the oasis of Tolga and M'Chouneche. Biofuel is also centered on waste of animals, as their waste is generally responsible for several pollution issues, but can be solved by producing renewable energy. Animal or plants waste gradually may be become a high-calorie source of energy[14].

Hydroelectric Energy in Algeria

In terms of popularity and industry in Algeria, One of the primary types of energy arrives after oil and natural gas, by producing 5% of the total energy generated . since then, Algeria has been able to produce itself a considerable amount of hydropower through the construction of power plants.

The Paradigm for Green Economy Policy in Algeria

The actions carried out by the government and its partners ensure the rational use and preservation of natural resources for the benefit of future generations, protecting agricultural land and the public hydraulic domain, by protecting ecosystems, combating desertification, pollution and nuisance, integrating waste management and investmenting in the green economy. In 2011, Algeria took one of the first actions towards green economy by fostering a capacity for renewable energy sources. This could diversify energy sources in Algeria . The energy sector of Algeria hold a predominant place in the Algerian economy: hydrocarbons alone account for 60% of the budget's revenue and 98% of export revenue. In 2015, Algeria is the 18th largest oil producer, the 10th largest natural gas producer and the 6th largest natural gas exporter in the world. More than 99% of energy production and consumption, including in the electricity sector, is extracted from hydrocarbons[15].

Nevertheless, the Algerian government is strating to consider ecological solutions by investing in new and renewable energies. According to the 2012 Algerian Renewable Energy and Energy Efficiency Development Program (PENREE), Algeria aims to achieve an installed capacity of 22,000 MW from renewable sources by 2030.

Figure 3. Division of program by technology sector
Source: CDER,(2016)

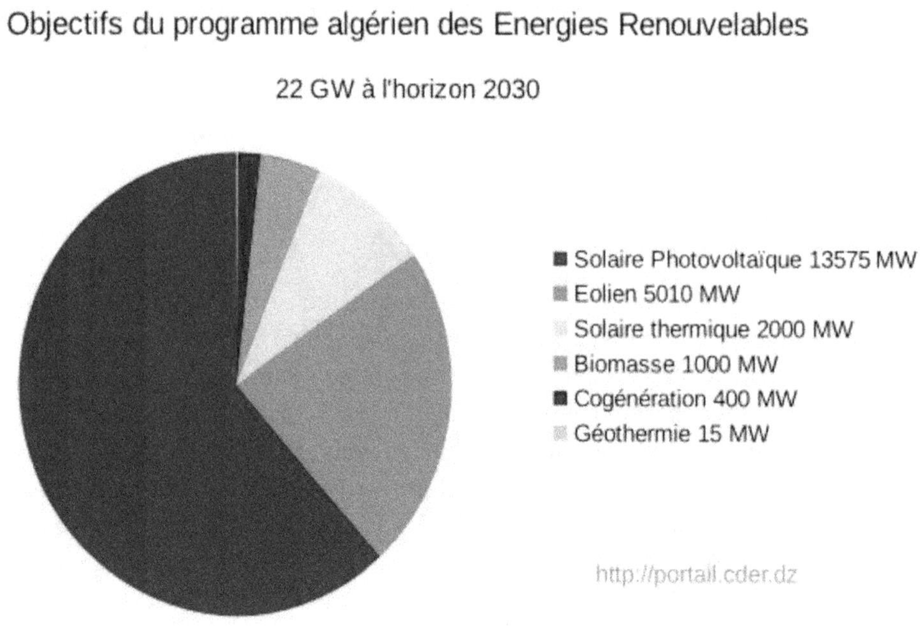

The Algerian program envisages the development of approximately 60 photovoltaic and concentrated solar power plants, wind farms and hybrid power plants by 2020. Photovoltaic Solar will be the main source of clean energy production as it would represent 60% of the total planned capacity by 2030.

The efficiency of the renewable energy plan to be introduced for the national market for the period 2015-2030 is 22,000 MW, separated by sector as follows:

Feed-n Tariff for PV-Solar: FIT is the most commonly used method for generating renewable energy (RE) worldwide[16]. A feed-in tariff is an option to pay a certain amount per unit of electricity sent to the grid to proprietors of distributed energy system (like solar). Mostly they are adjusted-price incentives that are trapped in over a contract period of 10 to 20 years, offering a long term, stable opportunity for property owners with distributed generation[17]. Depending on the size of the power station, the program divides the funding into two stages:

Figure 4. Phases of the new renewable energy program
Source: Renewable Energy Development Center Algeria,2016

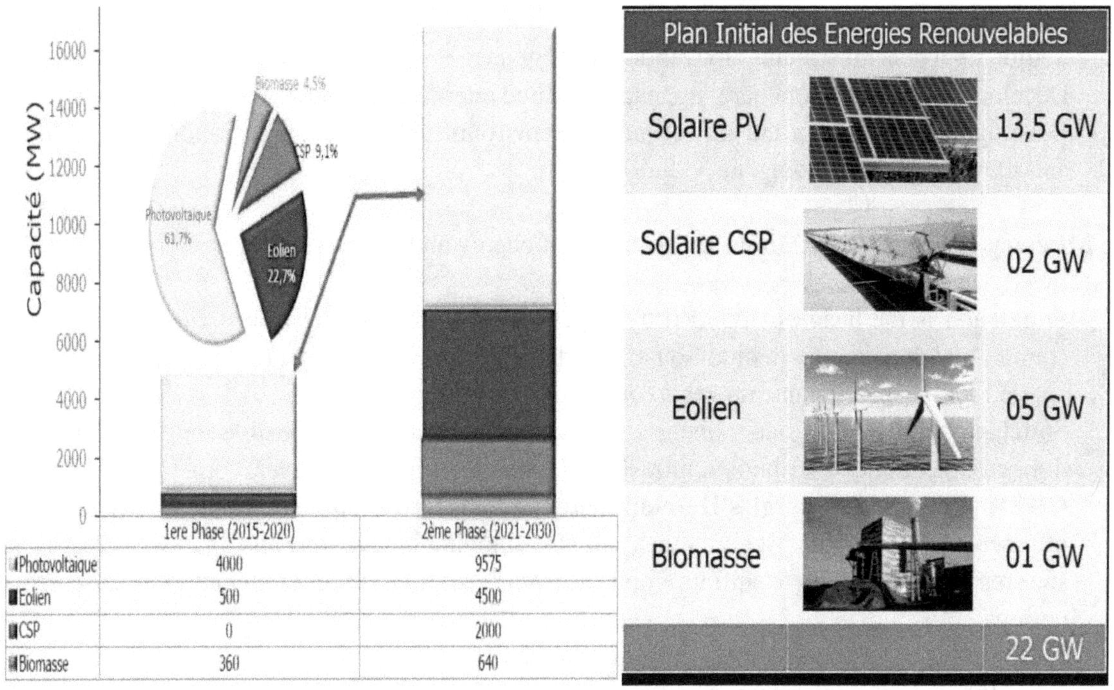

- **Phase 1:** Paid through 5 years with a limited number of functional hours.
- **Phase 2:** Paid through next 15 years.

*To be able to apply this law upon your private power plant, a power plant has to generate a minimum of 1 Mw[18].

Renewable Energy National Fund (RENF)

Fund was created in 2009 and developed in the framwork of sustainable development on the basis of renewable Energy promotion (Law04-90). Funds are given by 0.5% tax on oil revenues.

Renewable Energy Promotion in Framework of Sustainable Development

This policy was established and adopted in 2004, to address three key aspects of renewable energy:

- Full certification with at least the source of energy technology, so it can be marketed locally and internationally.
- A local renewable energy promotion observatory .
- Financial incentive system to support renewable energy development activities.

Law 99-09 on the Management of Energy: This law was enacted in 1999 to raise awareness of energy efficiency by:

a. Setting up a general structure for rational energy use.
b. Developing strategies for energy management and energy efficiency.
c. Developing green energy and protection of the environment by reducing emission of carbon dioxide and monoxide and avoiding air pollution.

Ministry of Energy and Mines: the Ministry of energy and Mines require

1. **Sonatrach** is the business founded by Algerian government to be in possession of mining the country's hydrocarbons. Behind Sonatrach, there are a number of other firms whose value can be applied to transport means or even to other fuel exploiting powers[19].
2. **Sonelgaz**: On the other side, Sonelgaz is the national business responsible for electricity generation, gas, electricity distribution, infrastructure maintenance, and serveral others[20].
3. **CDER (Renewable Energies Development Center),** carries out experimental work in the field of renewable ressources. Not only that, but also applying innovations and analysis to real-time issues regarding its ability to shift its emphasis from fossil fuels, to providing alternative ecofriendly options[21].
4. **NAFTEC** is a state owned subsidiary of Sonatrach for:
   d. Running oil refineries in Algerian region.
   e. Rafinery Management.
   f. Upgrade and extend the refinery to meet local and international business requirements[22].
5. **NAFTAL** is established in 1982, as a branch to Sonatrach, the primary mission of the organization is to supply and sell oil products and timber on the domestic market. The firm is even responsible for multiple of hydrocarbon pipelines[23].

CONCLUSION

In Algeria, the green economy is seen as a driving force for growth that can help adress two of the nation's major challenges: economic diversification and job growth. In particular, the government has initiated a set of reforms and programs aimed at diversifying the economy, boosting the business climat, improving energy security, protecting the environment, greening its industries and promoting land use.

Nonetheless, these policies need to be integrated and expressed better as part of a national program to support the green economy, with a focus for sustainable production and consumption, thus leading to the creation of jobs. A strategy should be supported by local stakeholder methods that address environmental sustainability challenges.

Algeria's energy efficiency potential may be difficult to accomplish by 2030, but it is not impossible. Let's think we are using 130,000 km2 of the Saharan desert and building solar panels everywhere, that would offer us enough electricity to power all of Europe and North Africa. Algeria has succeeded in establishing a strong power system for the fossil fuels, but it would take a bit of time to switch to green economy at this current level. But if we can find a method to change the pressurized supply for fossil fuels and adjust it with more innovative eco-friendly energy resources, then it should achieve the objectif of 100% renewable energy use progressively through time.

RECOMMENDATIONS

- Embracing a comprehensive green economy approach with specific and measurable goals and indicators, concentrating on green jobs, technological innovation, R&D, manufacturing integration and improvement of natural capital;
- Description of green jobs nomenclature for inclusion in the standard economic activity category;
- Adjusting the education system, professional training and continuing education to new jobs and improving partnerships between the universities, research centres, companies, trade and industry and professional associations;
- Setting up a more profitable national innovation system based on policies to promote innovation, especially for SMEs with limited financial funds and skills;
- Reshaping industrial policy to promote sustainable investment;
- Enhancing monitoring and evaluation processes, including the creation of green economy indicators;
- promoting the technology transfer in the form of cooperation (North-South and South-South)

REFERENCES

Al-shaleh, Y.M. (2011). An Empirical Insight into the Functionality of Emerging Sustainble Innovation Systems: The Case of Renewable Energy in Oil-Rich Saudi Arabia. *Int. J. Transit. Innov.*, 302–320.

Angelofd. (2012). Environmental innovation: in search of a meaning. *World Journal of Entrepreneurship, Management and Sustainable Development, 8*(2).

Bryson, J. R., & Lombardi, R. (2009). Balancing Product and Process Sustainability against Business Profitability: Sustainability as a competitive strategy in the Property Development process. *Business Strategy and the Environment, 18*(2), 97–107. doi:10.1002/bse.640

Calza. (2017). Types of green innovations: ways of implementation in a non-Green industry. *Sustainbility, 9*.

Chen, Y. S., Lai, S. B., & Wen, C. T. (2006). The influence of green innovation performance on corporate advantage in Taiwan. *Journal of Business Ethics, 67*(4), 331339. doi:10.100710551-006-9025-5

Driessen, P. H., & Hillebrand, B. (2002). Adoption and diffusion of green innovations. *Marketing for sustainability: towards transactional policy-making*, 343-355.

Emas, R. (2015). *The concept of Sustainable Development: definition and definig principles*. Florida International University.

Fussler, C., & James, P. (1996). *Eco-innovation: A Breakthrough Discipline for Innovation and Sustainability*. Pitman Publishing.

Hemosilla, C. (2010). Diversity of eco-innovations: Refections from selected case studies. *Journal of Cleaner Production, 10.*

Hopkins, M. (2004). *Corporate social responsibility: an issues paper.* Policy Integration Department, Working Paper No. 27 World Commission on the Social Dimension of Globalization International Labor Office Geneva.

Huber, M. (2008). Community for innovations: Developing an integrated concept for open innovation. In *IFIP Working Conference on Open IT-Based Innovation: Moving Towards Cooperative IT Transfer and Knowledge Diffusion* (pp. 503-510). Springer.

Ismail, T. N. T. (2011). Corporate Social Responsibility: The Influence of the Silver Book. *International Journal of Business And Management Studies, 3*(2), 371–383.

Jansson, J., Marell, A., & Nordlund, A. (2010). Green consumer behavior: Determinants of curtailment and eco-innovation adoption. *J. Consum. Mark.*

Kammerer, D. (2009). The effects of customer beneðt and regulation on environmental product innovation. Empirical evidence from appliance manufacturers in Germany. *Ecological Economics*, 68.

Kemp, R., & Arundel, A. (1998). *Survey indicators for environmental innovation.* Academic Press.

Koestoer, Y. T. (2007). Corporate Social Responsibility in Indonesia Building internal corporate values to address challenges in CSR Implementation. In Good Corporate and Social Governance in Promoting ASEAN's Regional Integration, Jakarta, Indonesia.

Kucukoglu, M., & Ibrahim, P. (2015). Positive influences of green innovation on company Performance. *Social and Behavioral Sciences, 195.*

Larsen, O. M. (2005). *Governing innovation for sustainable development: integration of environmental nad innovation policies in Norway.* ProSus Working Paper.

Ma, Y. (2018). The sources of green management innovation: Does internal efficiency demand pull or external knowledge supply push? *J. Clean. Prod.*

Oltra, V. (2008). *Environmental innovation and industrial dynamics: the contributions of evolutionary economics.* DIME Working Papers on Environmental Innovation.

Osterwalder, A., Pigneur, Y., & Clark, T. (2010). *Business model generation: A handbook for visionaries, game changers, and challengers.* Alexander Osterwalder & Yves Pigneur.

Ramus, C. A. (2002). Encouraging innovative environmental actions: What companies and managers must do. *Journal of World Business, 37*(2), 151–164. doi:10.1016/S1090-9516(02)00074-3

Rave, T., Goetzke, F., & Larch, M. (2011). *The determinants of environmental innovations and patenting: Germany reconsidered* (No. 97). Ifo Working Paper.

Rennings, K. (2000). Redefining innovation—Eco-innovation research and the contribution from ecological economics. *Ecological Economics, 32*(2), 319–332. doi:10.1016/S0921-8009(99)00112-3

Roh, T. (2019). Open for green innovation: from the perspective of green process and green consumer innovation. *Journal Sustainbility, 11.*

Schiederig, T. (2011). *What is green innovation? A quantitative literature review.* The XXII ISPIM Conference.

Stambouli, A. B., Khiat, Z., Flazi, S., & Kitamura, Y. (2012). A review on the renewable energy development in Algeria: Current perspective, energy scenario and sustainability issues. *Renewable & Sustainable Energy Reviews, 16*(7), 4445–4460. doi:10.1016/j.rser.2012.04.031

Zott, C. (2011). The Business Model: Recent Developments and Future Research. *Journal of Management, 37*(4), 1019–1042.

ENDNOTES

1 LEEC is a joint venture established in 1988 by the International Institute for Environment and Development (IIED) & the department of Economics of University College London (UCL).

2 https://www.medgreeneconomy.org/assets/downloads/greeneconomy-med-web.pdf

3 World Green Economy Organisation(2018), "World green economy report" university of Cambridge and UNDP

4 Competitiveness and Innovation Framework Programme (2007 to 2013), European Commission

5 Europa INNOVA Thematic Workshop, Lead Markets and Innovation, 29–30th June 2006, Munich, Germany.

6 Fussler, C. and James, P. (1996) Eco-innovation: A Breakthrough Discipline for Innovation and Sustainability (Pitman Publishing). • '[Eco-innovation is] Innovation which is able to attract green rents on the market.'

7 Emas, R, (2015), « the concept of Sustainable Development: definition and definig principles », Florida international university,online publication.

8 Oltra, V, (2008), « Environmental innovation and industrial dynamics: the contributions of evolutionary economics », DIME Working papers on environmental innovation.

9 Schiederig,T and al, (2011), «what is green innovation ? A quantitative literature review », the XXII ISPIM Conference.

10 OECD/European Commission/Nordic innovation joint workshop « the future of eco innovation: the role of business models in green transformation », 2012, copenhagen, danmark

11 OECd, op-cit

12 https://www.uneca.org/sites/default/files/uploaded-documents/SROs/NA/AHEGM-ISDGE/egm_ge-_algeria.pdf

13 Renewable Energy Development Center, Algeria, CDER. Accessed December 2019.

14 A. B. Stambouli, Z. Khiat, S. Flazi, Y. Kitamura, "A review on the renewable energy development in Algeria: Current perspective, energy scenario and sustainability issues", Renewable and Sustainable Energy Reviews, vol. 16, pp.4445-4460, 2012.

15 Ministry of Environment and renewable Energy, The National Conservatory for environmenal Education (CNFE), « Economie Verte et energie Renouvelables », (2019).

16 Al-shaleh, Y.M. An Empirical Insight into the Functionality of Emerging Sustainble Innovation Systems: The Case of Renewable Energy in Oil-Rich Saudi Arabia. Int. J. Transit. Innov. Syst. 2011, 1, 302–320. [

17 Fields, S. Feed-in Tariffs: A Primer on Feed-in Tariffs for Solar. Energysage. Available online: https://news.energysage.com/feed-in-tariffs-a-primer-on-feed-in-tariffs-for-solar/ (accessed on 12 December 2019).

18 "Ministère De L'Énergie - Algérie - Les Entreprises Du Secteur." Ministère De L'Énergie - Algérie - Les Entreprises Du Secteur. Accessed May 2019.

19 SONATRACH." SONATRACH. Accessed December 2019

20 "SONELGAZ." SONELGAZ. Accessed December 2019.

21 "Renewable Energy Development Center Algeria". December 2019

22 http://www.algerie-dz.com/article9048.html (December,2019)

23 http://www.naftal.dz/fr/

Chapter 7
Examining Causal Linkages for Sustainable Development:
A Literature Review

Ritu Rana
National Institute of Technology, Hamirpur, India

Manoj Sharma
National Institute of Technology, Hamirpur, India

Mir Sayed Shah Danish
University of the Ryukyus, Japan

ABSTRACT

This chapter aims to summarize the literature comprising several variables crucial for sustainable development and the examination of causal linkages existing among these variables. This chapter explores the existing literature related to the causality-testing techniques used by several researchers and academicians in the field of sustainable development. This study sums up with a wide variety of existing causality-testing techniques in the literature (both conventional and latest/improved ones). Still, researchers are making efforts for further studies and strive to continually look for examining these causal linkages in a better way. Although several variables have been used in previous studies to explore the causality relationships for sustainable development, it seems there is no limit to the number of such explanatory variables. Therefore, causality testing for sustainable development has a broader scope always for further research advancements in this area.

DOI: 10.4018/978-1-7998-4915-5.ch007

INTRODUCTION

Globalization has made the whole world a common marketplace for all. Nowadays, economic activities have crossed all the boundaries due to this globalization. These economic activities take place at country level to achieve a certain target of economic growth. The world nations are distinguished based on their level of economic growth. These nations are classified into developed, developing, and under-developed nations. In developing and under-developed countries, policy-makers are tried to frame policies that improve economic growth and encourage business activities on a large scale. These countries somehow relied on technology and financial support of developed countries for economic development. Foreign Direct Investment (FDI) is considered the best way to accumulate foreign capital in such countries. Developing countries provide relaxations in taxes and environmental regulations to attract foreign investors. Markets are made open and trade barriers are minimized for the sake of economic growth. Meanwhile, studies show that these economic activities have an adverse effect on environment and cause greenhouse gas (GHG) emissions. Researchers are claimed that these foreign investments contribute to environmental pollutions in these countries due to open-door investment policies and regulations, whereas these investors have stricter environmental regulations in their home country. Thus, there arises a debate on reducing these adverse environmental effects from highly polluting developing countries. This debate can be regarded as a debate relating to sustainable development.

Sustainable development is a principle wherein human development goals are achieved, whilst sustaining natural systems' ability to empower society within required natural resources and ecosystem services. It can be classified as development that meets the needs of the present without compromising the ability of future generations. Sustainable development is a continuous and unpredictable process demonstrated with variables. However, some studies are confirmed that sustainable development trends can be predictable to some extent. Several studies have been conducted to evaluate causal linkages between several determinants of economic growth and determinants of environmental performance. These causal linkages between determinants of growth and environment have been examined by researchers using various causality testing techniques. These studies differ from one another in many aspects. Difference among these studies is mainly based on applied variables that result in opposing findings with the same variables. Variables used in the study are time frame of the study, regional differences and, most importantly on the basis of the causality testing techniques applied to examine such causal linkages. Researchers are focused on new variables and new techniques and methodologies to define relationships more accurately than the previous ones. Thus, there is always a scope for finding something new and improving/ overcoming the previous inaccurate findings. The purpose of this study is to explore existing literature related to causality testing techniques and also to deal with various determinants of economic growth used by several researchers and academicians in the field of sustainable development.

BACKGROUND

Winchester and Salji (2016) emphasized that 'reviews' should be informative, personal (but unbiased synopsis of information), and it should provide a balanced view of conflicting findings and inconsistencies along with established and current thinking instead of just providing an exhaustive list of all that has been published. Nakano and Muniz Jr. (2018) are also of the opinion that 'reviews' provide a comprehensive analysis of academic production regarding a certain topic, and mention them as very useful for both new and experienced researchers. Thus, it is important to conduct a comprehensive review of studies available in the literature before conducting any research study. This section has been further divided into three sub-sections. The first sub-section contains the two basic concepts used by researchers that define the growth-environment relationship upon which the previous and current literature is based. The second sub-section lists out some important variables, which are addressed in literature to explain such relationship. The third sub-section mentions some important causality testing techniques used in the literature to find out the causal linkages among these variables. Although all the sections are interlinked, and literature survey includes a review of all the studies related to examination of causal linkages for sustainable development, but for the ease of reading, these separate sections are classified.

Basic Concepts

There are two famous concepts in the literature that form the basis of causal linkages between economic growth and environmental performance. Namely, the Pollution Haven Hypothesis (PHH) and the Environmental Kuznets Curve (EKC) Hypothesis.

The Pollution Haven Hypothesis (PHH)

The PHH concept states that 'the movement of investments in dirty industries is from developed countries to developing countries because of the lower environmental regulations and costs of complying in these countries to attract foreign investments for their economic growth as compared to the regulations and costs in the country of origin of such investments' (Elliot and Shimamoto, 2008; Johnson and Turner, 2010; Blanco et al., 2011; Shao, 2018). Developing countries have comparatively relaxed environmental laws and regulations than the developed countries to attract FDI. As per this hypothesis, the flow of FDI in dirty industries originates from developed countries. At the same time, this hypothesis is criticized on grounds. Strict environmental standards and regulations are comprising cost to the firm that can be caused to ignore tendency of multinational enterprises towards standardization of technology across all plants to yield cost benefits in the processes. Studies are reported various supporting and opposite views on the existence of this hypothesis. The researchers, they are in support of this hypothesis are Akbostanci, Tunç and Türüt-Asik (2004), Acharyya (2009), Blanco et al. (2011), Shahbaz et al. (2015), Shao (2018), etc. Blanco et al. (2011) only studied PHH in the dirty sector without finding robust evidence for other sectors that questioning its existence). Also, Shahbaz et al. (2015) finding is sensitive to different groups of income and regions. On the other hand, there are some studies that tried to examine the existence of PHH but failed to confirm it e.g. Dietzenbacher and Mukhopadhyay (2007), Elliott and Shimamoto (2008).

The Environmental Kuznets Curve (EKC) Hypothesis

In contrast to the PHH, the environmental Kuznets curve (EKC) hypothesis or the inverted U-shaped curve is another hypothesis. This hypothesis states that as economic growth increases income levels, environmental degradation grows at first, stabilizes at middle-income levels, and declines with higher income levels. EKC hypothesis supports the view that growth will ultimately also contribute to the solution for environmental problems in the developing world (Shafik and Bandyopadhyay, 1992; Panayotou, 1993; Kuznets 1995; Johnson and Turner, 2010). Grossman and Krueger (1995) found the reason behind this is due to the people becoming more concerned about the quality of their environment as their income levels rise. This inverted-U-shaped curve was first introduced in 1954 by Simon Kuznets that later became a benchmark to monitor the environmental performance with the increase in income levels. This hypothesis is also criticized on the grounds that it follows a pattern of 'pollute first and control pollution later.' Some studies related to EKC hypothesis include Grossman and Krueger (1991), Grossman and Krueger (1995), Shafik and Bandyopadhyay (1992), Panayotou (1993), Selden and Song (1994), Dinda (2004), Rana and Sharma (2019), etc. On the contrary, there are some studies that inferred that the inverted-U-shaped curve does not hold true every time and for all indicators of environmental degradation. Some of such studies are Lopez (1994), John et al. (1995), McConnell (1997), Stern (2003), etc.

Several studies tried to examine possible determinants of EKC hypothesis e.g., Rock (1996) and Rothman (1998) for trade; Torras and Boyce (1998) for political freedom; Kaufman et al. (1998) for density of economic activity; Suri and Chapman (1998) for economic structure; Hitam and Borhan (2012) and Shahbaz et al. (2015) for FDI, etc.

Important Variables

Previous studies and current literature comprise of several variables to evaluate causal linkages between economic growth and environmental performance. These variables can be regarded as the determinants of sustainable development. A description of such determinants of sustainable development used in the literature is given in this section.

FDI has been a major variable used in the literature to study the causal linkages for sustainable development. Literature contains studies on FDI and its effects on the economic growth of the host country. A number of researchers emphasize that FDI is the main driver of economic growth, especially in developing countries. It is evident in the literature review conducted by Almfraji & Almsafir (2014). Various studies in literature are examined FDI and its impacts on both economic growth (expressed in terms of GDP) and environmental pollution to analysis causal relationships hiring the EKC hypothesis e.g., Mehta (2002), Jorgenson et al. (2007), Baek and Koo (2008), Acharyya (2009), Choi et al. (2010), Alam et al. (2011), Hitam and Borhan (2012), and Shahbaz et al. (2015). Literature also linked FDI to environmental pollution, and tried to examine the existence of PHH and causal relationships among them (e.g., Jorgenson et al., 2007; Seker et al., 2015; Shahbaz et al., 2015; Bokpin, 2017). A wide range of literature devoted to analyses environmental pollution, which is represented by either carbon dioxide (CO_2), sulfur dioxide (SO_2) and nitrous oxide (N_2O) emissions (Seker et al., 2015).

Some indicators of global trade such as exports and imports are counted as causal linkages contributors for sustainable development e.g. Hitam and Borhan (2012), Chakraborty and Mukherjee (2013), Sawhney and Rastogi (2015), and Sawhney and Majumder (2015). Trade liberalization is used as a variable to examine such relationships (e.g. Choi et al., 2010; and Rana and Sharma, 2020).

Another crucial and most important determinant of sustainable development is 'energy consumption' that has been used in almost all the studies relating to causal linkages for sustainable development e.g. Choi et al. (2010), Alam et al. (2011), Seker et al. (2015), and Linh and Lin (2015).

One more important indicator that has been used in recent studies relating to causal relationships in sustainable development is technological advancements and/or technology gaps. Studies containing technology as a determinant of sustainable development include Shafik and Bandyopadhyay (1992), Borensztein et al. (1998), De Mello (1999), Xu (2000), Li and Liu (2005), and Schembri and Petit (2009).

Technology transfer is one of the major benefits that FDI is considered to bring to the host country (Xu, 2000; Li and Liu, 2005; and Schembri and Petit, 2009).

The above discussion shows that FDI is the most important factor for sustainable development. Still, there are authors that do not found FDI to be the real reason for environmental destruction in the host country e.g. Dietzenbacher and Mukhopadhyay (2007), Pao and Tsai (2011), and Kizilkaya (2017). Though there is a tendency to blame FDI for environmental degradation, there are studies that have also included the level of domestic investments (LDI) in a developing country to focus on its causal linkages with FDI for sustainable development. Such studies include Driffield and Hughes (2003), Farla, De Crombrugghe and Verspagen (2016), Pananond and Cuervo-Cazurra (2018), Firebaugh (1992), Yildirim (2014), Ullah et al. (2014), and Elmarzougui et al. (2016).

In different studies in literature, it proved that different sources of electricity generation also contribute to environmental degradation whilst providing benefits for economic growth and infrastructure development e.g. Rana & Sharma (2018), Planete-Energies (2016), and International Energy Agency (IEA) (2018). Some determinants or variables which are used in the previous studies are summarized as follows:

1. **Economic Growth:** It is an increase in the capacity of an economy to produce goods and services over a period of time. It is generally measured in terms of Gross Domestic Product (GDP). GDP is a monetary measure of market value of all final goods and services produced in domestic territory of a country over a period of time.
2. **Foreign Direct Investment:** As per International Monetary Fund (IMF), FDI is an investment that involves a long-term relationship reflecting a lasting interest of a resident entity in one economy (direct investor) in an entity resident in an economy other than that of the investor (host country). Well-managed FDI supports economic growth, non-debt financial resource (capital stock), employment generation (both direct and indirect), technical know-how (expertise), new managerial practices (capabilities), augmentation of existing stock of knowledge through labor training, creative skill acquisition and diffusion, taxes and revenues generation, better quality products, competitive prices of products, and many more.
3. **Carbon Dioxide (CO_2) Emissions:** As per World Bank, CO_2 is a naturally occurring, principal anthropogenic greenhouse gas. It is the reference gas against which other greenhouse gases are measured. It has the maximum global warming potential causing anthropogenic climate change. It is a major source of ocean acidification since it dissolves in water and forms carbonic acid. Stems out from burning oil, coal and gas for energy use, burning wood and waste materials, and from industrial processes such as cement production and etc.
4. **Level of Domestic Investment:** Domestic investments are also made in energy and electricity generation to facilitate foreign investments and to cope up with increasing demand that comes from economic growth. Investments mainly in energy projects with private participation represent domestic investments.
5. **Exports:** Exports are the goods and services produced in one country and purchased by residents of another country. It doesn't matter what the good or service is. It doesn't matter how it is sent. It can be shipped, sent by mail, or carried in personal luggage on a plane. If it is produced domestically and sold to someone in a foreign country, it is an export. Exports are one component of international trade.
6. **Imports:** They are the goods and services bought by a country's residents that are produced in a foreign country. Like exports, imports are also one component of international trade.

7. **Energy Consumption:** Energy consumption is considered a variable of utmost importance in the studies of causal relationships among environmental pollution and economic growth. Several studies have used energy consumption as an endogenous variable to examine such relationships (e.g. Alam et al., 2011; Choi et al., 2010; and Seker et al., 2015).

Trade Openness: Choi et al. (2010) measured trade liberalization at time t i.e., $OPEN_t$ as the sum of imports and exports as a percent of total GDP in a given year as follows:

$$OPEN_t = IMP_t + EXP_t \tag{1}$$

where, IMP_t is imports of goods and services (% of GDP) and EXP_t is exports of goods and services (% of GDP) at time t.

8. 8. **Technology Gap:** Li and Liu (2005) measured technology gap as follows:

$$TG_t = \frac{GDP_{max_t} - GDP_t}{GDP_t} \tag{2}$$

where, GDP_{max_t} is the GDP per capita (current US $) of the United States of America (USA) at time t.

9. **Governance and Institutional Quality:** There arises a moderating role of good governance and quality institutions on the FDI-environment relationship as suggested by Bokpin (2017) in the African context. As per Bokpin (2017), FDI promotion must be done in a restricted manner while keeping the environmental sustainability in consideration, and governance and institutions must be strengthened to regulate the firms' activities as the presence of quality institutions can assure the effects of FDI less harmful to environment.
10. **Electricity Generation:** Electricity produced from various sources of power generation has also been used in several studies to explore causal linkages for sustainable development. These include coal sources, hydroelectric sources, renewable energy sources, etc.

Therefore, it becomes clear now that there may be any number of determinants that can explain the causal linkages for sustainable development, and the literature is rich in such determinants. Researchers are still working on them to find out other important variables to facilitate sustainable development.

Causality Testing Techniques

There are two famous concepts in the literature that form the basis of causal linkages between economic growth and environmental performance. Namely, the Pollution Haven Hypothesis (PHH) and the Environmental Kuznets Curve (EKC) Hypothesis.

Literature consists of various causality testing techniques to examine the causal linkages among various variables. Similarly, while examining causal linkages for sustainable development, different causality testing techniques are used by researchers. Techniques used in the literature varies as per the objectives and purpose of conducted studies. Some of the causality testing techniques are summarized below. These causality testing techniques are not limited in this age of demand for economic growth. Researchers are constantly working on them and are trying to find out new ways to examine causal linkages for sustainable development in a better way.

Vector Auto Regressive (VAR) Models

In economics, it is quite common to have models where some variables are not only explanatory for a given dependent variable but are also explained by the variables that they are used to determine, which gives rise to the models of simultaneous models. Thus, a clear identification between endogenous and exogenous variables becomes necessary (Asteriou, 2007). Sims (1980) criticized any differentiation among endogenous and exogenous variables on the grounds that if there is simultaneity among a number of variables, then all these variables should be treated in the same way i.e., all variables are treated as endogenous. This means that in its general reduced form, each equation has the same set of regressors which leads to the development of Vector Auto Regressive (VAR) models. VAR is defined as a system of Auto Regressive Distributed Lag (ARDL) equations describing dynamic evolution of a set of variables from their common history, where 'vector' implies multiple variables involved. VAR allows all variables to interact with themselves and with each other in regression analysis without any theoretical structure on the estimators. VAR can explain dynamic structure of time-series better than the static Ordinary Least Squares (OLS) estimation method, but at the expense of complexity in interpretation of effects. Let Y_t and X_t be two series, where Y_t is affected by not only its past (or lagged) values but also by current and lagged values of X_t and vice-versa, then a simple bi-variate VAR model is given by:

$$Y_t = A_1 + B_1 X_t + \sum_{j=1}^{p} C_j Y_{t-j} + \sum_{j=1}^{p} D_j X_{t-j} + u_{1t} \tag{3}$$

$$X_t = A_2 + B_2 Y_t + \sum_{j=1}^{p} E_j Y_{t-j} + \sum_{j=1}^{p} F_j X_{t-j} + u_{2t} \tag{4}$$

where, it is assumed that u_{1t} and u_{2t} are uncorrelated error terms called impulses or shocks in the language of VAR. If there are I(1) variables, VAR is estimated using first differences of the variables to remove common trends and if the variables are cointegrated, the Error Correction Term (ECT) should be included in VAR.

Error Correction Mechanism (ECM)

In case of a bi-variate model, the purpose and nature of the structural modeling is to discover the dynamic causal relationship between Y_t and X_t i.e., to discover the short-run slope or long-run elasticity of Y_t with respect to X_t ($\partial Y_t / \partial X_t$). For simplifying the analysis, following (ARDL) (1,1) model is considered:

$$Y_t = A_0 + A_1 Y_{t-1} + B_0 X_t + B_1 X_{t-1} + u_t \tag{5}$$

From eq. 5, both short-run and long-run effects (slopes or elasticities) can be analyzed as follows for any expanded model ARDL (p,q):

a. a. Short-run or static effect:

$$\frac{\partial Y_t}{\partial X_t} = B_0 \tag{6}$$

b. b. Long-run or dynamic or equilibrium effect:

$$\frac{\partial Y_t}{\partial X_t} = \frac{B_0 + B_1 + B_2 \ldots + B_q}{1 - A_1 - A_2 \ldots - A_p} \tag{7}$$

By subtracting Y_{t-1} from both sides of eq. 5 and adding and subtracting B_0X_{t-1} to the RHS (right-hand side), we get what is widely known as the Error Correction Model (ECM), which is applicable with all ARDL models as follows:

$$\Delta Y_t = B_0 \Delta X_t - \pi \left(Y_{t-1} - \alpha - \beta X_{t-1} \right) + u_t \tag{8}$$

where, $\pi = \left(1 - A_1\right)$ i.e., speed of adjustment ($0 < \pi < 1$)

and $\left(Y_{t-1} - \alpha - \beta X_{t-1}\right)$ is the error correction term (ECT) where $\alpha = \frac{A_0}{\left(1 - A_1\right)}$ and $\beta = \frac{\left(B_0 + B_1\right)}{\left(1 - A_1\right)}$

Notice the negative sign of the πECT, which signifies that the result of the coefficient π, from the model generated through system, must have a negative sign along with being significant.

Therefore, ECM and ARDL are basically the same if the series are integrated of the same order [often I(1)] and are cointegrated. According to Asteriou (2007), when the two variables are cointegrated, the ECM incorporates not only short-run but also long-run effects.

Vector Error Correction Models (VECM)

Introduction of ECT in a VAR model makes the model a Vector Error Correction Model (VECM), which can be seen as a restricted VAR. VECM is an appropriate modeling strategy when the variables are cointegrated. It is useful when long-run forecast is desired, as VAR doesn't explicitly take into account the long-run relationship. VECM gives long-run structural relations plus information on adjustment, which provides better insight in economic processes.

A VAR (p) can be reformulated in a VECM. Suppose we have three variables; Y_t, X_t and Z_t, which can all be endogenous i.e., we have that (using matrix notation for $W_t=[Y_t, X_t, Z_t]$):

$$W_t = A_1 W_{t-1} + A_2 W_{t-2} + \ldots + A_p W_{t-p} + u_t \tag{9}$$

The general case of the VECM includes all the various options that can happen is as follows:

$$\Delta W_t = \Pi W_{t-1} + \Gamma_1 \Delta W_{t-1} + \Gamma_2 \Delta W_{t-2} + \ldots + \Gamma_{p-1} \Delta W_{t-p+1} + u_t \tag{10}$$

where, Δ is the difference operator, W_t is a vector with all the variables under study, $\Gamma_1,\ldots,\Gamma_{p-1}$ represents the matrix of the short-run dynamics, Π represents the long-run dynamics and $\Pi = \pi\beta'$ (Johansen, 1988; and Johansen and Juselius, 1992) where π will include the speed of adjustment to equilibrium coefficients and β' will be the long-run matrix of coefficients. Therefore, the $\beta' W_{t-1}$ term is equivalent to error correction term ($Y_{t-1} - \alpha - \beta X_{t-1}$) in the single-equation case (eq. 8), except that now $\beta' W_{t-1}$ contains up to *p-1* vectors in a multivariate framework.

The possible ranks of matrix Π and their respective implications are given in Table 1 as under (Binh, 2013), when there are *k* numbers of variables:

Table 1. Rank of matrix Π and its implications

Rank of Π	Implications
r = 0	There is no cointegration. No stable long-run relations between variables. VECM is not possible (only VAR in first differences).
0 < r < k	There are r cointegrating vectors. These vectors describe the long-run relationships between variables. VECM is introduced.
r = k	All variables are already stationary. No need to estimate the model as VECM. VAR on untransformed data can be done.

Johansen Cointegration Test

Cointegration is an important requirement for any economic model using non-stationary time series data. Because if the variables do not cointegrate, a case of spurious regression may arise making econometric work almost meaningless. Suppose that if there really is a genuine long-run relationship between two variables, then although the variables will rise over time (because they are trended), there will be a common trend or cointegrating vector that links them together (Binh, 2013). Engle and Granger (1987) also proposed that even though all the variables in the model are non-stationary, the series may be stationary if the variables have a linear combination. For an equilibrium, or long-run relationship to exist, a linear combination of the variables (which is a stationary variable i.e., an I(0) variable) is required. From economic perspective, two variables will be cointegrated in case of a long-run relationship (or equilibrium). If we have more than two variables in a model, then there is a possibility of having more than one cointegrating vector (Asteriou, 2007) i.e. Variables in the model can result from several equilibrium relationships. Generally, in case of *k* number of variables, only up to *k-1* cointegrating vectors can exist. Johansen's methodology is used to find out how many cointegrating relationships exist among variables.

In order to identify the number of cointegration vectors, Johansen (1988) proposes the trace and maximum eigenvalue statistics. The trace statistic is based on a likelihood ratio test about the trace of the matrix. The trace of a matrix is a very simple operation defined only for square matrices. For any $n \times n$ matrix A, the trace of a matrix A, is the sum of its diagonal elements (Wooldridge, 2013). The maximum eigenvalue statistic is designed to test the null hypothesis that Rank (Π) = *r*, i.e., *r* cointegration relationships against the alternative of *r+1* as follows:

$$\lambda_{max}\left(r, r+1\right) = -T\ln\left(1-\hat{\lambda}_{r+1}\right) \quad (11)$$

where, T represents the number of observations and $\hat{\lambda}_i$ shows the estimated values of the roots.

The trace statistic considers whether the trace is increased by adding more eigenvalues beyond the rth eigenvalue. Here, the null hypothesis is that the number of cointegrating vectors is less than or equal to *r*.

$$\lambda_{trace}\left(r\right) = -T\sum_{i=r+1}^{n} ln\left(1-\hat{\lambda}_i\right) \quad (12)$$

ARDL Bounds Test

Johansen cointegration test suffers a limitation that it requires all time-series to have the same order of integration. It cannot be performed when some of the variables are I(0) and some are I(1). Having same order of integration is not always possible because of different unit root tests providing different orders of integration even for exactly the same time-series. Another cointegration testing technique is the Auto Regressive Distributed Lag (ARDL) bounds test. This methodology of Pesaran and Shin (1999) and Pesaran et al. (2001) is preferred over Johansen cointegration test because it can be used with a mixture of I(0) and I(1) data. A general ARDL model is written as follows:

$$\Delta Y_t = \beta_0 + \sum_{i=1}^{k}\beta_i \Delta y_{t-i} + \sum_{i=0}^{k}\delta_i x_{t-i} + \varphi_1 y_{t-1} + \varphi_2 x_{t-1} + \mu_t \quad (13)$$

where, β_i and δ_i are short-run coefficients; φ_1 and φ_2 are ARDL long-run coefficients; μ_t is the disturbance term and *k* is the number of lags.

Modified Wald (MWALD) Test

Modified Wald (MWALD) test was proposed by Toda and Yamamoto (1995). This Toda-Yamamoto (TY) multivariate approach for testing the causality has its advantages over others in the sense that it can be used even when the variables are integrated of different orders. It requires the estimation of an augmented VAR irrespective of whether the time series is I(0) or I(1) and whether it is cointegrated or not. Also, if the system has a unit root, the conventional OLS of VAR in level-based Wald statistics have non-standard asymptotic distribution that may involve annoying parameters (Toda and Phillips, 1993).

TY procedure put restrictions on the parameters of VAR (l) from an augmented VAR ($l+i_{max}$) model, where l is the optimal lag length and i_{max} is the maximum order of integration of variables (Alam et al., 2011). A general VAR ($l+i_{max}$) model in TY approach is written as:

$$y_t = \alpha + \beta_1 y_{t-1} + \ldots + \beta_{l+i_{max}} y_{t-(l+i_{max})} + \varepsilon_t \tag{14}$$

where y_t consists of K endogenous variables, α is a vector of intercept terms, β are coefficient matrices and ε_t are white noise residuals. The null hypothesis in TY causality is based on zero restriction on first l parameters (H_0: $\beta_1 = \ldots = \beta_l = 0$) of the *kth* element of y_t. Until l^3 i_{max}, the model is valid (Kurozumi and Yamamoto, 2000). The coefficients for the 'extra' lag are not included while performing the Wald tests. They are there just to fix up the asymptotics.

Causality Testing in Time and Frequency Domain

This causality testing technique is the latest in its field and the researchers are now applying this scientific technique to study the causal linkages for sustainable development. This is an attempt to take into account the structural shifts for getting robust results, which is missing in Toda-Yamamoto (TY) approach. TY approach suffers this limitation of assuming not to have structural shifts in data as explained by Nazlioglu et al. (2016) and later by Gormus et al. (2018) in their studies related to oil prices and real estate investment trusts and high yield bonds and energy markets, respectively. This study applied this technique that argued to provide robust and accurate results.

CONCLUSION

This review throws light on various causality testing techniques (both conventional and the latest improved techniques) that paves the way for further advancement in the field. The conventional causality testing techniques include Granger (non-) causality, Sims Causality, Johansen and Juselius causality, etc. The latest ones consist of Modified Wald (MWALD) test, causality testing in time, and frequency domain. Several previous studied variables in the literature are discussed in this chapter, which linked with causal relationships for sustainable development. These variables are outlined such as foreign direct investment (FDI), domestic investments, energy consumption, technology gap, imports, exports, trade openness, etc. This study implies that though a wide variety of causality testing techniques exist in the literature that examines several determinants of sustainable development. Still, researchers are making efforts for further studies and strive to constantly look for examining these causal linkages in a better way. They constantly try to find out any advancement and the key variables affecting sustainable development goals. To conclude, sustainable development and the ways to test causal relationships among its different variables is a wide scope, which needs further discussion.

REFERENCES

Acharyya, J. (2009). FDI, growth and the environment: Evidence from India on CO_2 emission during the last two decades. *Journal of Economic Development*, *34*(1), 43–58. doi:10.35866/caujed.2009.34.1.003

Akbostanci, E., Tunç, G. I., & Türüt-Asik, S. (2004). *Pollution haven hypothesis and the role of dirty industries in Turkey's exports*. ERC Working Papers 0403, ERC-Economic Research Center, Middle East Technical University, revised Feb 2004.

Alam, M. J., Begum, I. A., Buysse, J., Rahman, S., & Huylenbroeck, G. V. (2011). Dynamic modeling of causal relationship between energy consumption, CO_2 emissions and economic growth in India. *Renewable & Sustainable Energy Reviews*, *15*(6), 3243–3251. doi:10.1016/j.rser.2011.04.029

Almfraji, M. A., & Almsafir, M. K. (2014). Foreign direct investment and economic growth: Literature review from 1994 to 2012. *Procedia: Social and Behavioral Sciences*, *129*, 206–213. doi:10.1016/j.sbspro.2014.03.668

Asteriou, D., & Hall, S. G. (2007). *Applied econometrics: A modern approach using Eviews and Microfit* (Revised Edition). Palgrave Macmillan.

Baek, J., & Won, W. K. (2008). *A dynamic approach to the FDI-environment nexus: The case of China and India.* No 6508, 2008 Annual Meeting, American Agricultural Economics Association (New Name 2008: Agricultural and Applied Economics Association), Orlando, FL.

Binh, P. T. (2013). *Unit root tests, cointegration, ECM, VECM and causality models*. Topics in Time Series Econometrics.

Blanco, L., Gonzalez, F., & Ruiz, I. (2011). *The impact of FDI on CO_2 emissions in Latin America.* Pepperdine University, School of Public Policy Working Papers. Paper 28.

Bokpin, G. A. (2017). Foreign direct investment and environmental sustainability in Africa: The role of institutions and governance. *Research in International Business and Finance*, *39*, 239–247. doi:10.1016/j.ribaf.2016.07.038

Borensztein, E., De Gregorio, J., & Lee, J.-W. (1998). How does foreign direct investment affect economic growth? *Journal of International Economics*, *45*(1), 115–135. doi:10.1016/S0022-1996(97)00033-0

Chakraborty, D., & Mukherjee, S. (2013). How do trade and investment flows affect environmental sustainability? Evidence from panel data. *Environmental Development*, *6*, 34–47. doi:10.1016/j.envdev.2013.02.005

Choi, E., Heshmati, A., & Cho, Y. (2010). *An empirical study of the relationships between CO_2 emissions, economic growth and openness.* IZA Discussion Papers, No. 5304, 1-27.

De Mello, L. (1999). Foreign direct investment-led growth: Evidence from time series and panel data. *Oxford Economic Papers*, *51*(1), 133–151. doi:10.1093/oep/51.1.133

Dietzenbacher, E., & Mukhopadhyay, K. (2007). An empirical examination of the pollution haven hypothesis for India: Towards a green Leontief paradox? *Environmental and Resource Economics*, *36*(4), 427–449. doi:10.100710640-006-9036-9

Dinda, S. (2004). Environmental Kuznets curve hypothesis: A survey. *Ecological Economics*, *49*(4), 431–455. doi:10.1016/j.ecolecon.2004.02.011

Driffield, N., & Hughes, D. (2003). Foreign and domestic investment: Regional development or crowding out? *Regional Studies*, *37*(3), 277–288. doi:10.1080/0034340032000065433

Elliott, R., & Shimamoto, K. (2008). Are ASEAN countries havens for Japanese pollution-intensive industry? *World Economy*, *31*(2), 236–254. doi:10.1111/j.1467-9701.2007.01088.x

Elmarzougui, E., Larue, B., & Tamini, L. T. (2016). Trade openness, domestic and foreign investments, and the environment. *Modern Economy*, *7*(05), 591–605. doi:10.4236/me.2016.75065

Engle, R., & Granger, C. W. J. (1987). Cointegration and error correction: Representation, estimation and testing. *Econometrica*, *55*(2), 251–276. doi:10.2307/1913236

Farla, K., De Crombrugghe, D., & Verspagen, B. (2016). Institutions, foreign direct investment, and domestic investment: Crowding out or crowding in? *World Development*, *88*(C), 1–9. doi:10.1016/j.worlddev.2014.04.008

Firebaugh, G. (1992). Growth effects of foreign and domestic investment. *American Journal of Sociology*, *98*(1), 105–130. doi:10.1086/229970

Gormus, A., Nazlioglu, S., & Soytas, U. (2018). High-Yield Bond and Energy Markets. *Energy Economics*, *69*, 101–110. doi:10.1016/j.eneco.2017.10.037

Grossman, G. M., & Krueger, A. B. (1991). *Environmental impacts of a North American free trade agreement.* NBER Working Paper Series, 3914.

Grossman, G. M., & Krueger, A. B. (1995). Economic growth and the environment. *The Quarterly Journal of Economics*, *110*(2), 353–377. doi:10.2307/2118443

Hitam, M. B., & Borhan, H. B. (2012). FDI, growth and the environment: Impact on quality of life in Malaysia. *Procedia: Social and Behavioral Sciences*, *50*, 333–342. doi:10.1016/j.sbspro.2012.08.038

IEA. (2018). *Global energy and CO_2 status report 2017*. IEA.

Johansen, S. (1988). Statistical analysis of cointegration vectors. *Journal of Economic Dynamics & Control*, *12*(2), 231–254. doi:10.1016/0165-1889(88)90041-3

Johansen, S., & Juselius, K. (1992). Testing structural hypothesis in a multivariate cointegration analysis of the PPP and the UIP for UK. *Journal of Econometrics*, *53*(1-3), 211–244. doi:10.1016/0304-4076(92)90086-7

John, A., Pecchenino, R., Schmmelpfennig, D., & Schreft, S. (1995). Short- lived agents and the long-lived environment. *Journal of Public Economics*, *5*(1), 12–20. doi:10.1016/0047-2727(94)01459-2

Johnson, D., & Turner, C. (2010). *International business: Themes and issues in the modern global economy* (2nd ed.). Routledge, Taylor & Francis Group. doi:10.4324/9780203861615

Jorgenson, A. K., Dick, C., & Mahutga, M. C. (2007). Foreign investment dependence and the environment: An ecostructural approach. Society for the Study of Social Problems, 54(3), 371-394.

Kaufman, R. K., Davidsdottir, B., Garnham, S., & Pauly, P. (1998). The determinants of atmospheric SO_2 concentrations: Reconsidering the environmental Kuznets curve. *Ecological Economics*, *25*(2), 209–220. doi:10.1016/S0921-8009(97)00181-X

Kizilkaya, O. (2017). The impact of economic growth and foreign direct investment on CO_2 emissions: The case of Turkey. *Turkish Economic Review*, *4*(1), 106–118.

Kurozumi, E., & Yamamoto, T. (2000). Modified lag augmented autoregressions. *Econometric Reviews*, *19*(2), 207–231. doi:10.1080/07474930008800468

Kuznets, S. (1995). Economic growth and income inequality. *The American Economic Review*, *45*, 1–28.

Li, X., & Liu, X. (2005). Foreign direct investment and economic growth: An increasingly endogenous relationship. *World Development*, *33*(3), 393–407. doi:10.1016/j.worlddev.2004.11.001

Linh, D. H., & Lin, S. M. (2015). Dynamic causal relationships among CO_2 emissions, energy consumption, economic growth and FDI in the most populous Asian countries. *Advances in Management and Applied Economics*, *5*(1), 69–88.

Lopez, R. (1994). The environment as a factor of production: The effects of economic growth and trade liberalisation. *Journal of Environmental Economics and Management*, *27*(2), 163–184. doi:10.1006/jeem.1994.1032

McConnell, K. E. (1997). Income and the demand for environmental quality. *Environment and Development Economics*, *2*(4), 383–399. doi:10.1017/S1355770X9700020X

Mehta, P. S. (2002). The Indian mining sector: Effects on the environment & FDI inflows. *CCNM Global Forum on International Investment, Conference on Foreign Direct investment and the Environment.*

Nakano, D., & Muniz, J. Jr. (2018). Writing the literature review for an empirical paper. *Production*, *28*(0), e20170086. doi:10.1590/0103-6513.20170086

Nazlioglu, S., Gormus, A., & Soytas, U. (2016). Oil Prices and Real Estate Investment Trusts (REITs): Gradual-Shift Causality and Volatility Transmission Analysis. *Energy Economics*, *60*(1), 168–175. doi:10.1016/j.eneco.2016.09.009

Pananond, P., & Cuervo-Cazurra, A. (2018). The complementarity of foreign and domestic investment by emerging market multinationals. In Contemporary Issues in International Business: Institutions, Strategy and Performance. doi:10.1007/978-3-319-70220-9_7

Panayotou, T. (1993*). Empirical tests and policy analysis of environmental degradation at different stages of economic development.* Working Paper WP238, Technology and Employment Programme, International Labour Office (ILO).

Pao, H. T., & Tsai, C. M. (2011). Multivariate Granger causality between CO_2 emissions, energy consumption, FDI (foreign direct investment) and GDP (gross domestic product): Evidence from a panel of BRIC (Brazil, Russian Federation, India, and China) countries. *Energy*, *36*(1), 685–693. doi:10.1016/j.energy.2010.09.041

Pesaran, M. H., & Shin, Y. (1999). An autoregressive distributed lag modelling approach to cointegration analysis. In S. Strom (Ed.), *Econometrics and Economic Theory in the 20th Century: The Ragnar Frisch Centennial Symposium*. Cambridge University Press. 10.1017/CCOL521633230.011

Pesaran, M. H., Shin, Y., & Smith, R. J. (2001). Bounds testing approaches to the analysis of level relationships. *Journal of Applied Econometrics*, *16*(3), 289–326. doi:10.1002/jae.616

Planete-Energies. (2016, December). *Understanding electricity. Feature Report*. https://www.planete-energies.com/

Rana, R., & Sharma, M. (2018). On the causality between electricity generation, energy consumption, investment patterns and CO2 emissions in India. *Proceedings, 2nd European Conference on Electrical Engineering & Computer Science (EECS 2018)*. 10.1109/EECS.2018.00023

Rana, R., & Sharma, M. (2019). Dynamic causality testing for EKC hypothesis, pollution haven hypothesis and international trade in India FDI. *The Journal of International Trade & Economic Development*, *28*(3), 348–364. doi:10.1080/09638199.2018.1542451

Rana, R., & Sharma, M. (2020). Dynamic causality among FDI, economic growth and CO_2 emissions in India: With open markets and technology gap. *International Journal of Asian Business and Information Management*, *11*(3), 3. doi:10.4018/IJABIM.2020070102

Rock, M. T. (1996). Pollution intensity of GDP and trade policy: Can the world be wrong? *World Development*, *24*(3), 471–479. doi:10.1016/0305-750X(95)00152-3

Rothman, D. S. (1998). Environmental Kuznets curve- Real progress or passing the buck? A case for consumption-based approaches. *Ecological Economics*, *25*(2), 177–194. doi:10.1016/S0921-8009(97)00179-1

Sawhney, A., & Majumder, P. (2015). *Is India becoming a waste haven of metal scrap? MPRA Paper 67186*. University Library of Munich.

Sawhney, A., & Rastogi, R. (2015). Is India specialising in polluting industries? Evidence from US-India bilateral trade. *World Economy*, *38*(2), 360–378. doi:10.1111/twec.12164

Schembri, P., & Petit, O. (2009). Clean technology transfers and North-South technological gap: An important issue for environmental policies. *Inter Economics*, *120*, 109–130.

Seker, F., Ertugrul, H. M., & Cetin, M. (2015). The impact of foreign direct investment on environmental quality: A Bounds testing and causality analysis for Turkey. *Renewable & Sustainable Energy Reviews*, *52*, 347–356. doi:10.1016/j.rser.2015.07.118

Selden, T. M., & Song, D. (1994). Environmental quality and development: Is there a Kuznets curve for air pollution emissions? *Journal of Environmental Economics and Management*, *27*(2), 147–162. doi:10.1006/jeem.1994.1031

Shafik, N., & Bandyopadhyay, S. (1992). Economic growth and environmental quality: Time-series and cross-country evidence. World Bank Publications.

Shahbaz, M., Nasreen, S., Abbas, F., & Anis, O. (2015). Does foreign direct investment impede environmental quality in high, middle and low-income countries? *Energy Economics*, *51*, 1–35. doi:10.1016/j.eneco.2015.06.014

Shao, Y. (2018). Does FDI affect carbon intensity? New evidence from dynamic panel analysis. *International Journal of Climate Change Strategies and Management*, *10*(1), 27–42. doi:10.1108/IJCCSM-03-2017-0062

Sims, C. A. (1980). Macroeconomics and reality. *Econometrica*, *48*(1), 1–48. doi:10.2307/1912017

Stern, D. I. (2003). *The environmental Kuznets curve*. International Society for Ecological Economics Internet Encyclopaedia of Ecological Economics.

Suri, V., & Chapman, D. (1998). Economic growth, trade and the environment: Implications for the environmental Kuznets curve. *Ecological Economics*, *25*(1), 195–208. doi:10.1016/S0921-8009(97)00180-8

Toda, H. Y., & Phillips, P. C. B. (1993). Vector autoregressions and causality. *Econometrica*, *61*(6), 1367–1393. doi:10.2307/2951647

Toda, H. Y., & Yamamoto, T. (1995). Statistical inferences in vector autoregressions with possibly integrated processes. *Journal of Econometrics*, *66*(1-2), 225–250. doi:10.1016/0304-4076(94)01616-8

Torras, M., & Boyce, J. K. (1998). Income inequality and pollution: A re-assessment of the environmental Kuznets curve. *Ecological Economics*, *25*(2), 147–160. doi:10.1016/S0921-8009(97)00177-8

Ullah, I., Shah, M., & Khan, F. U. (2014). Domestic investment, foreign direct investment, and economic growth nexus: A case of Pakistan. *Economics Research International*, 1-5.

Winchester, C. L., & Salji, M. (2016). Writing a literature review. *Journal of Clinical Urology*, *9*(5), 308–312. doi:10.1177/2051415816650133

Wooldridge, J. M. (2013). *Introductory econometrics: A modern approach* (5th ed.). South-Western Cengage Learning.

Xu, B. (2000). Multinational enterprises, technology diffusion, and host country productivity growth. *Journal of Development Economics*, *62*(2), 477–493. doi:10.1016/S0304-3878(00)00093-6

Yildirim, E. (2014). Energy use, CO_2 emission and foreign direct investment: Is there any inconsistence between causal relations? *Frontiers in Energy*, *8*(3), 269–278. doi:10.100711708-014-0326-6

Zhang, Z., & Meng, X. (2019). Internet penetration and the environmental Kuznets curve: A cross-national analysis. *Sustainability*, *11*(1358), 1–12. doi:10.3390u11051358

Chapter 8
Sustainable Urban Agro Ecology and Its Implications With Food Systems

José G. Vargas-Hernández
https://orcid.org/0000-0003-0938-4197
University Center for Economic and Managerial Sciences, University of Guadalajara, Mexico

ABSTRACT

This chapter has the aim to analyze the implications and interrelationships between a sustainable urban agro ecology and the food system. The beginning assumption of this analytical review considers that sustainable urban agro ecology has positive implications in the development of a sustainable urban food system. The analysis is based on the theoretical and empirical literature review confronted with common spatial-functional observations of urban development and configurations. The analysis concludes that the sustainable urban food system based on agro ecology is growing as an alternative movement towards the building and maintenance of a fairer and healthier urban sustainable environmental development.

INTRODUCTION

The aim of this paper is develop a framework to guide research and action in urban agro ecology and its relationship with sustainable food systems. Spatial-functional analysis can be used to assess the ecological and production functions of the configuration (Lovell et al. 2010). There is a geographical colonialism of knowledge in the agro ecological literature (Gómez et al. 2013) in the publication of agro ecological knowledge between the place where the topic takes place and where it is published. Agro ecology gained ground in academia disseminating information and knowledge sharing with farmer and producers' cooperation on the alternative agriculture production. Agro ecology is a new high knowledge used to diversify farming systems aimed to produce their own fertilizers and pest controls diminishing dependence on external inputs such as chemical fertilizers and pesticides (Altieri and Nicholls 2004).

The current food crisis is driven by human activity causing a natural catastrophic event in climate change with serious consequences in agriculture activities linked to the biodiversity crisis. It is the ecol-

DOI: 10.4018/978-1-7998-4915-5.ch008

ogy of food systems (Francis, *et al.* 2003). Cities are recognizing the relevance of urban agro ecology using biological diversity rather than chemical inputs. The shift of local food economies towards agro ecological small scale and diverse production limited by access to land, affordability and regulations on land use and housing, is a challenge. Agro ecology production is in debate with food sovereignty questioning the food insecurity due to the stresses of the existing modes of food production based on commodity fluctuations of prices and claiming for a new food system.

Urban agro ecology and farming contribute to sustainable food systems change, healthy nutrition, food security and sovereignty, health, wellbeing, community development, economic efficiency, social inclusion and equity, political involvement, environmental sustainability, etc., and the benefits have been associated with food justice to benefit the disadvantaged urban communities. However, the urban agro ecological movement has not received a rigorous analytical scrutiny.

URBAN AGRO ECOLOGY

Agro ecology requires a fundamental cultural and philosophical approach to productive and efficient agriculture. Farming land had been incorporated by ancient cities. The new colonies of Greeks and Romans were settled around the agricultural activities (Kostof, 1991). Mediterranean and Middle Eastern societies have cultivated gardens for aesthetic and production purposes (O'Brien, 2010a). During the medieval times, agricultural farming land was available around the city walls (Cockrall-King, 2012; Howe et al., 2005; Steel, 2008). Steel (2008) and Van der Schans and Wiskerke (2012). During the pre-industrial times, intensive farming disappeared from cities and agriculture was relocated in an urban-rural divide.

Some former industrial cities in developed countries, now in shrinkage processes have a large supply of vacant lots where a large range of people are engaging in urban agro ecology (Colosanti & Hamm 2010). Urban agro ecology is a strategy for regenerating shrinking cities and leveraging a treatment for vacant land. The shrinking cities context of urban agro ecology in community garden projects have an influence on the benefits to community members contributing to the stability of (Tranel & Handlin, 2006).

The green agricultural revolution has increased the crop production but has proved to be unsustainable due to the side effects that damage the environment, biodiversity losses, etc. A cropping system in traditional agro ecology contributes to the stability and productivity of peasant farming in stressful conditions of the environment. Urban agro ecology practices are questioned if they can overcome the structural drivers, income and health disparities and poverty (Raj, Raja, & Dukes, 2016).

The agro ecological practices promote the reintroduction of biodiversity in farming structure and management through agroforestry, multiple cropping, ecologically-based rotations, and the integration of animals with crops. Agro ecological practices emphasize cover cropping that increase diversity and crop rotations. Crop rotation is an organic practice applied for weed control. Globalized industrial agriculture based on export crops, transgenic and biofuel crops is reshaping the world's food supply with economic, social and environmental impacts.

However, agro ecological practices may be fully compatible with gradual mechanizing farms. Agro ecological and organic methods and practices produce equal and most if the times substantially increased yields to elaborate enough food compared to conventional methods in many units of production. Biodynamic and organic agro ecology farming employ alternative practices such as the use of nitrogen-fixing cover crops, rotations, natural controls of pests and diseases, organic composts for soil fertility and organic matter management. Thus, agro ecological and organic practices outperform traditional agriculture practices which contributes to increase the farm-level productivity, food security and household income.

Urban agro ecology practices as greening agendas are benevolent with the potential to provide food quality and affordability, community resilience and cohesion, promote sustainability and reduce ecological footprints, etc. Agro ecological practices includes diversity of species and farm techniques and tasks developed and tested in many regions (Pretty 2008). Urban agro ecology in practice has interwoven elements between agricultural, gardening and farming practices in various forms of public or private indoor, vertical, rooftop, peri urban, and other urban forms and its management (Tornaghi, 2011).

Land used for industrial agriculture exploitation has as consequences the degradation, desertification and deforestation, as well as contamination, water consumption and global warming. Reliance on fossil fuel-based agricultural inputs is related to water capture in soil and carbon sequestration. Industrial agriculture is based on production systems characterized by large-scale farming and economies of scale, vertical integration and corporate concentration. Small agro ecological farming is being adversely impacted by modern industrial agribusiness practices, technology and policy, which have an impact on the biodiversity, environmental wellbeing, food security, etc.

The intensive use of industrial agriculture has brought social and environmental concerns over negative impacts of urbanization processes. Urbanization processes are exacerbating the problems of agro ecological land decreasing and soil degradation (UNEP 2012, Chappell and LaValle, 2011). In European contexts, the establishment of urban agro ecological sites, such as allotments, are the legacy of wartime food shortages.

Urban agro ecology counteracts specific mechanisms of neoliberal urbanism localization (Brenner and Theodore, 2002: 368–375) characterized by regional and intra-national collaboration of new food systems to support food sovereignty. The publication of the International Assessment of Agricultural Knowledge, Science and Technology for Development (IAASTD) in McIntyre et al. (eds.) (2009) advocated the use of agro ecology in sustainability initiatives. In 2010, Special Rapporteur on the Right to Food published by the United Nations (UN) (De Schutter, O. 2014, 2010). considered agro ecology as the basis for food security. New agro ecological technologies are making contributions to food security at the household, communities, regional and national levels.

Agro ecology is a form of resistance to an economic capitalist system based on the rationality of profits. Agro ecology flourishes on movements, collectives, communities, organizations, families, etc. Agro ecology may be adopted as a set of technologies for production rather than a form of resistance against the dominant industrial food production system that exacerbates socio-economic inequalities and is environmentally unsustainable. Traditional agro ecology is the source of promising models to promote biodiversity and sustain year-round yields without agrochemicals. By 2050 it will be needed 70% increase in agricultural production to meet consumption (Burney et al. 2010). Cereal production is required to feed animal and for agro fuel production.

Agro ecology was addressed as relevant subject at the UN Conference on Sustainable Development (Rio+20) in June 2012. The agro ecology concept must engage producers and citizens in a dialogue of critical understanding, collectively shared consciousness and mutual learning. Agro ecology is a concept that has been considered as a science of agricultural development, which is rejected by social movements on the basis that is elite scientific knowledge and demanding to be supported knowledge and experience of local food producers.

The agro ecological science is blending with indigenous knowledge systems to enhance food security and conserve natural resources. Gliessman (2007) contends that agro ecology is the science that applies ecological concepts and principles to the design and management of sustainable food systems. Agro ecology applies ecological concepts and principles to the design, implementation and management of sustainable food systems. Agro ecological science focus on more socially and environmentally sustainable agricultural crop production.

Agro ecology is an emergent and evolving concept used to develop a common framework for principles based on biodiversity enrichment to provide healthy adequate food. Agro ecology is a scientific discipline and sustainable farming approach for healthier food production systems. Agroecologist is an emerging alternative and relational approach building on farmer's knowledge and promising to meet human needs by offering healthy food, human dignity and environmental sustainability.

Agro ecology is defined as the ecology of food systems which fosters broader research questions based on interdisciplinary solutions based on sustainable agro ecology and sustainable food systems. Agro ecology is also defined by the engaged participants including local governments, policy-makers, food producers, social movements, researchers, mainstream institutions. Urban agro ecology efforts of government and social organizations have been relevant to avert the shortfall of food availability for urban populations (Companioni et al., 1997).

Resource needs to achieve the goals of urban agro ecology projects are beyond the material land, soil and financial needs, money, technical knowledge and expertise, etc. Place-based projects offer localized opportunities to develop food democracy through other agro ecology alternatives to the industrial corporate food system (Hassanein, 2003). Agro ecology place-based projects have limitations to overcome the structural inequities in the food system due to demographic, socioeconomic and political power asymmetries (Broad, 2016).

In the context of urban sustainable development, cities can approach a multifunctional urban food forestry project combining elements of urban agro ecology and agroforestry which may contribute to food security and justice in the context of climate change. Urban food agro forestry is a strategy for improving urban food security through free nutrient-dense produce and the multi functionality of urban agriculture. Urban ecological agroforestry overcomes the lack of water and land, water imbalances, low solar radiation, soil compaction, soil and air pollution, etc., and secures land tenure (Lovell 2010; De Zeeuw et al. 2011; Blanco and Lal 2010; Erickson-Hamel and Danso 2010; Pearson 2010).

Urban food forestry has limitations in meeting food security, poverty, malnutrition and in providing a balanced diet and in the use of urban land use. Urban agro ecology may support living-wage jobs in contexts where there are high land prices and where the consumer market is weak and cannot pay to cover costs of production and wages (Daftary-Steel, Dignity, Herrera, & Porter, 2015). Another limitation is the precariousness of land access for urban agro ecology for disadvantaged communities. The agro ecology based food system counters the global corporate industrial agro-food system with short-term yields of crops linked global environmental sustainability problems that compromises the socio ecological foundations of food systems. Agro ecological and organic management practices used by communities would be more focused on selecting high-yielding crops and improved quality.

There are no comprehensive data on the demographics of urban agro ecology practitioners, only in some specific case-based studies. However, there is a demographic increase on urban agro ecology practitioners to gentrification of low-income neighborhoods (Reynolds, 2015) who benefit from public investments in urban agro ecological programs.

Urban agro Ecology is a mechanism used to appropriate urban space (Thibert, 2012), creating commons (Eizenberg, 2012; Roman-Alcalá, 2015) and the right can be demanded to the city (Purcell & Tyman, 2014). The urban agro ecology projects are urban spaces for political community dialogue and cultural resistance actions, while at the same time provide personal benefits, motivation and satisfactions to individuals to work and interact with others for growing their own food and the common enjoyment of eating it. Urban agro ecology is associated with improved mental well-being and reduced stress (Armstrong, 2000).

Urban agro ecology has the potential to improve health tied to socioeconomics and environmental contexts. Urban agro ecology is a tool for health promotion, environment and income generation, contributing to strengthen individual and community leadership and empowerment to facilitate knowledge multiplication (Maria Ribeiro, Maria Bógus, and Wada Watanabe 2015).

Common principles of an agro ecological system include using recycled nutrients, energy and integrating crops and livestock on farming, diversifying genetic, species and natural resources by biological interacting synergies among the components of agro bio ecological diversity to achieve productivity. Urban agro ecology includes a wide range of activities, such as growing, processing, marketing and distribution of vegetables, herbs, fruit, grains and raising domestic animals, fish (aquaculture), bees, etc. Sustainable farming includes the social dimension that does not include the organic farming, which also increases the gas emissions (Kremen et al. 2012). Agro ecology is the science that supports sustainable agriculture and provides solutions to the environmental crises based on sustainability, productivity, resilience, and equity.

Agro ecology is a place-based and community-based experimentation science with sustainability context dependent economic, social and environmental impacts aimed to deliver pro-poor development. Community supported agro ecology schemes are grounded in food re localization of direct relationship arrangements shared in market networks. Popular agro ecology is heterogeneous crop mixes and sizes targeting to increase the food production of food to be supplied to the communities and specially to the tourist sector.

Agro ecology as an agricultural model provides a productive and economically efficient, environmentally sustainable and socially inclusive way to increase food security and justice to alleviate poverty. The agro ecological workforce of producers and farmers have strong agro ecological backgrounds in knowledge and experience although may lack the social and financial capital for land access. Migrant workers employed on agro ecological activities are gaining spaces as urban growers in cities where production is widespread and the state led support is minimal.

Urban agro ecology tends to be fragmented, dispersed and structurally weak mainly due to insufficient land spaces to cultivate, resources and lack of organization and management. Lack of access to land and water management, insufficient infrastructure, investment and technology, insufficient livestock in agro ecological systems lead to inadequate growth in urban agro ecological production and discriminatory multilateral trading system that contributes to food insecurity. The urban geography of food insecurity is growing due to changes in the supply chain mainly affected by the urbanization processes and the economic crisis resulting in the widening of socioeconomic and ecological inequalities.

Agro ecology is a component of agricultural production that has the potential to achieve food sovereignty supported by farmers, agricultural institutions and social movements. Food sovereignty principles require the control of food systems policy, practice and research, food producers. Food sovereignty movements link economic, socio-political and environmental demands with agro ecological food system initiatives with producers and consumers. The agro ecology science, practice and movement is a mode of production that requires to be supported to improve indigenous household and community food sovereignty and security to create autonomy and yield stability in alternative spaces out of the political and economic phenomena.

Many national, regional and local organizations have adopted agro ecology initiatives and developed partnerships and close ties with other proponents as part of a food sovereignty approach. Partnerships should be participative and inclusive of marginalized and vulnerable groups contributing to the appropriate design and implementation of solutions. Social movements efforts addressing the broader agro ecological conditions such as land rights, agro ecology policies and programs, etc. make it possible in specific spaces and localities. The agro ecology movement supports land small holdings with land rights and social equity.

The current spatial concentration of agricultural production in operation has an impact on ecological and socio political forces that has reduced the demand for land dedicated to crops and freeing urban green vacant spaces and marginal lands leading to agro ecological cultivations. Urban and suburbs areas can adapt to agro ecological production and resilience challenges derived from urban expansion in which the produce is channeled through regional and local markets.

Small-scale sustainable farmers producing food tends to protect livelihoods, biodiversity, environment, food security and health, jobs, etc., which cannot be under the control of large multinational agribusiness. Co-constitutive forces shape tensions as inherent characteristics of the complexity of urban agro ecology places which may be compressible as a business model to analyze economic and social organization performance, competitiveness and profitability.

Despite the existing evidences that agro ecology enhances the resilience and productivity of small-scale farming, it is arguable the performance of traditional crops, requiring other intensification approaches such as the adaptation of animal combinations. Small scale farming can be redesigned using ecological principles and using local natural resources to enhance the habitat with healthy plant growth and beneficial organisms and stressing pests. More labor-intensive model of agro ecology supported by strategies in managing people, plants and animals through farmer to farmer systems and cooperatives, may increase the economic and social protection for livelihoods.

The agro ecological system structure and functions require to recognize the problems and make adjustments and appropriate changes in design and management to be implemented. Agro ecologically-based management supports natural resource conservation and sustainable pest management. Public space management can develop agro ecology and gardening for urban food production and for leisure in urban green spaces for therapeutic and educational benefits (Bhatti and Church, 2000; Davis and Middleton, 2012; Jamison, 1985; Jones, 2005; O'Brien, 2010b; Wakefield et al., 2007). A design strategy has been implemented using urban green infrastructure for agriculturally retrofit, or agrifit, by planting urban fruit species in the urban forest as for example with then technique guerrilla grafting mixing low productive fruit trees with high quality cultivars.

Agro ecology management in urban and peri-urban areas with its spatial urban planning contributes to develop new typologies of spatial, environmental, societal, productive and economic identities inter related in new forms of networks and governance. Coalitions among small farmers, civil society organizations, consumers, local governments research organizations and social movements. all in coordinated action are committed to foster sustainable agro ecology toward to more economic efficiency, social inclusion and justice and environmental sustainability. Governance coordination of their relationships between urban and rural actors and their connectivity for food exchange nodes is a relevant policy and strategy to achieve better food security.

Traditional management of agro ecology in the form of raised fields, terraces, polycultures agroforestry systems, etc., using creativity of local farmers has been a successful indigenous agro ecological strategy. The concept of Eco belts is closely related to urban food agroforestry (Bentrup et al. 2001; Bentrup and

Kellerman 2003) which stems from the multi functionality of food tress (MacDaniels and Lieberman 1979) and heavy yields of nutrient-dense fruits (Christaki 2012). Urban agro forestry reduces disasters and contributes to adapt to climate change by reducing urban temperature, runoffs, floods, dust, etc.

Agro ecology in developing countries is built upon the local natural resources of land, water, local varieties of seeds and indigenous knowledge. Traits relevant to indigenous local farmers such as resistance to drought, performance in poly crop systems, competitive ability, storage quality, etc., are replaced by transgenic crops affecting biodiversity, increasing the risk of losing farmer´s abilities to produce with natural inputs under changing environments. Small indigenous farms rained agro ecologists are more resilient to damages of climate change and seem through the development of adaptation strategies allowing them to be able to cope with weather and climate fluctuations.

Land uses abundance and distributed contributes to farm sustainability thought the employment of tools for farm spatial designed processes (Benoit et al. 2012; Lovell et al. 2010; Sterk et al. 2006). Land use functions depend on spatial and topographic relationships, windbreaks, soil and water conservation, runoff filtration, habitat provision, contour cultivation, nitrogen fixation in poly cropping, among other functions (Ajayi 1987; Fujita et al. 1992, Tacio 1993; Bunch 2002). Agro ecology enhances the soil by promoting beneficial organisms to produce healthy food.

Unused and vacant urban land, private or public, can be available for agro ecological farming for food systems through policies and incentives for land access and water, such as tax breaks, low leasing, donation of inputs, agro ecological training programs, etc. Less use of higher productive land for people needing agro ecological food may leave more land available for conservation and regeneration of natural habitat. Land uses in permaculture design requires an analysis of in-site heterogeneity, topography, existing vegetation, and microclimate (Mollison 1988; Jacke and Toensmeier 2005). Agricultural and agro ecological microclimate is an ongoing debate in the scientific literature (Orlandini et al. 2006).

The agro ecological matrix model is an alternative to the agriculture intensification in land-sparing within a framework of small-scale agro ecological production. The urban population serviced by agro ecological products but not involved in agro ecology activities should be considered in terms of land units required and land maintainable by the agro ecological system to support them. The agro ecological land needed may be less than the land maintainable leading to an over-production in agro ecological healthy food. However, the agro ecological urban land needed to sustain the population is greater than the available urban land.

In a complex and interwoven agro ecological system, the value of land and water require of local government, private and community-level investments. Functional agro biodiversity has a negative impact in soil and water decomposition rates, pest abundance, etc. Carbon content levels in agro ecological soils are critical in lost topsoil and decreasing number of birds (Donald et al. 2006). 11.5 percent of total atmospheric emissions comes from chemical pests and fertilizers released by agricultural activities (FAOSTAT, 2016). Industrial agriculture is carbon intensive and contributes 15% to greenhouse gas emissions (Kasterine and Vanzetti 2010) and could rise by 40% by 2030 (Smith et al. 2007).

Waste water can be reused in agro ecology food production through proper participatory management policies. Maintenance of the wetlands, agro ecology and aquaculture can be combined with treatment and reuse waste water. Urban wastewater only should be optimally re used in agro ecology if it is ensured its low cost treatment at facilities using biotechnologies.

Agro ecology is a development model that would lessen dependence on industrial chemical inputs such as pesticides and fertilizers, and genetically modified (GMO) seeds. Agro ecology must be free an isolated from contamination from GMO crops, cross fertilization and genetic pollution to maintain genetic diversity and act as a safeguard against environmental failure. Crop auxiliaries increases agricultural productivity and decreases health and financial costs of pesticides (Östman et al. 2003; Weisenburger 1993). Mixed state and seeds are planted in nursery then transplanted into fertilized soil in unsuited areas.

The agro ecological movements support further research of the economic, social and environmental aspects of permaculture and like other, it requires an institutional and normative framework for holistic and positive action, on personal empowering and responsibility (Smith 2002). Community supported agro ecology movements close the relationships between the producers and the consumers of agro ecological food at the rural-urban food interface.

Urban food strategies can be relocated in local/urban contexts supported by urban agro ecology and community growing and agricultural production in relation to both food security and sustainability objectives. Urban food strategies support access to land for agro ecology, food manufacturing, storage, distribution and marketing. A strategy of food systems connects spaces, actors and politics on sustainable procurement initiatives by embracing all the involved public and private stakeholders, economic development agencies, advocating organizations, etc.

Political pressure from organized small agro ecological farmers and social movements to develop more responsive policies will result in more economic agro ecological efficiency, social equity and inclusion, environmental sustainability, preservation of natural resources, food sovereignty, health and security.

Eco agriculture and organic agriculture for new forms of plant nutrition. The term agro ecology is being used and at the risk of being co-opted for monoculture and organic food approaches and sustainable intensification and climate-smart agriculture trends. Organic, permaculture, biodynamics, etc., are some trajectories towards agro ecology. Agro ecology and permaculture interrelationships have potential for a dialog to integrate the scientific and practical knowledge which entails adaptation of mixed-methods triangular between quantitative and qualitative analysis to address the challenge of a heterogeneous field.

Permaculture is derived from the term permanent agriculture and used to contrast continuous agriculture (Rasul and Thapa 2003; Geist and Lambin 2002). A permanent agriculture connotes agricultural systems incorporating perennial species (Smith 1929) redefined later as permanent culture as the scope of permaculture encompassed human settlement more broadly (Mollison 1988). Permaculture increases harvest labor and structural complexity (Reich 2010).

Permaculture's principles with movements focusing on energy-intensive agro ecological systems (Mollison and Holmgren 1978). Permaculture offers core premises for proposals in agro ecology and agroforestry (Mollison and Holmgren 1978; Mollison 1979, 1988; Holmgren 2004; Shepard 2013). The International permaculture movement is less institutionalized and organized than other international agro ecological movements (Borras et al. 2008; Rosset et al. 2011; Martínez-Torres and Rosset 2010; q.v. Grayson 2010) limiting the potential for mobilization and political support (de Molina 2012).

The adaptation strategies implemented to minimize crop failure are based on the increased use of local varieties more tolerant to drought, mixed cropping, agroforestry, wild plant gathering, etc. Agroforestry is economically efficient by lowering inputs such as water and labor and increasing the crops (Lassoie and Buck 2000; Olson et al. 2000; Bentrup and Kellerman 2003; Lal 2012). Agro forestry food as a large-scale project incorporates multi-functional trees and introduce biodiversity into agro ecological production systems to achieve sustainable and diversified food production. Large-scale urban agro ecological production are enabled by biophysical interactions with energy flows and socially mediated ecological processes of production in a geographical urban configuration.

Cultivating agro ecological and agroforestry food in urban green areas is a growing trend in civic culture that has an advantage the involvement of communities, local food movements, schools, businesses, social organizations, etc. and eager volunteer labor to develop new models of production, distribution and consumption, agritourist, education, leisure, and other activities. Urban agro ecology production occurring in urban green areas and gardens within cities and peripheries benefits urban population and receive direct influence from cities (Companioni et al., 1997). A system of urban agro ecology is composed of urban green areas, farms and gardens developed along agro ecological principles, emphasizing diversification, the use of local resources, eliminating the use of synthetic chemical pesticides and fertilizers, recycling, etc.

Transdisciplinary trends in agro ecological food and urban food forestry include some models of production, distribution and consumption, such as the community supported agriculture, pick up orchards, farmers markets, consumer cooperatives, organic food stores and restaurants, which provide for profit and charity opportunities, at little costs of management, transport, storage and marketing.

Exchange of urban agro ecology produce can be through barter and other informal channels selling at discounted prices to low-income consumers, or donate to food banks and shelters for poor consumers (Levkoe, 2011). Many urban agro ecology farmers and participants grow more vegetables, herbs and fruits beyond their own consumption needs and have to share their surplus with other local consumers, community members and food banks (Balmer et al., 2005). There is mounting consumer interest in local organic foods grown agro ecologically by farmers with whom consumers can interact.

In some countries, the State supports comprehensive agro ecological policies and provides investment for producers. Investments in urban food agro ecology is needed to reduce urban poverty and to meet the growing need of urban food. Government dual policy design can encourage and stimulate agro ecological-based production, offering some tax incentives, and increasing the price of agro ecological and organic products above the market average without excluding the needed poor people by providing special access. In this sense, the dual agro ecological policy approach can provide economic support to production while protecting the access to food consumption for the most vulnerable people.

Some local governments and municipalities provide economic incentives for urban agro ecology (Huang & Drescher, 2015). The beneficiaries of incentives may be the land private property owners who get lower taxes and not the disadvantaged communities who experience inequities in food security (Havens & Roman-Alcalá, 2016. The incentives can have regressive effects for food justice concerns.

The agro ecological approach supports policy and legislation to change the industrial agro ecological change and compromises advocating the principles to be applied as a strategy to local farming in all sectors. Agro ecology is emerging in several dimensions on efforts to expand initiatives despite the precarious funds and restrictive government policies. Agro ecological productivity is being limited by the climate change effects such as flooding, droughts, excessive heat, etc.

Export-oriented agro ecology based on economies of scale to target international markets without compromising indigenous food sovereignty and security. Export costs of agro ecology products are risen creating tensions in production prices for the domestic consumption market and in achieving self-sufficiency with the subsequent worsening of living conditions. Agro ecological food and certified organic products are relevant to promotion, sovereignty and security of sustainable food production.

URBAN SUSTAINABLE FOOD SYSTEMS

The food system based on industrial agriculture has tremendous health and environmental costs, contributing to accelerate the pace of climate change with the increase of carbon dioxide emissions. Industrial agriculture failed to deliver satisfactory food system affecting human health and contaminating agro ecological systems. The agro ecological food system has to prove to be efficiently productive, sustainable resource-conserving and socially inclusive.

The dismantling of the industrial agro food complex have be replaced with restoring local agro ecological alternatives opposing corporate control over production and consumption and supporting local producers to create local food systems. Changes in the global and local food systems expand environmental sustainability awareness and extends to sustainable social relationships aimed to reduce the ecological footprint. Urban agro ecological radical alternatives of space connect to other emerging forms of urban natural environment of re-communing urban land.

In 2006, several cities, among which were London and Oakland pioneered the urban food strategies and since then, more cities around the world joined the initiative called as plan, charter or strategy. The food plan of New City has moved from food system insecurity to opportunity. New York City has become the leader in food system's change (New York City Council 2010, 3). A health-focused food system is aimed to nourish the environment, protect against climate change, promote social justice, create local and diverse economic development and builds community (Toronto Public Health Department 2010, 6).

Current food systems have become highly dependent on oil. Food system emissions from production to consumption, contribute with more than 20% of global greenhouse gas emissions (Vermeulen et al, 2012). Dependence of the food system on fossil fuels has a high environmental impact on human diets. Urban agro ecology is connected to food ethics, food justice and sovereignty movements (Gottlieb and Joshi, 2010; Haiven, 2009; Tornaghi, 2011).

The concept of food justice as an attribute of food systems is being used by government agencies, community and social organizations and some academics (Gottlieb and Joshi 2010; Cadieux and Slocum 2015). A well-crafted definition of food justice which considers the distributive and procedural justice and the structural change of conditions of socioeconomically disadvantaged groups affected by the current food system. This definition of food system is "the right of communities everywhere to produce, process, distribute, access, and eat good food regardless of race, class, gender, ethnicity, citizenship, ability, religion, or community." (Institute for Agro ecology and Trade Policy, 2012).

Urban agro ecology is a strategy of food cultivation to foster food justice. The environmental and social benefits of urban agro ecology are associated with the promotion of food justice to assuming that benefits everyone rather than determine who benefits most. Food justice advocates engage in urban agro ecology with the aim of expanding access to healthy food (Rajan & Duncan, 2013; Mares & Alkon, 2011; Reynolds & Cohen, 2016). The food justice movement advocates granting access to urban and local land for agro ecology practices in food production and recreational activities while confronting and resisting economic, social and environmental inequities embedded in dominant and traditional farming and food systems (Gottlieb and Joshi, 2009: 149).

Some local government initiatives of urban agro ecology projects in urban sites are motivated by ethics and food justice (Nordahl 2009). The emerging food system policy is disconnected with the ethics values of sustainable environment, land access, production, distribution and consumption incorporating commoditized versions of urban agro ecological and socio environmental exclusion practices.

Food justice movement brings awareness on the disparities embedded in any food system, advocates policies, strategies and practices such as place-based projects in urban agro ecology and tries to change the food system through the use of politics to achieve more fair food system (Santo, Palmer, & Kim, 2016). The food justice movement advocates the principles of sustainability, equity, justice, and food security. Food justice movement calls all consumers to grow their own food and to those who have access to land used for agro ecology production that food system workers earn fair livable wages.

Food justice movement calls warm about the disparities and dysfunctionalities perpetuated in the dominant and traditional food systems and the alternative food systems as well as the disparities that exist from production, distribution and consumption in any food system (Alkon & Agyeman, 2011). Growing local sustainable food for better food systems must prioritize food justice by supporting strategies of food certification and labeling, targeted more explicitly to local food justice (Allen, 2010; Born & Purcell, 2006). Food justice movement demands that disadvantaged communities benefit from any initiative to strengthen local, healthy agro ecology projects to support urban food systems.

The complexity of the current food system of cities is facing socio political ecology challenges for the need for radical change to achieve sustainability of food system and food security in a constructive way. The agro ecology scientific community benefits from funding on the agro ecology informed by indigenous and local food systems such as agro ecology of fisheries.

Urban agro ecology improves the wellbeing and health of people with a fairer food system which increases food access and security, generates income, enhances community development, build skills and reduces inequalities. Urban agro ecology contributes to develop skills and capacity building and education of participants who learn about the natural environment in cultivating food (Okvat & Zautra, 2011).

Urban agro ecology is becoming a source of fresh produce and food for the urban and suburban populations. Food agro ecological production is interconnected and interdependent human, nature and ecological processes more focused on conservation and food affordability. It is possible to feed the growing population taking into account the interactions, interdependencies and feedbacks of the technological, economic, social and environmental dimensions of a sustainable agro ecological systems (de Schutter 2014, The Economics of Ecosystems and Biodiversity TEEB 2015). Low-input agro ecology techniques, based on sustainable urban inputs, crop diversification, and organic pests' control and resource conserving can contribute to the agro ecological food production being socially inclusive and environmentally healthy.

Urban agro ecology is interested on healthy food production taking care from land cultivation to supply and consumption of urban agro ecological food. Urban agro ecology as a means of food production within city for more sustainability environment and consumption of fresh vegetables and fruits, reduce dependency on external resources and imports from rural areas making them less vulnerable to food crisis. Rather than chemical inputs, urban agro ecology uses biological diversity to build soil quality, manage water uses, increase crop yields, tackle health crises and reduce climate change and poverty. Urban agro ecology also promotes the amelioration of human health.

Some agro ecology's principles have been adopted by some social movements to make food sovereignty as the long term strategic vision of agro ecology. Agro ecology can optimize the sustainability and productivity of agro systems to give form to food sovereignty defined as the right of people to develop and maintain their capacity to produce their basic food crops based on their own productive and cultural diversity.

Social movements make agro ecological food sovereignty a core part of both short-term tactics and a longer-term strategy. Climate-Smart Agro ecology (CSA) is an agro ecology-based food system is an efficient strategy facing climate change. An eco-socio-logical approach to CSA is a participative, collective an inclusive development initiative under climate change to manage food systems. CSA initiatives contribute to the agro ecology food system supported by the principles of sustainability to create the environmental sustainability and social safeguards.

The concept of agro ecologically food production sovereignty systems is becoming more relevant in Latin America involving farmers in active participation of technological innovation to strengthen local capacities. The agro ecological revolution in Latin America (Altieri & Toledo, 2011) is concerned with issues of agro ecology-based agricultural production, food sovereignty, agro exports and biofuels expansion.

Agro ecological food producers supported by organizations at all space levels embrace the principles of agro ecology. The different meanings of agro ecology are contested by citizens and other stakeholders involved in social movements using some strategies to advance agro ecology. Agro ecological food producers combine efforts with other participants and stakeholder into a veritable social movement. Urban and peri-urban smallholder agro ecology food producers connecting with consumers and feeding urban residents, while taking care of maintaining equitable and resilient food value chains for sustainable urban eco systems and protects the biodiversity integration.

Empirical and longitudinal data are needed at the policy implementation stage to configure the relationships between urban food strategies and food system actors. Urban food strategies are contextually constrained subject to economic and market forces and changing consumer behavior. New urban food strategies and policies features socioeconomic, political and environmental dynamics to identify and take into consideration food security (Nally, 2014; Unger and Wooten 2006).

Producer-consumer interactions and participation in food networks with learning channels and contents contribute for learning about agro ecology production and food. Agro ecological producers are required to establish minimum requirements to enter the formal markets and being recognized by developing and building capacities and developing inclusive and participatory methods to create and guarantee organic products.

Agro-food may be sourced in a geographical area based on agro-ecological methods used and sustainability considerations (Feagan, 2007). Neither small business nor governments invest but large agro food business invest to achieve higher productivity and lower costs to ensure supply. Fostering food agro ecological systems requires the agreement and linkage involvements in collaborating with networks of all the stakeholders, landowners, farmers, producers, consumers, market organizations, communities, civil society, non-governmental organizations, business, public institutions, research institutions, universities, etc.

Establishment of localized resilient agro ecological food systems and circular economy production and consumption models in local economies must be promoted by ecological units of local governments (Pimbert, 2009). Agro ecology needs the support from political commitment aimed to food sovereignty while taking care of natural resources. Food sovereignty is supported by access to land, water and seeds focusing on local autonomy of energy and technological sovereignty, production-consumption cycles, markets, and farmer-to-farmer networks. Other organizations are heavily involved in fostering agro ecological seed security.

Urban agro ecology contributes to healthier and sustainable livelihood conditions of farmers and consumers involved in grounded-based activities, and to better food sovereignty and sustainability based on economic, social, political and environmental dimensions. To implement an agro ecological food system, the state, social and private actors and academic and research institutions must be proactive.

Food sovereignty requires a commitment of smallholder farmers to the principles of agro ecology and their participation in natural food production. Urban agro ecology for food production in urban spaces restrained by environmental conditions, represents a challenge for innovation, such as intensified technologies for plant growing, use of soilless for production such as idle and underutilized land, roofs, walls, vacant lots, brownfields, underground under artificial lighting (Smit and Nasr, 1992; Solon, 2014; Osborne, 2015; Opitz et al., 2016; Mok et al., 2014).

Food sovereignty advocates the control of agro ecological food production alternatives in small-scale sustainable fisher folk and to include fishing, gathering, hunting and farming supported by food movements. Food sovereignty is the right that human beings have to define their healthy, ecologically and culturally food produced through sustainable methods. Food sovereignty has been defined as the right of peoples to healthy and culturally appropriate food produced through ecologically sound and sustainable methods, and their right to define their own food and agriculture systems (Declaration of Nyeleni of the Forum for Food Sovereignty, February 2007).

The cause of food sovereignty serves social movements as the leaver to transform the traditional industrial agriculture mode of production into a healthier and nutritious food system. Non-governmental organizations (NGOs) and transnational agro ecological movements after ignoring local knowledge and needs, and after steady decline of traditional agriculture have started to support sustainable agro ecology embracing principles of food security and sovereignty.

Some urban food system initiatives as mechanisms for enhancing resilience and improving food security, may increase food availability. Food security, physical and financial access to food are interrelated issues that have a strong urban dimension. Urban agro ecology cultivation increases food access and food security to all household of participants, practitioners, consumers and recipients (Algert, Baameur, & Renvall, 2014).

Urban agro ecology is one way for some households to meet their food needs and it is a limited intervention to address food insecurity and access to fresh and healthy food (Alaimo, Packnett, Miles, & Kruger, 2008). Critics argue that urban agro ecology as strategy is limited to increase food security if the consumers lack access to land, physical capacities and skills needed in these activities (Ghose & Pettygrove, 2014).

Access to quality food is a moral imperative and a matter of *national* security (Morgan and Sonnino 2010). Agro ecological development has long-lasting policies with positive effects to increase a sustainable food security system for local populations. Food security is threatened by food losses and waste occurring at different stages of the supply chain affecting the amount of food production. Improving both global and local management of food systems contributes to food security.

Urban agro ecology has the ability to contribute to food security for any household in any city depending on land available, climate and skills of farmers (Grewal & Grewal, 2012). The concept of food citizenship is related with a change on food system meaning that local consumers of food must value locally grown agro products and processed to support the farmers. Urban agro ecology addresses urban food insecurity and food of bad quality due to urbanization developing and its implications in increasing poverty of urban population. Urban agro ecology is becoming relevant by its limited potential in reduction of urban poverty and food insecurity by providing a substantial share of income and important source of livelihood for the urban poor groups and their dietary adequacy.

Converting land to urban agro ecology leads to community food security and self-sufficiency in any city (Colasanti, 2010) although urban agro ecology farming has limits to supply adequate food for the city (Thibert, 2012). Urban agro ecology movement focuses on social organization of agro ecological features aligned to the socio-economic issues, environmental sustainability and food security implications. Food security is addressed in spatial terms to provide access to nutritional food and meet the welfare needs of urban populations (Whelan, Wrigley, Warm and Canning, 2002). Urban food strategies and policies are related to the new governance mechanisms that involves civil society (Blay-Palmer, 2017, FAO 2011; Viljoen and Wiskerke, 2012).

Public city bodies of different sizes have created or reconfigure urban food systems. Local governments are promoting actions to become sustainable food systems innovators as a response for a new geography of food security. In most urban food systems, food security strategies focus on the issues of nutrition and public health. Urban agro ecology increase the nutrition and fresh food knowledge of participating households' in community faming and gardening (Alaimo et al., 2008). Urban agro ecology young participants like more to taste the vegetables they grew themselves (Allen et al., 2008).

The urban agro ecology movements supported by spontaneous popular groups, international organizations, government organizations, and civil society efforts are relevant to make available and supply urban food for the urban populations (Companioni et al., 1997). The food sovereignty movement is supporting agro ecological producers to resist the industrial food system by controlling food production, distribution, marketing and consumption within domestic and local communities. The Food Sovereignty movement is defined as the "right of people to define their own food and agriculture" (Peoples Food Sovereignty Network, 2002; Patel, 2009; Sage, 2014).

Urban and peri-urban food policies are enforced by local governments to overcome critical impacts on environment, natural resources and ecological connectivity aimed to produce fresh and healthy food. The people food policy (PFP) it is a process of citizen-based policy engaged on agro ecological food movement adopted in other places. Urban consumers are not only of agro ecological food products produced in the rural and urban realms (Morgan, 2009).

CONCLUSION

Agro ecology is a sustainable alternative of oil dependent food agro ecology production system aimed to preserve the capabilities of future generations to feed themselves. Sustainable urban food strategies are a shift in the traditional food system. Urban food strategies must define and delimit their local food system and the benefits to deliver. Urban agro ecology farmers and producers may contribute to save money by supplementing to the households some of their produce expenditures.

Agro ecology and sustainable agro-food systems have to scale up a lot opportunities and challenges. The agro ecological food system has policy implications to encourage support and always represents some challenges to the state and society, including the investments, use of social capital to create organization network process in participatory environment.

Food agro ecology is deeply implicated in the loss of biodiversity, climate crisis, erosion of live-hoods, contamination of water and land, etc. Agro ecology expansion is contingent upon farmer production practices and social relations of exchange. Economic, social and environmental costs of agro ecological production systems should be evaluated and compared. These models increase the access and availability of nutrient-dense foods.

Agro ecology and bio economics are green alternatives which are being put into practice for healthy food systems but require setting up new institutional arrangements to build and share knowledge. The food system integrates production, consumption and consumer practices into the broader economic and political systems in which occur. Agro ecology redesigns local food systems to achieve economic efficiency, social inclusion and ecological sustainability. An unsustainable food system has consequences for the health and wellbeing, economic performance, environmental sustainability, social injustice, etc.

However, urban agro ecology strategy cannot resolve the fundamental causes of disparities, impoverishment and food injustice of socioeconomically disadvantaged communities, sometimes stigmatized by structural racism. Urban agro ecology is a limited solution to food injustice and food insecurity resulting from systemic conditions of poverty, low wages, and income disparity. It is not clear whether urban agro ecology may lead to economic and job opportunities for disadvantaged communities.

The food system is related with economic efficiency, social equity, environmental sustainability, the health and wellbeing people and fair trade. A food system makes healthy food affordable and gives priority to the wellbeing contributing to create a thieving economy, enhancing the urban environment and protecting biodiversity and natural resources.

An autarchic approach to food system change as the result of scaling political action does not create opportunities. Some cities place a strong emphasis on localized sustainable food system to support more environmental benefits derived from local production and consumption of food taking into consideration other factors such as energy use and storage. Growing fresh food in local urban agro ecology spaces reduces energy in transportation, cooling, processing, packaging, etc. Any sustainable food system has the inherent attributes of health and wellbeing interacting and shaping with time, spaces and places.

A reduction of food system's distance also reduces the impacts on greenhouse gas emissions and that rising fuel costs have on food prices. The food system may be limited by the intermediary actors, stakeholders and activities essential for the development of relationships between the food producers and consumers in local markets. Most residents of urban contexts are not directly engaged in food production but food consumers that have to rely on financial access to purchase nutritious food with health implications for the urban poor who cannot afford to buy healthy food. Food systems are vulnerable when one or more of the availability, access, utilization and stability components are jeopardized or uncertain.

Local authorities should exercise their political will and provide financial, technical and practical support to stimulate urban producers to develop the sustainable agro ecological production. Policy reforms supported by food sovereignty proponents may contribute to the production systems of agro ecological food. Agro ecology will last with a backing up food sovereignty policy. Sustainable urban food strategies are a shift in the traditional food system. Urban food strategies must define and delimit their local food system and the benefits to deliver. Urban agro ecology farmers and producers may contribute to save money by supplementing to the households some of their produce expenditures (Brown & Carter, 2003).

Measures should be recommended by the coordination of environment, agro ecology and health units to prevent risks eventually associated. This data can be used by urban producers, urban city planners and policymakers, governments and all the stakeholders involved.

Mechanisms of cooperation between agro ecology, health, environment and waste management departments should be created to assess, design strategies to prevent and mitigate risks. Urban agro ecology poses some specific health risks in communities where there is soil, water, and air pollution (Evans & Kantrowitz, 2002). Environmental pollution is present in all the agricultural production, food transportation, storage and retailing processes. The potential of a food strategy strengthens the agricultural sector, improving public health, the environment and economic growth, protecting soil and water resources, encouraging diversity, innovation and collaboration.

REFERENCES

Ajayi, M. T. (1987). Effects of spacing/number of plants per stand and fertilizer placement on performance of two maize varieties intercropped with cowpea in the forest zone of Nigeria. *Ghana Journal of Agricultural Science*, *20*, 39–46.

Alaimo, K., Packnett, E., Miles, R. A., & Kruger, D. J. (2008). Fruit and vegetable intake among urban community gardeners. *Journal of Nutrition Education and Behavior*, *40*(2), 94–101. doi:10.1016/j.jneb.2006.12.003 PMID:18314085

Algert, S., Baameur, A., & Renvall, M. (2014). Vegetable output and cost savings of community gardens in San Jose, California. *Journal of the Academy of Nutrition and Dietetics*, *114*(7), 1072–1076. doi:10.1016/j.jand.2014.02.030 PMID:24751664

Alkon, A., & Agyeman, J. (2011). *Cultivating food justice: Race, class, and sustainability*. MIT Press. doi:10.7551/mitpress/8922.001.0001

Allen, J. O., Alaimo, K., Elam, D., & Perry, E. (2008). Growing vegetables and values: Benefits of neighborhood-based community gardens for youth development and nutrition. *Journal of Hunger & Environmental Nutrition*, *3*(4), 418–439. doi:10.1080/19320240802529169

Allen, P. (2010). Realizing justice in local food systems. *Cambridge Journal of Regions, Economy and Society*, *3*(2), 295–308. doi:10.1093/cjres/rsq015

Altieri, M. A., & Nicholls, C. I. (2004). *Biodiversity and Pest Management in Agroecosystems*. Haworth Press. doi:10.1201/9781482277937

Altieri, M. A., & Toledo, V. M. (2011). The agroecological revolution in Latin America: Rescuing nature, ensuring food sovereignty and empowering peasants. *The Journal of Peasant Studies*, *38*(3), 587–612. doi:10.1080/03066150.2011.582947

Armstrong, D. (2000). A survey of community gardens in upstate New York: Implications for health promotion and community development. *Health & Place*, *6*(4), 319–327. doi:10.1016/S1353-8292(00)00013-7 PMID:11027957

Balmer, K., Gill, J., Kaplinger, H., Miller, J., Peterson, M., Rhoads, A., … Wall, T. (2005). *The diggable city: Making urban agro ecology a planning priority* [Student project]. Retrieved from https://pdxscholar.library.pdx.edu/usp_murp/52/

Benoit, M., Rizzo, D., Marraccini, E., Moonen, A. C., Galli, M., Lardon, S., Rapey, H., Thenail, C., & Bonari, E. (2012). Landscape agronomy: A new field for addressing agricultural landscape dynamics. *Landscape Ecology*, *27*(10), 1385–1394. doi:10.100710980-012-9802-8

Bentrup, G., & Kellerman, T. (2003). Agroforestry and GIS: achieving land productivity and environmental protection. *Proc. of the 8th North American Agroforestry Conference*, 15–25.

Bentrup, G., Schoeneberger, M., Josiah, S., & Francis, C. (2001) Ecobelts: reconnecting agriculture and communities–case studies. In *Proceedings of the Ecospheres Conference*. University of Nebraska.

Bhatti, M., & Church, A. (2000). I never promised you a rose garden: Gender, leisure and home-making. *Leisure Studies, 19*(3), 183–197. doi:10.1080/02614360050023071

Blanco, H., & Lal, R. (2010). *Principles of Soil Conservation and Management.* doi:10.1007/978-1-4020-8709-7

Blay-Palmer, A., Levkoe, C. Z., Mount, P., & Nelson, E. (Eds.). (2017). *Nourishing Communities: From Fractured Food Systems to Transformative Pathways.* Springer.

Born, B., & Purcell, M. (2006). Avoiding the local trap: Scale and food systems in planning research. *Journal of Planning Education and Research, 26*(2), 195–207. doi:10.1177/0739456X06291389

Borras, S. M. Jr, Edelman, M., & Kay, C. (2008). Transnational agrarian movements: Origins and politics, campaigns and impact. *Journal of Agrarian Change, 8*(2-3), 169–204. doi:10.1111/j.1471-0366.2008.00167.x

Brenner, N., & Theodore, N. (2002). Cities and the geographies of 'actually existing neoliberalism'. *Antipode, 34*(3), 349–379. doi:10.1111/1467-8330.00246

Broad, G. (2016). *More than just food: Food justice and community change.* University of California Press. doi:10.1525/california/9780520287440.001.0001

Bunch, R. (2002). Increasing productivity through agroecological approaches in Central America: experiences from hillside agriculture. In N. Uphoff (Ed.), *Agroecological innovations increasing food production with participatory development* (pp. 162–172). Earthscan.

Burney. (2010). Greenhouse gas mitigation by agricultural intensification. Proceedings of the National Academy of Sciences.

Cadieux, K. V., & Slocum, R. (2015). What does it mean to do food justice? *Journal of Political Ecology, 22*(1), 1–26. http://jpe.library.arizona.edu/

Campanioni, N., Rodriguez, A. A., & Carrion, M. (1997). La Agricultura Urbana en Cuba: su participación en la seguridad alimentaria. In *Proceedings III Encuentro Nacional de Agricultura Organica* (pp. 9–13). Villa Clara, Cuba: Academic Press.

Chappell, M.J.; Lavalle, L.A. (2011). Food security and biodiversity: can we have both? *Agro ecology and Human Values, 28*, 3-26.

Christaki, E. (2012). Hippophae rhamnoides L. (Sea Buckthorn): A potential source of nutraceuticals. *Food and Public Heath, 2*(3), 69–72. doi:10.5923/j.fph.20120203.02

Cockrall-King, J. (2012). *Food and the City.* Prometheus Books.

Colasanti, K. (2010). Assessing the local food supply capacity of Detroit, Michigan. *Journal of Agro ecology, and Community Development, 1*(2), 41–58. doi:10.5304/jafscd.2010.012.002

Colasanti, K. J. A., & Hamm, M. W. (2010). Assessing the Local Food Supply Capacity of Detroit, Michigan. *Journal of Agriculture, Food Systems, and Community Development, 1*(2), 41–58. doi:10.5304/jafscd.2010.012.002

Daftary-Steel, S., Dignity, F., Herrera, H., & Porter, C. M. (2015). The unattainable trifecta of urban agro ecology. *Journal of Agro ecology. Systems, and Community Development*, *6*(61), 19–32. doi:10.5304/jafscd.2015.061.014

Davis, L., & Middleton, J. (2012). The perilous road from community activism to public policy: Fifteen years of community agriculture in Sandwell. In Sustainable Food Planning: Evolving Theory and Practice. Wageningen: Wageningen Academic Publishers.

De Molina, M. G. (2012). Agro ecology and politics. How to get sustainability? About the necessity for a political agroecologist. *Agroecology and Sustainable Food Systems*, *37*, 45–59. doi:10.1080/10440046.2012.705810

De Schutter, O. (2010). *Report submitted by the special rapporteur on the right to food*. United Nations Human Rights Council.

De Schutter, O. (2014). *Report of the Special Rapporteur on the right to food. Final report: the transformative potential of the right to food.* Human Right Council, General Assembly, United Nations.

De Zeeuw, H., Van Veenhuizen, R., & Dubbeling, M. (2011). The role of urban agriculture in building resilient cities in developing countries. *J Agric Sci*, *149*(S1), 153–163. doi:10.1017/S0021859610001279

Declaration of Nyeleni. (2007). Available online: https://nyeleni.org/IMG/pdf/DeclNyeleni-en.pdf

Donald, P. F., Sanderson, F. J., Burfield, I. J., & van Bommel, F. P. (2006). Further evidence of continent-wide impacts of agricultural intensification on European farmland birds, 1990–2000. *Agriculture, Ecosystems and Environment*, *116*, 189–196.

Eizenberg, E. (2012). Actually existing commons: Three moments of space of community gardens in New York City. *Antipode*, *44*(3), 764–782. doi:10.1111/j.1467-8330.2011.00892.x

FAO. (2011). *Food, agro ecology and cities: Challenges of food and nutrition security, agro ecology and ecosystem management in an urbanizing world*. http://www.fao.org/fileadmin/templates/FCIT/PDF/FoodAgriCities_Oct2011.pdf

FAOSTAT. (2016). *Emissions by sector*. Accessed December, 2016. 7, 2016. http://www.fao.org/faostat/en/#data/GT/visualize

Feagan, R. (2007). The place of food: Mapping out the 'local' in local food systems. *Progress in Human Geography*, *31*(1), 23–42. doi:10.1177/0309132507073527

Francis, C., Lieblein, G., Gliessman, S., Breland, T. A., Creamer, N., Harwood, R., Salomonsson, L., Helenius, J., Rickerl, D., Salvador, R., Wiedenhoeft, M., Simmons, S., Allen, P., Altieri, M., Flora, C., & Poincelot, R. (2003). Agroecology: The ecology of food systems. *Journal of Sustainable Agriculture*, *2003*(22), 99–118. doi:10.1300/J064v22n03_10

Fujita, K., Ofosu-Budu, K. G., & Ogata, S. (1992). Biological nitrogen fixation in mixed legume-cereal cropping systems. *Plant and Soil*, *141*(1-2), 155–175. doi:10.1007/BF00011315

Geist, H. J., & Lambin, E. F. (2002). Proximate causes and underlying driving forces of tropical deforestation. *Bioscience*, *52*(2), 143–150. doi:10.1641/0006-3568(2002)052[0143:PCAUDF]2.0.CO;2

Ghose, R., & Pettygrove, M. (2014). Actors and networks in urban community garden development. *Geoforum*, *53*(May), 93–103. doi:10.1016/j.geoforum.2014.02.009

Gliessman, S. R. (2007). *Agroecology: The ecology of sustainable food systems* (2nd ed.). CRC Press.

Gómez, L. F., Ríos-Osorio, L., & Eschenhagen, M. L. (2013). Agroecology publications and coloniality of knowledge. *Agronomy for Sustainable Development*, *33*(2), 355–362. doi:10.100713593-012-0109-6

Gottlieb, R., & Joshi, A. (2009). Food Justice. Cambridge, MA: MIT Press.

Gottlieb, R., & Joshi, A. (2010). *Food justice*. MIT Press. doi:10.7551/mitpress/7826.001.0001

Grayson, R. (2010). The permaculture papers 3: childhood. *Pac. Edge*. https://pacific-edge.info/2010/09/the-permaculture-papers-2/

Grewal, S. S., & Grewal, P. S. (2012). Can cities become self-reliant in food? *Cities (London, England)*, *29*(1), 1–11. doi:10.1016/j.cities.2011.06.003

Haiven, M. (2009). Food, finance, crisis, catalyst: Global capitalism and the revolutionary value of food sovereignty. *Política y Cultura*, *2*. https://www.politicsandculture.org/2009/11/03/max-haiven-food-finance-crisis-catalyst-global-capitalism-and-the-revolutionary-value-of-food-sovereignty

Hassanein, N. (2003). Practicing food democracy: A pragmatic politics of transformation. *Journal of Rural Studies*, *19*(1), 77–86. doi:10.1016/S0743-0167(02)00041-4

Havens, E., & Roman-Alcalá, A. (2016). *Land and sovereignty land for food justice? AB 551 and structural change*. Food First. Retrieved from https://foodtank.com/news/2017/01/14707/

Holmgren, D. (2004). *Permaculture: principles and pathways beyond sustainability*. Holmgren Design, Hepburn.

Howe, J., Bohn, K., & Viljoen, A. (2005). Food in time: The history of English open urban space as a European example. In A. Viljoen (Ed.), *Continuous Productive Urban Landscapes: Designing Urb Agriculture for Sustainable Cities* (pp. 95–107). Architectural Press.

Huang, D., & Drescher, M. (2015). Urban crops and livestock: The experiences, challenges, and opportunities of planning for urban agriculture in two Canadian provinces. *Land Use Policy*, *43*(February), 1–14. doi:10.1016/j.landusepol.2014.10.011

Institute for Agro Ecology and Trade Policy. (2012). *Draft principles of food justice*. Minneapolis, MN: Author. Retrieved from https://www.iatp.org/documents/draft-principles-of-food-justice

Jacke, D., & Toensmeier, E. (2005). *Edible forest gardens: ecological design and practice for temperate climate permaculture* (Vol. 2). Chelsea Green Publishing Company.

Jamison, M. S. (1985). The joy of gardening: Collectivist and bureaucratic cultures in conflict. *The Sociological Quarterly*, *26*(4), 473–490. doi:10.1111/j.1533-8525.1985.tb00239.x

Jones, L. (2005). The explosion of garden visiting in France. In T. Richardson & N. Kingsbury (Eds.), *Vista: The Culture and Politics of Gardens* (pp. 52–65). FrancesLincoln.

Kasterine, A., & Vanzetti, D. M. (2010). The Effectiveness, Efficiency and Equity of Market Based and Voluntary Measures to Mitigate Greenhouse Gas Emissions from the Agri-Food Sector (February 24, 2010). *UNCTAD Trade and Environment Review, 2010*. Available at SSRN: https://ssrn.com/abstract=1768948

Kostof, S. (1991). *The City Shaped: Urban Patterns and Meanings Through History*. Thames and Hudson.

Kremen, C., Bacon, C., & Iles, A. (2012). Diversified farming systems: An agro-ecological, systems-based alternative to modern industrial agriculture. *Ecology and Society*, *17*(4), 44. doi:10.5751/ES-05103-170444

Kremen, C., & Miles, A. (2012). Ecosystem services in biologically diversified versus conventional farming systems: Benefits, externalities, and trade-offs. *Ecology and Society*, *17*(4), 40. doi:10.5751/ES-05035-170440

Lal, R. (2012). Urban ecosystems and climate change. In R. Lal & B. Augustin (Eds.), *Carbon sequestration in urban ecosystems* (pp. 3–19). Springer Science. doi:10.1007/978-94-007-2366-5_1

Lassoie, J. P., & Buck, L. E. (2000). Development of agroforestry as an integrated land use management strategy. In H. E. Garrett, W. J. Rietveld, & R. F. Fisher (Eds.), *North American agroforestry: an integrated science and practice* (pp. 1–29). American Society of Agronomy.

Levkoe, C. Z. (2011). Towards a transformative food politics. *Local Environment*, *16*(7), 687–705. doi:10.1080/13549839.2011.592182

Lovell, S. (2010). Multifunctional urban agro ecology for sustainable land use planning in the United States. *Sustainability*, *2*(8), 2499–2522. doi:10.3390u2082499

Lovell, S. T., DeSantis, S., Nathan, C. A., Olson, M. O., Méndez, V. E., Hisashi, C., Kominami, H. C., Erickson, D. L., Morris, K. S., & Morris, W. B. (2010). Integrating agroecology and landscape multifunctionality in Vermont: An evolving framework to evaluate the design of agroeco-systems. *Agricultural Systems*, *103*(5), 327–341. doi:10.1016/j.agsy.2010.03.003

MacDaniels, L. H., & Lieberman, A. S. (1979). Tree crops: A neglected source of food and forage from marginal lands. *Bioscience*, *29*(3), 173–175. doi:10.2307/1307798

Mares, T., & Alkon, A. (2011). Mapping the food movement: Addressing inequality and neoliberalism. *Environment and Society*, *2*(1), 68–86. doi:10.3167/ares.2011.020105

Maria Ribeiro, S., Maria Bógus, C., & Wada Watanabe, H. A. (2015). Agroecological urban agro ecology from the perspective of health promotion. Saúde e Sociedade, 24(2), 730-743.

Martínez-Torres, M. E., & Rosset, P. M. (2010). La Vía Campesina: The birth and evolution of a transnational social movement. *The Journal of Peasant Studies*, *37*(1), 149–175. doi:10.1080/03066150903498804

McIntyre, & ... (Eds.). (2009). *The International Assessment of Agricultural Knowledge, Science and Technology for Development: Summary for Decision-Makers Report of the Global Report*. Island Press. M. A.

Mok, H. F., Williamson, V. G., Grove, J. R., Burry, K., Barker, F., & Hamilton, A. J. (2014). Strawberry fields forever? Urban agriculture in developed countries: A review. *Agronomy for Sustainable Development*, *34*(1), 21–43. doi:10.100713593-013-0156-7

Mollison, B. (1979). *Permaculture: a designer's manual*. Tagari Publications.

Mollison, B. (1988). *Permaculture: a designer's manual*. Tagari.

Mollison, B., & Holmgren, D. (1978). *Permaculture one: a perennial agricultural system for human settlements*. Tagari, Tyalgum.

Morgan, K. (2009). Editorial: Feeding the city: The challenge of urban food planning. *International Planning Studies*, *14*(4), 341–348. doi:10.1080/13563471003642852

Morgan, K., & Sonnino, R. (2010). The urban foodscape: World cities and the new food equation, Cambridge Journal of Regions, Economy and Society, Cambridge Political. *Economy and Society*, *3*(2), 209–224.

Nally, D. (2014). Governing precarious lives: Land grabs, geopolitics, and 'food security'. *The Geographical Journal*, *181*(4), 340–349. doi:10.1111/geoj.12063

New York City Council. (2010). *FoodWorks: a vision to improve NYC's food system*. The New York City Council, New York City. Obtained from http://ngfn.org/resources/ngfndatabase/knowledge/foodworks_fullreport_11_22_10.pdf

Nordahl, D. (2009). *Public Produce: The New Urban Agriculture*. Island Press.

O'Brien, D. (Ed.). (2010a). *Gardening: Philosophy for Everyone*. Wiley-Blackwell. doi:10.1002/9781444324563

O'Brien, D. (2010b). Cultivating our garden: David Hume and gardening as therapy. In D. O'Brien (Ed.), *Gardening: Philosophy for Everyone* (pp. 192–203). Wiley-Blackwell. doi:10.1002/9781444324563.ch14

Okvat, H., & Zautra, A. (2011). Community gardening: A parsimonious path to individual, community, and environmental resilience. *American Journal of Community Psychology*, *47*(3), 374–387. doi:10.100710464-010-9404-z PMID:21222153

Olson, R. K., Schoeneberger, M. M., & Aschmann, S. G. (2000). An ecological foundation for temperate agroforestry. In H. E. Garrett, W. J. Rietveld, & R. F. Fisher (Eds.), *North American agroforestry: an integrated science and practice* (pp. 31–61). American Society of Agronomy.

Opitz, I., Specht, K., Bergers, R., Sieber, T. R., & Piorr, A. (2016): Toward Sustainability: Novelties, Areas of Learning and Innovation. *Urban Agriculture. Sustainability*, *8*, 1–18.

Orlandini, S., Marta, A. D., & Mancini, M. (2006). The agroclimatic analysis at farm scale. *Meteorological Applications*, *13*(S1), 87. doi:10.1017/S1350482706002593

Osborne, M. (2015). *Urban farming's growing importance*. TechBerlin. Available at: https://techberlin.com/articles/urban-farmings-growing-importance/

Östman, Ö., Ekbom, B., & Bengtsson, J. (2003). Yield increase attributable to aphid predation by ground-living polyphagous natural enemies in spring barley in Sweden. *Ecological Economics*, *45*(1), 149–158. doi:10.1016/S0921-8009(03)00007-7

Patel, R. (2009). What does food sovereignty like? *The Journal of Peasant Studies*, *36*(3), 663–706. doi:10.1080/03066150903143079

Pearson, C. J. (2010). Guest editorial: Challenging, multidimensional agriculture in cities. *International Journal of Agricultural Sustainability*, *8*(1-2), 3–4. doi:10.3763/ijas.2009.c5008

Peoples Food Sovereignty Network. (2007). *Statement on peoples' food sovereignty*. Available at: https://www2.world-governance.org/article70.html

Pimbert, M. (2009). *Towards food sovereignty*. London: International Institute for Environment and Development. https://pubs.iied.org/pdfs/14585IIED.pdf

Pretty, J. (2008). Agricultural sustainability: Concepts, principles and evidence. *Philosoph Trans Royal Soc. Botanical Sciences*, *363*(1491), 447–465. PMID:17652074

Purcell, M., & Tyman, S. K. (2014). Cultivating food as a right to the city. *Local Environment*, *20*(10), 1132–1147. doi:10.1080/13549839.2014.903236

Raj, S., Raja, S., & Dukes, B. (2016). Beneficial but constrained: Role of urban agro ecology programs in supporting healthy eating among youth. *Journal of Hunger & Environmental Nutrition*. Advance online publication. doi:10.1080/19320248.2015.1128865

Rajan, S. R., & Duncan, C. A. M. (2013). Ecologies of hope: Environment, technology and habitation: Case studies from the intervenient middle. *Journal of Political Ecology*, *20*(20), 73–79. doi:10.2458/v20i1.21758

Rasul, G., & Thapa, G. B. (2003). Shifting cultivation in the mountains of South and Southeast Asia: Regional patterns and factors influencing the change. *Land Degradation & Development*, *14*(5), 495–508. doi:10.1002/ldr.570

Reich, L. (2010). Why I'm not a permaculturist. *Gard. Rant.* https://gardenrant.com/2010/03/why-im-not-a-permaculturist.html

Reynolds, K. (2015). Disparity despite diversity: Social injustice in New York City's urban agro ecology system. *Antipode*, *47*(1), 240–259. doi:10.1111/anti.12098

Reynolds, K., & Cohen, N. (2016). *Beyond the kale: Urban agro ecology and social justice activism in New York City*. University of Georgia Press.

Roman-Alcalá, A. (2015). Broadening the land question in food sovereignty to northern settings: A case study of Occupy the Farm. *Globalizations*, *12*(4), 545–558. doi:10.1080/14747731.2015.1033199

Rosset, P. M., Sosa, B. M., Jaime, A. M. R., & Lozano, D. R. Á. (2011). The campesino-to-campesino agroecology movement of ANAP in Cuba: Social process methodology in the construction of sustainable peasant agriculture and food sovereignty. *The Journal of Peasant Studies*, *38*(1), 161–191. doi:10.1080/03066150.2010.538584 PMID:21284238

Sage, C. (2014). The transition movement and food sovereignty: From local resilience to global engagement in food system transformation. *Journal of Consumer Culture, 14*(2), 254–275. doi:10.1177/1469540514526281

Santo, R., Palmer, A., & Kim, B. (2016). Vacant lots to vibrant plots: A review of the benefits and limitations of urban agro ecology. Baltimore, MD: Johns Hopkins Center for a Livable Future.

Shepard, M. (2013). *Restoration agriculture: real world permaculture for farmers.* Acres U.S.A.

Smit, J., & Nasr, J. (1992). Urban agriculture for sustainable cities: Using wastes and idle land and water bodies as resources. *Environment and Urbanization, 4*(2), 141–152. doi:10.1177/095624789200400214

Smith. (2007). *Agriculture, in Climate Change 2007: Mitigation.* Contribution of WG III to the Fourth Assessment Report of the Intergovernmental Panel on Climate Change.

Smith, C. (2002). Learning for the environment: An examination of personal empowerment through learning permaculture. *Post Scr, 3*, 12–25.

Smith, J. R. (1929). *Tree crops: a permanent agriculture. Quinn and Boden Company, Inc.* Rahway.

Smith, J. R. (1929). *Tree crops: a permanent agriculture.* New York Brace & Co.

Solon, O. L. (2014): Vast underground bomb shelter reappropriated by urban farmers. *Wired.* Available at: https://www.wired.co.uk/article/underground-farm-zero-carbon-food

Steel, C. (2008). *Hungry City: How Food Shapes Our Lives.* Vintage.

Sterk, S., van Ittersum, M. K., Leeuwis, L., Rossing, R., van Keulen, H., & Van de Ven, G. W. J. (2006). Finding niches for whole-farm design models—Contradictio in terminis? *Agricultural Systems, 87*(2), 211–228. doi:10.1016/j.agsy.2004.11.008

Tacio, H. D. (1993). Sloping agricultural land technology (SALT): A sustainable agroforestry scheme for the uplands. *Agroforestry Systems, 22*(2), 145–152. doi:10.1007/BF00705143

The Economics of Ecosystems and Biodiversity (TEEB). (2010). *Mainstreaming the economics of nature: a synthesis of the approach, conclusions and recommendations of TEEB.* TEEB.

Thibert, J. (2012). Making local planning work for urban agro ecology in the North American context: A view from the ground. *Journal of Planning Education and Research, 32*(3), 349–357. doi:10.1177/0739456X11431692

Tornaghi, C. (2011) *Steal this veg. The Red Pepper.* Available at: https://www.redpepper.org.uk/steal-this-veg

Toronto Public Health Department. (2010). *Annual report.* Toronto Public Health.

Tranel, M., & Handlin, L. B. Jr. (2006). Metromorphosis: Documenting change. *Journal of Urban Affairs, 28*(2), 151–167. doi:10.1111/j.0735-2166.2006.00265.x

UNEP. (2012). *The Emissions Gap Report 2012. A UNEP Synthesis Report, United Nations Environment Programme.* UNEP.

Unger, S., & Wooten, H. (2006). *A food systems assessment for Oakland, CA: Toward a sustainable food plan.* Oakland, CA: Mayor's Office of Sustainability Van der Schans, JW, Wiskerke, JSC (2012) Urb agriculture in developed economies. In A. Viljoen & J. S. C. Wiskerke (Eds.), *Sustainable Food Planning: Evolving Theory and Practice* (pp. 245–258). Wageningen Academic Publishers.

Vermeulen, S. J., Campbell, B.M., & Ingram, J. S. I. (2012). *Climate Change and Food Systems.* Academic Press.

Wakefield, S., Yeudall, F., Taron, C., Reynolds, J., & Skinner, A. (2007). Growing urban health: Community gardening in South-East Toronto. *Health Promotion International, 22*(2), 92–101. doi:10.1093/heapro/dam001 PMID:17324956

Weisenburger, D. D. (1993). Human health effects of agrichemical use. *Human Pathology, 24*(6), 571–576. doi:10.1016/0046-8177(93)90234-8 PMID:8505035

Whelan, A., Wrigley, N., Warm, N., & Canning, S. E. (2002). Life in a 'Food Desert'. *Urban Studies (Edinburgh, Scotland), 39*(11), 2083–2100. doi:10.1080/0042098022000011371

KEY TERMS AND DEFINITIONS

Agro Ecology: Is the discipline that is responsible for administering the ecological principles of the production of food, fuels, fibers, and pharmaceuticals. This encompasses a wide range of approaches and they consider it a science and a way of seeing life, whether organic, conventional, intensive, or extensive.

Ecosystem Services: They are resources or processes of natural ecosystems (goods and services) that benefit human beings. It includes products such as clean drinking water and processes such as waste decomposition.

Environmental Development: An economic and social development that respects the environment. The objective of sustainable development is to define viable projects and reconcile the economic, social, and environmental aspects of human activities; It is about making progress in these areas without having to destroy the environment.

Food System: Refers to food produced, processed, distributed, and consumed locally.

Transition: Step or change from one state, way of being, etc., to another. Intermediate state between an older one and another that is reached in a change.

Urban Agro Ecology: Is called agroecological practices that are developed in or near cities. Urban agroecology usually develops on the roofs of buildings, on the walls of houses and on the balconies and terraces of buildings.

Chapter 9
Green Energy Technologies as the Road Map to Sustainable Economic Growth in Kenya

Joan Mwihaki Nyika
https://orcid.org/0000-0001-8300-6990
Technical University of Kenya, Kenya

ABSTRACT

Kenya, just like other developing countries of the world, is energy deficient. Power supply is characterized by flaws including inaccessibility by marginalized rural communities, high cost of electricity, power outages, and over-reliance on biomass and hydropower sources. Despite these trends, the country has great potential to produce green energy including wind, biofuels, solar, and geothermal power in excess of the entire country's demand. This chapter explores the status of each of these green energy sources and their production potential using in Kenya. Findings show that the production capacity for all of the sources is high though limited tapping and harnessing of these energy resources is done. This could be attributable to lack of exploitation technology and expertise, huge capital investments involved and the uncertainty on the effects of these renewable energy sources to food and water security. For optimal exploitation, Kenya must invest in the energy sector hugely to enhance technical know-how on its development and cater for the associated financial opportunity costs.

INTRODUCTION

Natural resources such as geothermal heat, tides, rain, wind and sunlight provide green energy. Green energy is a derivative of natural processes whose replenishment is constant. Approximately 16% of energy consumption worldwide comes from these green sources, 10% is sourced from traditional biomass, which is important in heating processes and 3.4% is hydropower (UNEP, 2011). In 2013, renewables with exception of hydropower accounted for 8.5% of the generated electricity worldwide and this figure rose up to 9.1% in 2014 and was equivalent to $270.3 billion investment in these power sources (McCrone, 2015). Approximately 19% of electricity sources is from green energy with 16% being from hydropower

DOI: 10.4018/978-1-7998-4915-5.ch009

and 3% from other new renewable sources in developing countries such as Kenya (Kimuyu et al., 2015). Equally, the investment on these green technologies has risen by 32% and as of 2010, more than US$211 billion had been invested on these technologies and processes worldwide. In Africa, investment on green technologies rose from US$750 million to US$3.6 billion, largely because of investments by Kenya and Egypt (Kimuyu et al., 2015). The increased awareness on the need for sustainability due to high pollution tendencies, climate variation and change, increased energy demands, global warming and natural resources depletion have prompted these use of renewables (Mahmure et al., 2015). The increased uptake of green technologies has facilitated the growth of these energy sources all over the world. The World Economic Forum ascertained this supposition and suggested that green energy sources have been critical in the last decade in bridging the energy deficiency gap for about 1.3 billion people worldwide in the wake of emerging markets, which require huge energy inputs (McCrone, 2015).

Kenya in particular is endowed with a variety of green energy sources including bio-fuel, geothermal, hydropower, wind, solar and biomass although their use is limited. In recent times, the expansion of the green energy sector is on rise in the country with the growing need and rising cost of electricity, gas prices, global oil and the pressure to be environmentally stable (Kiplagat et al., 2011). In Kenya, the three main sources of energy are electricity, petroleum and biomass accounting for 5.5, 19.1 and 74.6% of total consumption, respectively (Kiplagat et al., 2011). The sector is characterised by over-reliance on biomass, inaccessibility to modern green technologies, power outages, inability to meet existent energy demand especially in marginalised areas and over-dependence on hydropower. Green energy technologies are foreseen bridges to close these gaps and are potentials to sustainable economic growth, which is environmentally sensitive. This book chapter therefore explains the status of the energy sector and the role of existent green technologies in promoting sustainable economic growth using Kenya as the case study.

The Progress on Green Energy in Developing Countries

In developing countries and transitioning economies, global warming is a growing challenge in addition to food scarcity and financial crises. With this trend, countries, which are poor will become poorer while the rich will be richer unless the adoption of a green economy is implemented (Saidmamatov et al., 2014). A green economy translates to greening of job opportunities (creation of jobs that are friendly environmentally), generation of renewable energy (use of solar photovoltaics (PV), wind and biofuels) and transportation systems (use of hybrid cars) (McCollum et al., 2018). Some countries have launched stimulus packages for financial support to green energy projects. A case example is the USA that has allocated US$ 32 billion on renewable energy projects (McCollum et al., 2018). Similarly, US$ 194 billion green stimulus funding was set aside for clean energy projects around the world including energy-smart technologies, renewable energy projects, carbon storage, transport and capture initiatives according to the UNEP SEFI (2011). Among developing countries, the largest investors of green energy are India, China and Brazil that collectively account for 90% or US$ 60 billion of available investment (Saidmamatov et al., 2014). The other developing countries share the remaining 10% of the total investments and are experiencing a surge in green energy projects. According to UNEP SEFI (2011), investments in Latin America, Asia and Africa have grown by at least two folds since 2010. The investments are concentrated in a few countries only. Saidmamatov et al. (2014) noted the trend and highlighted the need to upscale energy infrastructure to do with its distribution and transmission and improve funding on such projects to expand coverage and maintenance capacity by many developing nations.

Contrary to the views of Saidmamatov et al. (2014), Arndt et al. (2019) claimed that many developing countries are well positioned to adopt a variety of green energy technologies. This is because they are endowed with natural renewable energy sources such as sunshine and adequate wind speed; they are experiencing a surge in energy demands; their populations comprise of dispersed rural populations whose access to electricity and other modern forms of energy are limited; and they lack legacy energy systems unlike developed countries. Due to these reasons, these countries can leapfrog to green/ renewable technologies that are low cost, environmentally benign and reliable. Adopting to these technologies such as wind and solar has an integration challenge where the available supply does not meet the demand. Therefore, institutional reorganisation to enhance integration capacity of green energy and existent electricity infrastructure is essential (Arndt et al., 2019).

Technologies and Costs of Green Energy

Improvements in technology and the rising costs of electricity have pushed producers and consumers to seek alternative energy measures. Consequently, significant investments on wind and solar PV are ongoing. According to the Energy Information Agency, EIA (2017) report, the trend to adopt green energy technologies such as wind and solar PV reduced the levelized cost of electricity (LCOE) by 62 and 81%, respectively while gas and coal reduced by 37 and 15%, respectively. The green energy technologies are now cost competitive compared to conventional technologies such as gas, oil and coal. The LCOE of renewable energy is also on a reducing trend contrary to projections by EIA over time (Lazard, 2016). Other projections that have overestimated the LCOE of solar and wind technologies include those by Cho (2010), Tomabechi (2010) and Eurek et al. (2017). Overestimation of these costs and the underestimation of green energy potential are attributable to the electricity production mixes, which do not appreciate green energy's role in bridging the demand gap. A case example of this trend is from the intergovernmental panel on climate change (IPCC) report that supposed that fossil fuels would remain the dominant power sources despite noting that investments in solar PV had tripled compared to those of coal (Arndt et al., 2019). Therefore, intensive research is needed to model the costs associated with adopting green energy compared to those of expanding and maintaining existent conventional energy sources.

Investment Capacity and Volumes of Green Energy

The reduced costs affiliated with renewable energy in developing countries combined with enactment of supportive policies, conventional energy sources are becoming less popular. According to the IEA (2018), global investments in green energy were estimated at $US 300 billion in 2017, which was about 70% of total energy investments. According to Frankfurt school of finance and management (2018), global investment on renewable energy was estimated at $USD 2.2 trillion owing to reduced costs of these technologies and developing countries have had a considerable share of these developments. The countries accounted 63% of these global investments in green energy by 2017 with Brazil, India and China dominating in these projects. These investments have opened avenues for power mixes to complement existent conventional energy sources.

The Target for Renewable Energy in Developing Countries

The greatest advantage associated with green energy use is reduction in emissions and improved energy efficiency. To minimize emissions to levels that retain greenhouse gas emissions at 450 parts per million (ppm) by 2050, renewable energy investments should translate to 27% of carbon dioxide reductions according to the IEA (2015, 2018). Developing countries are targeting to minimize negative impacts on ecosystems and biodiversity through adoption of green energy technologies according to the UNEP (2012). By responding to these challenges, the countries will be better positioned to respond to climate change effects. To achieve these targets by the end of 2020, the lowest cost for solar grid expansion has been applied in densely populated rural and urban areas. Developing countries such as Kenya, South Africa, Vietnam and China have realized the initiative successfully (Saidmamatov et al., 2014). According to the World Bank (2009), regional grid expansion in developing countries of Africa could promote hydropower trading and the subsequent supply of cheap power. The overall effect would be reduced vulnerability to fluctuating oil prices and the ever-increasing carbon emissions. To adopt these technologies effectively, institutional reorganization and international collaboration and cooperation in research and development on these technologies is indispensable (McCollum et al., 2018).

Description of Kenya's Geography

Kenya is a country located in Africa's East coast, on the equator, bordered by Uganda, Ethiopia and Sudan on the west, north and northwest, respectively. It covers an area of 582, 646 km^2 and is number 47 in terms of size worldwide (Kiplagat et al., 2011). Its main physical features are the Indian Ocean, Mount Kilimanjaro and Lake Victoria situated at the southeast, south near Tanzania and south west, respectively.

Climate in the country is tropical with annual average temperatures being 22°C. At the coast, climate is relatively humid and hot, while the northeast and northern parts are very dry and the inland is temperate. The central highlands experience equatorial climate. Early mornings and nights are usually cool though sunshine is experienced throughouOt the year. July - August are the coolest months while January - March are the hottest. Rainfall occurs in two seasons, the long and short rains experienced in April to June and October to December, respectively (Kiplagat et al., 2011). The highlands are agriculturally rich while the Lake Victoria region has the Kano plains suited for irrigated agriculture unlike the northern parts that are arid and the main economic activities of the region is agro-pastoralism. The country's vast physical features including the plains, Great Rift Valley, mountains, lakes and rivers are major tourist attractions.

In Eastern Africa, Kenya is the most developed economy and is a logistical, commercial and economic hub of East Africa. It has a population of more than 45 million people and a gross domestic product of more than $28 billion (Boampong & Phillips, 2016). The country's economy heavily relies of agriculture and agro-based industries.

Energy Status in Kenya

Energy Production and Consumption Trends

Kenya as a growing economy has put the country's power supply and in particular, electricity under increasing pressure in recent times. For instance, the country's electricity demand increased by 19% annually in the period between 2004 and 2013 (Mokveld & Eije, 2018). It is from this growing demand that the country initiated the least cost power development plan whose focus is on developing hydroelectric and geothermal power plants. This initiative has resulted to a positive impact since electricity tariffs reduced by 30% in 2015 following the commissioning of a 280 MW geothermal plant in the country (Mokveld & Eije, 2018).

The existing capacity though can barely match up with the growing demand because of overreliance on hydropower to meet 39% of power demands in the Kenya as of 2015. The situation is worse during dry spells that have reduced water levels and expensive alternatives such as fossil fuel-based thermal generation are used to bridge capacity gaps. The domestic energy demand is additionally under pressure due to the rising population and the diversification of economic activities skewed to higher consumption. Energy imports are also used to substitute the domestic production capacity but they slow the nation's economic growth due to their exorbitant nature. For instance, 25% of the country's import budget is spent on crude petroleum imports. In events, imports are not adequate, unreliable supply characterised by blackouts and inadequate power infrastructure supplement high-energy costs. These opportunity costs have implications on productivity as Mokveld and Eije (2018) noted claiming that Kenyan companies are 10% less productive because of power fluctuations and outages. Therefore, reliable, affordable and sustainable energy supply for all citizens should be a priority during national policy formulation.

Overall Energy Production and Consumption Trends

Kenya has three main sources of energy that include electricity, petroleum and biomass accounting for 5.9, 19.1 and 74.6%, respectively. A summary of the sources and their supply and consumption capacities in kilotonnes of oil equivalent (ktoe) is shown in Table 1 (International Energy Agency, IEA, 2015). The energy is consumed by industrial, commercial, residential and transport sectors to the tune of 4229, 1153, 2544 and 3100 Gigawatt hours (GWh), respectively (IEA, 2015).

Table 1. Energy supply and consumption in Kenya (IEA, 2015)

Energy Source	Ktoe supply capacity	Ktoe consumption capacity
Coal	349	349
Crude oil	709	-
Oil products	3659	3951
Natural gas	-	-
Nuclear	-	-
Hydro	326	-
Wind, solar, geothermal	3855	-
Wastes and biofuels	16208	10771
Electricity	2	682
Total	25100	15753

Transmission and distribution of power in the country is controlled by a national grid, which operates as an integrated system linked by 66 kilovolts (kV) transmission lines and 132 and 220 Kv transmission systems. Grid system is attributable to the supply challenges characterised by outages and inaccessibility that result from deficiencies, technical or non-technical losses at the system. According to Mokveld and Eije (2018), only 56% of the country's population had access to electricity with 78% and 39% of this figure being urban and rural dwellers, respectively. This is a blatant inequity in energy supply and accessibility for rural and urban population. The main grid operators Kenya Power Limited Company (KPLC) and Kenya electricity transmission company limited (KETRACO) in cognition of this challenge have introduced smart grid technology to enhance system reliability. This initiative has seen the extension of power distribution and transmission lines and the construction of 4, 200 megavolt amperes (MVA) substations in the main cities of Kenya: Nairobi and Mombasa.

Off grid energy supply is being expanded to rural areas to promote accessibility through grid extension programs to marginalised area. Through the Kenya rural electrification authority, the Greenfield photovoltaic mini-grids are being extended and existent mini-grids that are diesel-powered are being retrofitted (Mokveld & Eije, 2018). Through these initiatives, the country has become one of the best off-grid served population in Africa evident from the rise of electricity access from 26% in 2012 to 55% by 2016 (Mokveld & Eije, 2018).

Organisation of the Energy Sector

Organisation of the energy sector in Kenya constitutes of the oversight bodies such as the rural electrification authority (REA), energy regulatory commission (ERC) and the ministry of energy (Kiplagat et al., 2011). The sector also has electricity producers such as Kenya electricity generating company (KenGen) among other independent power producers (IPPs). The latter contribute to 24% of effective power generating capacity while the former 74%. The ministry of energy facilitates the soliciting, release and use of resources to enhance the power sector, prepares a cost efficient power plan and formulates policies in the sector. Some of these policies relevant to the energy sector and their functions are summarised in Table 2 (Owino et al., 2016). The ministry has a green energy sector with ten outstations in the country's major ecological zones to disseminate knowledge regarding renewable energy and collect data on its adoption by citizens. REA was established in 2007 to fast track power accessibility in rural areas while ERC, which was promulgated under the energy act of 2006 provides environmental, economic and technical regulation and oversight to the sector. KPLC offers distribution and transmission services after buying power from the producers; KenGen and IPPs while KETRACO extends and constructs existent and new transmission lines, respectively (Kiplagat et al., 2011).

Financing the Energy Sector in Kenya

Investment and financing of energy related projects particularly those that are renewable is on the rise in recent years. According to Donastorg et al. (2017), the introduction of new and improved renewable energy systems supported by research and development as well as policy support has enhanced energy investments worldwide. For instance, in 2015, $286 billion was investment in energy projects with developed and developing countries accounting for $156 and $130 billion of this share, respectively (Donastorg et al., 2017). The main funding mechanisms for these projects were crowd funding, tax equity loans, tax incentives, feed-in tariffs, tradable green certificates, private and public sector funding.

Table 2. Policies aimed at improving the energy sector in Kenya (Owino et al., 2016)

Policy/ Plan	Year	Function/s
Sessional paper 4	2004	Specified the measures the government should take to promote adoption of renewable energy
2008-2012 energy plan		The plan aimed at forming a geothermal production company, transforming KPLC to a private-sector led but government-owned energy transmission company, establishment of energy testing laboratories and promoting the use of green energy
Energy act	2006	The act aimed at consolidating and amending legislations related to energy by giving powers to REA and ERC to conduct their mandates. The act outlined protocols to attaining flexible licenses for IPPs and approaches to promote green energy use.
Feed-in-tariffs policy	2008	The policy aimed at attracting private investors to produce electricity from green sources including hydro, biomass resources, solar and wind by defining their fixed tariffs.
National energy and petroleum policy and energy bill 2015		The policy aimed at establishing a competitive market system for IPPs around the country to promote its dominance.
National energy policy	2018	The policy aimed at enforcing the use of clean and eco-friendly energy sources accessible to all Kenyans.

In Kenya, the private sector, government and development partners finance energy projects. The private sector uses equity and debt to fund such projects while the government feeds in its finances through the national budget. The government has recently increased its funding for the energy sector through KenGen to develop the Geothermal Development Company (GDC) while the private sector has enhanced its financial facilitation in small hydro projects and in wind and solar energy exploitation (Kimuyu et al., 2015). Although the country does not have updated information on energy financing, the financial year 2007-2008 saw Kshs, 18 billion investment on energy development, which was significantly low, compared to Kshs. 68 billion invested in the 2011/2012 fiscal year (Ministry of Finance, 2011). Tariffs charged on generated power following an agreement by the distributor and generator are key financiers of the Kenyan energy sector. A number of feed-in-tariffs enforced by a policy of 2008 (as shown in Table 2) are also being used to finance the sector. These tariffs are summarised in Table 3 and are aimed at encouraging investment in green energy. The use of mixed financial systems for the energy sector incorporating both public and private sector has promoted the use of green energy in Kenya particularly geothermal and wind energy as Eberhard et al. (2018) noted. Development finance institutions (DFIs) such as Iber Africa power, Kipevu diesel power, Rabai, Tsavo, Gulf and Thika power all based in Kenya have provided loans to KenGen to finance power supply improvements (Eberhard et al., 2018)

GREEN ENERGY PROCESSES AND TECHNOLOGIES IN KENYA

Statistics on Renewable Energy Availability

Kenya is endowed with a variety of green energy resources including wind, solar, biomass, geothermal and hydropower. According to Ayieko (2011), the country has more than 7,000 MW of undeveloped geothermal energy resources and intends to develop 5,500 MW by 2030. The location of the Kenya at the equator makes it endowed with solar insolation that has the potential to produce power though it is largely untapped. A number of renewable energy sources in the country, their power potential and areas

Table 3. Feed-in- tariffs used in Kenya and their rates in $ / kWh at the interconnection point (Ministry of Energy, 2008).

Type of green energy	Plant capacity (MW)	Highest firm power tariff ($/ kWh)	Highers non-firm power tariff ($/kWh)
Solar	0.5 - 10	0.2	0.1
Biogas	0.5 – 40	0.08	0.06
Small hydro	5.1 – 10 1.0 – 5.0 0.5 – 0.99	0.08 0.1 0.12	0.06 0.08 0.12
Biomass	0.5 - 100	0.08	0.06
Wind	0.5 - 100	0.12	0.12
Goethermal	≤70	0.085	-

where they are endowed are summarised in Table 4. With the implementation of the 2015 national energy and petroleum policy in the country, adoption of these renewable energy sources will be up-scaled to improve power accessibility countrywide. The following sections discuss these renewable energy sources and their development status in the country for better insight.

Table 4. Renewable energy sources in Kenya and their power potential (Ayieko, 2011).

Renewable Energy source	Production potential	Areas
Large hydro	6,000 MW	Five drainage basins
Small hydro	3,000 MW	
Biomass - cogeneration	300 MW	Sugarcane belt
Other biomass including biodiesel, power ethanol and biogas	300 MW	High and medium potential regions
Wind	346 watts/m^2	North eastern, Coast and Nairobi
Solar	4-6 kWh/m^2/day	80% of the country
Geothermal	7,000 MW	Rift valley

Hydropower

Hydropower generation in Kenya dates back to the 1940s when Masinga dam was constructed along the Tana River basin (Kimuyu et al., 2015). The hydropower potential of the country is approximately 3000-6000 MW. However, about 1,558 MW out of these is untapped despite being of economic significance. Large hydro projects with a capacity greater than 30 MW are shown to be capable of tapping 1,310 of the 1,558 MW (Kiplagat et al., 2011). The major drainage basins of Kenya including the Tana River, Ewaso Ng'iro north River, Athi River, Rift Valley and Lake Victoria basins are the main locations of the hydropower potential. Table 5 shows the drainage basins used in hydropower generation and their power production capacities. Small, mini and micro hydropower projects also contribute 3,000 MW of the country energy supply and are located at various river sites of rural areas (Mbaka & Mwaniki,

Table 5. Hydropower potential from Kenya's drainage basins

Drainage basin	Potential Capacity (MW)	Mean annual energy (GWh/year)
Athi River	84	460
Rift Valley	345	630
Ewas Ng'iro North	155	675
Lake Victoria	355	1680
Tana River	570	2490
Total	1509	5935

2016). They supply communities at their vicinity with power independently. Table 6 shows the various large and small hydropower projects that have been exploited in the country. These projects contribute to at least 70% of the country's electricity production annually. Tea processing companies, KenGen and non-governmental organisations own them. Through the feed-in tariff policy, the Kenyan government has enhanced the exploitation of small hydro resources in the past decade.

Table 6. Examples of small and large hydropower implementations in Kenya and their capacities in MW (Mbaka & Mwaniki, 2016)

Plant	Year of Installation	Installed Capacity (MW)
Large Hydros		
Sondu	2008	60
Turkwel	1991	106
Kiambere	1988	144
Masinga	1981	40
Gitaru	1978	145
Gitaru unit 3	1999	80
Kamburu 1 and 2	1974	63
Kamburu 3	1976	31
Kindaruma	1968	40
Examples of small hydros		
Ndula	1925	2
Mesco	1933	0.3
Sagana falls	1955	1.5
Wanjii 1 and 2	1952	5.4
James Finlay 5	1999	1.1
Broke Bond 1-4	2000s	0.6
Imenti	2009	0.9
Tungu-Kabiru	2000	0.01

Geothermal Energy

Kenya is a market leader in geothermal energy production in Africa and is richly endowed with this resource. According to the government of Kenya (2008), Olkaria in Kenya's Rift Valley is the prospective region for construction and drilling of geothermal power stations with an average capacity of 15 MW. Explorations in other areas is also ongoing in collaborations by the government and the private sector as shown in Figure 1.

Figure 1. Prospective areas for geothermal exploration in Kenya (Mangi, 2018).

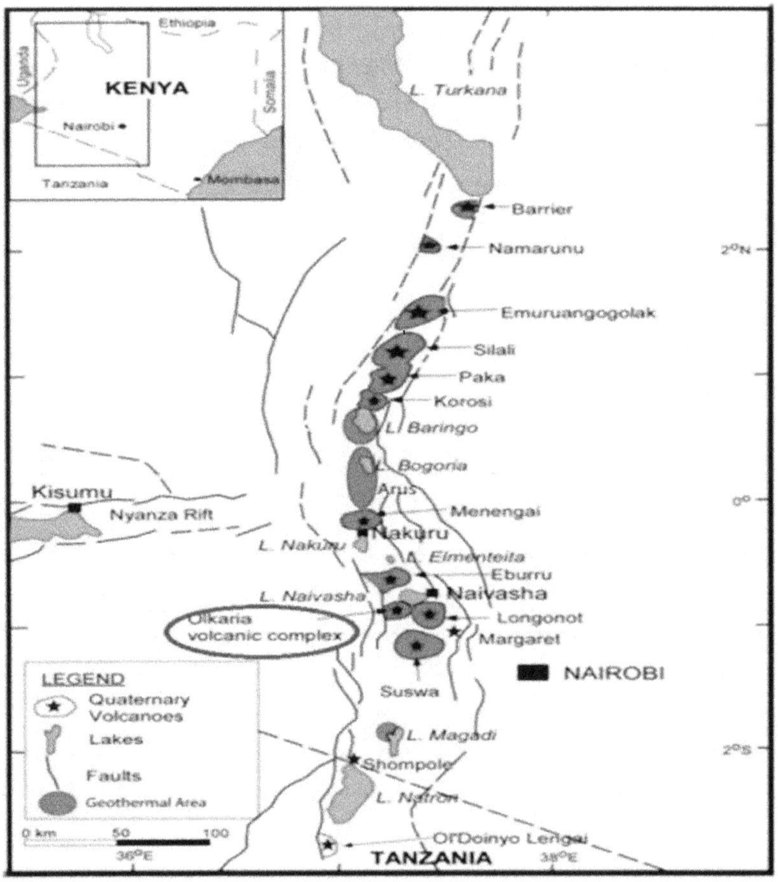

The geothermal capacity that was in operation in 2017 was 690 MW while 358 MW was under construction. This value of operational capacity rose to 745 MW by July 2019 following government efforts under the GDC (Kahlen et al., 2019). The country's projected use of geothermal energy by 2020 is 2,765 MW. The preference to geothermal power in Kenya is attributable to the low operating cost of its power stations compared to hydro ones, the ubiquitous and benign nature of this resource and the potential for high temperature geothermal energy in the country that is yet to be exploited (Micale et al., 2015). Unexploited geothermal energy in the country is quantified between 4,000 and 7,000 MW (Kiplagat et al., 2011; Mangi, 2018). To exploit this resource optimally, Kenya will have to overcome the huge capital investment, technological flaws in steam technologies used to process geothermal energy and the lack of human capacity specialised in processing the resource (Kimuyu et al., 2015; Kahlen et al., 2019).

Biofuels

Liquid energy sources such as bioethanol and biodiesel also occur in Kenya. Using plant feedstock such as jatropha (*Jatropha curcas*), castor (*Ricinus communis*), sunflower (*Helianthus annuus*) and croton (*Croton megalocarpus*), biodiesel is made through transesterification (Ndegwa et al., 2011). Other potential feedstock for biodiesel production include rapeseed (canola), oil palm, cotton and coconut palm that are grown at various regions of the country. Bioethanol is produced by fermenting sweet sorghum, sugarcane and cassava (Ndegwa et al., 2011). Examples of the feedstock are shown in Figure 2.

Compared to petroleum-based fuels, these energy sources are biodegradable and do not pollute the environment through the release of greenhouse gases. Using 50,000 hectares of land, Kenya has the potential to produce 32 million litres equivalent to 27, 400 metric tonnes of biodiesel which could save the country $71 million by substituting this resource with its conventional gasoline and diesel imports (Kimuyu et al., 2015). The government of Kenya under the national biofuel policy of 2010, the bioethanol strategy of 2009 to 2012 and the biodiesel development strategy of 2008-2012 has pushed for the expanded production of biofuel feedstock as Ndegwa et al. (2011) noted. Evidence of this is the commitment of 500,000 acres for cultivation of jatropha by the national government according to Kiplagat et al. (2011). To optimise on the production of biofuels, careful economic analyses to clarify uncertainty regarding their profitability and application without compromising on food security and natural resources (land and water) use is required (Ndegwa et al., 2011).

Wind Energy

Kenya has high wind energy potential compared to other sub-Saharan Africa countries. This potential is rated at 346 W/m^2 and could be used to meet the entire country's power demands (Kazimierczuk, 2019). This potential is favoured by the country's varied nature of surfaces and complex topography, which enable the availability of wind resource throughout the year. High wind speeds occur at the coastal, Rift valley and North West regions (Kenya Renewable Energy Association, KEREA, 2017). Wind energy production in the country began in the 20th century by colonialists who used it in water lifting. However, these projects were abandoned post-colonialism due to limited funding, poor planning and feasibility assessment. Competitive non-subsidised wind energy generation came into being in 2005 following the global increase in oil prices. As of 2017, wind energy made a 1.1% contribution to the country's energy demand with a capacity of 26 MW through the installed wind grid at Nairobi metropolitan at Ngong hills (Kazimierczuk, 2019). The Ministry of Energy (2016) expressed optimism for an upward trend in production as in 2016, 4 wind IPPs with a production capacity of 221 MW acquired operating licences through the feed-in tariffs, another 6 were awaiting licensing and had a capacity of 390 MW while another 9 were conducting feasibility studies for the same purpose. In addition to these is the Lake Turkana wind power project (Figure 3) at Marsabit County of Kenya's northeastern region that has a capacity of 310 MW and is now operation (REVE, 2018).

These milestones are notable compared to other developing countries around the globe. Although Kenya has no wind energy policies, its enhanced adoption to meet power demands of the country were

Figure 2. Examples of feedstock used in biofuel production including a) jatropha, b) castor, c) sunflower, d) rapeseed, e) cotton and f) coconut (GTZ/ Ministry of Agriculture, 2008).

envisioned by the least cost power development planning (LCPDP) committee of Kenya's vision 2030 that projected a 2036 MW wind power generation capacity by 2030 in line with sustainable green energy development (Tigabu et al., 2017). The feed-in tariffs on wind energy in the country are also the cheapest in the continent. The Kenyan government has also introduced zero tax rates on imported wind energy- generation equipment and abolished value added tax on such equipment and their accessories to encourage private investment in the sector. Kazimierczuk (2019) noted these gains as positive efforts towards increasing the wind energy capacity to 2500 MW by 2035.

Although wind energy is a promising technology towards green energy adoption in Kenya, its full exploitation faces a number of challenges (KEREA, 2017; Tigabu et al., 2017). First, the processing of power purchase agreements (PPA) in the country is slow and does not cater for extreme economic problems such as inflation. Secondly, existent feed-in tariffs are limited to a production capacity of 50 MW although most of the operational projects have a higher capacity. As such, there is a need to amend the tariffs to cater for specific production capacities. Lastly, the country must contend with the requirements for wind energy projects such as high wind speed, large wind farms and high capital and infrastructural investments to make the new projects compatible with the existent grid.

Solar Energy

The average irradiance received in Kenya is between 4 – 6 kWh/m^2/ day and the annual average is 250 million tonnes of oil (mToe) equivalent to 5 kWh/m^2/ day (Kiplagat et al., 2011). With such high insolation rates amounting to 5 hours of peak sunshine, amount of energy potential ranges between 700 and 2650 kWh in mountainous regions and arid regions, respectively. However, only a small amount of this energy is harnessed. Kenya has a vibrant solar energy market that has grown exponential over the last decade due to reduced solar photovoltaic prices in addition to the cost effectiveness of the resource

Figure 3. Lake Turkana wind energy project (REVE, 2018).

(Muok & Palit, 2015). The country has the highest household solar ownership in the world rated from the more than 30,000 sales of small solar systems of between 2 to 100 watts every year (Kimuyu et al., 2015). Additionally, at least 7,000 are installed for water heating and drying agricultural harvest (Kiplagat et al., 2011).

The collaboration of the government and the private sector is attributable to these developments. Through KPLC, the government operates off-grid thermal generators that are separate from the national grid and utilise solar energy. Examples of existent stations are at Hola, Merti, Lodwar, Habasueini, Elwak and Mandera with a capacity of 60, 10, 60, 30, 50 and 300 kWp, respectively (Muok & Palit, 2015). The government is also providing solar lanterns in counties such as Isiolo and Wajir that were previously marginalised in regard to energy accessibility through collaborations between the ministry of energy and KPLC. Additionally, institutions and schools in marginalised northern Kenya are being supplied with solar photovoltaic from the national grid as part of promoting green energy adoption (Muok & Palit, 2015). Although difficult to quantify, the private sector has played a crucial role in making solar panels and lanterns affordable and installed in more than 300,000 Kenyan homes by 2014 as Newell et al. (2014) noted. This is made possible through access to credit with low interest rates to purchase solar energy- related equipment and accessories.

Linking Green Energy to Sustainable Economic Growth in Kenya

Kenya's energy sector is the most attractive in Africa with an electricity capacity of 2298 MW in 2015 up from 1600 MW in 2009. The government projects to increase this capacity of 23,000 MW by 2030 (Blue & Green Tomorrow, 2015). Most notable milestones in this strive is the adoption of renewable energy as detailed in the previous sections of this chapter. With an energy mix of geothermal, hydro, thermal, wind, potential coal and nuclear sources, investment in the energy sector will grow and translate to economic development by providing affordable energy to enhance manufacturing and production activities. Through the adoption of renewable energy, 70% of Kenyans have access to power and this has improved their quality of life by facilitating access to improved health services and livelihood improvement particularly in the agricultural sector. According to Kimuyu et al. (2015), renewable energy has improved general welfare of Kenyan people and increased individual household income.

Initiatives on renewable energy have attracted foreign investors such as Google, which contributed $200 billion to the Lake Turkana wind project and the UBBINK (UK-based company), which collaborated with Chloride Exide to set up a solar panel assembly plant in Nakuru County of Kenya (Blue & Green Tomorrow, 2015). Skypower (a North American firm) also collaborated with Kericho county government to set up a solar plant in the region. Similarly, US- based companies such as Envirofit and Burn have invested in Kenya to produce clean burning stoves and prevent the overuse of biofuels and biomass related energy sources including wood and charcoal, which threatens environmental sustainability. Renewable energy is also influencing technological innovation as was the case with Kenya's M-kopa that aids poor communities to purchase solar lamps, radios and mobile phone charging kits as well as panels affordably. Through the program, 200,000 households across the country have accessed energy (Blue & Green Tomorrow, 2015). Foreign investment and technological innovations will drive Kenya's economy by allowing increased flow of finances and tech-savvy knowledge to substitute conventional energy sources with green energy. These sentiments are in agreement with a study by Ahmed and Shimada (2019) that reported a positive correlation between the adoption of green energy and economic development in developing countries including Kenya.

Renewable energy particularly solar has been known to promote socio-economic grown in Kenya's off-grid areas (Owino, 2018). Many counties that have been off-grid for more than half a century today boast of having solar floodlights in shopping centres, along paths and streets. These include remote areas of Siaya County, where such lights have improved security creating a conducive environment for vendors to trade for longer hours (Owino, 2018). The use of portable lights in addition has substituted main electricity supply in events of power outages in addition to creating small employment opportunities in small and medium enterprises such as small firms using solar pumps to lift water, salons, barbershops and welding ventures (Moksnes et al., 2017). The access to renewable energy is synergistic of development in the education sector as learners have access to light for remedial studies even in less-developed regions of Kenya. The UN educational, scientific and cultural organisation, UNESCO (2019) shared similar sentiments highlighting that renewable energy eliminated pollution associated with conventional energy sources such as kerosene, has promoted the emergence of micro-finance investments aimed at empowering women and improved agricultural productivity for small-scale farmers resulting to improved socio-economic welfare of poor Kenyans.

According to Day and Kurdziel (2018), green energy adoption in Kenya has contributed directly or indirectly to economic development in a number of ways apart from providing affordable energy resources. For every 1 MW addition to existent grid, at least 800 fulltime jobs are created, which is a roadmap to poverty eradication (Yadoo & Cruickshank, 2012). Solar photovoltaics have been significant in influencing employment positively in the country compared to other renewable energy sources (Opiyo, 2016). The adoption of green energy has reduced the dependency on fossil fuel imports in the country. This has led to $ 5.5 to 17 million savings of related investment by the government, money that could be reinvested towards development according to Day and Kurdziel (2018). The improved access to power has facilitated enhanced health services in rural areas as advanced healthcare equipment are operational, which translates to ability to steer up economic growth through improved human welfare. The ability of renewable energy to promote water pumping, rainwater harvesting and storage has promoted water security, which translates to improved agricultural productivity for economic growth. Furthermore, it has improved resilience to overcome climate variation and change through minimal greenhouse gas emissions and efficient natural resources use (Moksnes et al., 2017). These trends have positive implications to economic growth now and in the future.

CONCLUSION

Energy accessibility remains a huge challenge worldwide despite the environmental effects of pollution affiliated with conventional energy sources and resulting to global warming and climate change. In cognition of this impediment, Kenya has made great strides in adopting green energy including solar photovolataics, biofuels, wind, hydro and geothermal power sources. These efforts are enabled by solid policies and plans and the collaboration of the private and public sector. Consequently, economic benefits in the form of creation of employment, attraction of foreign investments, efficient use of natural resources, better education and healthcare services and general improvement of quality of life among Kenyans are evident. Therefore, these trends could be emulated in other countries of the world for sustainable development. However, huge capital investments and the lack of technical knowledge to optimize development of these renewable energy sources is a limitation.

REFERENCES

Ahmed, M., & Shimade, K. (2019). The effect of renewable energy consumption on sustainable economic development; Evidence from emerging and developing economies. *Energies*, *12*(15), 2954. doi:10.3390/en12152954

Arndt, C., Arent, D., Hartley, F., Merven, B., & Mondal, H. (2019). Faster than you think: Renewable energy and developing countries. *Annual Review of Resource Economics*, *11*(1), 149–168. doi:10.1146/annurev-resource-100518-093759

Ayieko, Z. (2011). *World Bank Resource*. Available from: http://siteresources.worldbank.org/EXTAFRREGTOPENERGY/ Resources/717305-1327690230600/8397692-1327691237767/Rural_Electrification_in_Kenya_presentation_ Final_11thNov2011.pdf

Blue & Green Tomorrow. (2015). *Renewable energy as a catalyst of economic development in Kenya.* Available from: https://blueandgreentomorrow.com/energy/renewable-energy-as-a-catalyst-of-economic-development-in-kenya/

Boampong, R., & Phillips, M. (2016). *Renewable energy incentives in Kenya: Feed-in tariffs and rural expansion.* National Science Foundation's Sustainable Energy Pathways Initiative.

Cho, A. (2010). Energy's tricky tradeoffs. *Science*, *329*(5993), 786–787. doi:10.1126cience.329.5993.786 PMID:20705848

Day, T., & Kurdziel, M. (2018). *The role of renewable energy mini-grids in Kenya's electricity sectors*. New Climate Institute.

Donastorg, A., Renukappa, S., & Suresh, S. (2017) Financing renewable energy projects in developing countries: A critical review. *IOP Conference Series: Earth and Environmental Science*, *83*, 012012. 10.1088/1755-1315/83/1/012012

Eberhard, A., Gratwick, K., & Kariuki, L. (2018). Kenya's lessons from two decades of experience with independent power producers. *Utilities Policy*, *52*, 37–49. doi:10.1016/j.jup.2018.04.002

Educational, U. N. Scientific and Cultural Organisation (UNESCO). (2019). *Solar energy: changing rural lives in Kenya*. Available from: https://en.unesco.org/courier/2019-3/solar-energy-changing-rural-lives-kenya

Energy Information Administration (EIA). (2017). *Levelized cost and levelized avoided cost of new generation resources in the annual energy outlook 2017*. Available from: https://www.eia.gov/outlooks/archive/aeo17/pdf/electricity_generation.pdf

Eurek, K., Sullivan, P., Gleason, M., Hettinger, D., Heimiller, D., & Lopez, A. (2017). An improved global wind resource estimate for integrated assessment models. *Energy Economics*, *64*, 552–567. doi:10.1016/j.eneco.2016.11.015

Frankfurt School of Finance and Management. (2018). Global trends in renewable energy investment 2018. Frankfurt, Germany: Author.

German Technical Cooperation-GTZ Kenya/ Ministry of Agriculture, Government of Kenya. (2008). *A roadmap for biofuels in Kenya: opportunities and obstacles*. Available from: http://kerea.org/wp-content/uploads/2012/12/A-Roadmap-for-Biofuels-in-Kenya_Opportunities-and-Obstacles.pdf

Government of Kenya. (2008). *Least Cost Energy Development plan: 2009–2029*. Ministry of Energy.

IEA. (2018). *Status of power system transformation 2018: advanced power plant flexibility*. IEA.

International Energy Agency (IEA). (2015). *Kenya energy outlook*. Available from: https://www.iea.org/articles/kenya-energy-outlook

Kahlen, L., Kurdziel, M., Day, T., & Schiefer, T. (2019). *The role of geothermal and coal in Kenya's electricity sector and implications for sustainable development*. Ambition to Action.

Kazimierczuk, A. H. (2019). Wind energy in Kenya: A status and policy framework review. *Renewable & Sustainable Energy Reviews*, *107*, 434–445. doi:10.1016/j.rser.2018.12.061

Kenya Renewable Energy Association (KEREA). (2017). *Wind Energy. Renew Sources*. Available from: http://kerea.org/renewablesources/wind-energy/

Kimuyu, P., Mutua, J. & Wainaina, J. (2015). *Role of renewable energy in promoting inclusive and sustainable development in Kenya*. European Report on Development.

Kiplagat, J., Wang, R., & Li, T. (2011). Renewable energy in Kenya: Resource potential and status of exploitation. *Renewable & Sustainable Energy Reviews*, *15*(6), 2960–2973. doi:10.1016/j.rser.2011.03.023

Lazard, M. (2016). *Lazard's levelized cost of energy analysis—version 10.0.* Available from: https://www.lazard.com/media/438038/levelized-cost-of-energy-v100.pdf

Mahmure, O., Selam, A., Firat, S., Kara, M., & Ozel, S. (2015). A comparative analysis of renewable energy use and policies: Global and Turkish Perspectives. *Sustainability*, *7*(12), 16379–16407. doi:10.3390u71215820

Mangi, P. (2018). Geothermal development in Kenya-country updates. *Proceedings, 7th African Rift Geothermal Conference.*

Mbaka, J., & Mwaniki, M. (2016). Small hydro-power plants in Kenya: A review of status, challenges and future prospects. *Journal of Renewable Energy and Environment*, *3*(4), 20–26.

McCollum, D., Zhou, W., Bertram, C., Boer, H., & Bosetti, V. (2018). Energy investment needs for fulfilling the Paris agreement and achieving sustainable development goals. *Nature Energy*, *3*(7), 589–599. doi:10.103841560-018-0179-z

McCrone, A. (2015). *Frankfurt School-UNEP Centre/BNEF. Global Trends in Renewable Energy Investment*. Frankfurt School of Finance & Management.

Micale, V., Trabacchi, C., & Boni, L. (2015). *Using Public Finance to Attract Private Investment in Geothermal: Olkaria III Case Study, Kenya (2015).* Climate Policy Initiative, Venice, Italy. Available from: http://climatepolicyinitiative.Org/wpcontent/uploads/2015/06/150601_Final_Olkaria_ForWeb.Pdf

Ministry of Energy. (2008). *Strategy for the Development of the Biodiesel Industry in Kenya 2008-2012, Final Draft*. Government Printer.

Ministry of Energy. (2016). *Active developers of feed-in tariffs*. Author.

Ministry of Finance. (2011). Physical Infrastructure Sector Medium Term Expenditure Framework (MTEF) Report, 2011/12 – 2013/14. Nairobi: Government Printer.

Moksnes, N., Korkovelos, A., Mentis, D., & Howells, M. (2017). Electrification pathways for Kenya-linking spatial electrification analysis and medium to long term energy planning. *Environmental Research Letters*, *12*(9), 095008. doi:10.1088/1748-9326/aa7e18

Mokveld, K., & Eije, S. (2018). Final energy report Kenya. Netherlands Enterprise Agency, RVO.

Muok, B., & Palit, D. (2015). *Solar photovoltaic for enhancing electricity access in Kenya: what policies are required?* Policy Brief.

Ndegwa, G., Moraa, V., Jamnadass, R., Mowo, J., Nyabenge, M., & Iiyama, M. (2011). *Potential for biofuel feedstock in Kenya*. ICRAF Working Paper No. 139 Nairobi: World Agroforestry Centre.

Newell, P., Phillips, J., & Pueyo, A. (2014). *The political economy of low carbon energy in Kenya*. Institute of Development Studies. doi:10.1111/j.2040-0209.2014.00445.x

Opiyo, N. (2016). A survey informed PV-based cost-effective electrification options for rural sub-Saharan Africa. *Energy Policy*, *91*, 1–11. doi:10.1016/j.enpol.2015.12.044

Owino, H. (2017). *Kenya: solar energy drives socio-economic growth in off-grid regions*. Available from: https://scienceafrica.co.ke/kenya-solar-energy-drives-socio-economic-growth-in-off-grid-regions/

Owino, T., Kamphof, R., Kuneman, E., & Tilburg, X. (2016). Towards a green trajectory of economic growth and energy security in Kenya? Netherlands Institute of International Relations 'Clingendael', The Hague.

REVE. (2018). *Lake Turkana Wind Power in Loyangalani, Marsabit, is the largest wind energy project on the continent. REVE 2018*. Available from: https://www.evwind.es/2018/11/ 11/lake-turkana-wind-power-in-loyangalani-marsabit-is-the-largest-wind-energyproject-on-the-continent/65167

Saidmamatov, O., Eshchanov, B., Salaev, S., & Shimin, L. (2014). Renewable energy potential of developing countries; the drivers towards a green economy (a case study from Uzbekistan). *International Journal of Green Economics*, *8*(2), 134–143. doi:10.1504/IJGE.2014.065850

Tigabu, A., Kingiri, A., Odongo, F., Margrethe, R., Andersen, M. & Lema, R. (2017). *Capability development and collaboration for Kenya's solar and wind technologies: analysis of major energy policy frameworks*. Academic Press.

Tomabechi, K. (2010). Energy resources in the future. *Energies*, *3*(4), 686–695. doi:10.3390/en3040686

UNEP. (2012). *Towards a Green Economy: Pathways to Sustainable Development and Poverty Eradication*. Available online at: http://www.unep.org/greeneconomy/Portals/88/documents/ger/GER_synthesis_en.pdf

UNEP-Sustainable Energy Finance Initiative (SEFI). (2011). Global Trends in Renewable Energy Investment 2011. United Nations Environment Programme (UNEP) Sustainable Energy Finance Initiative (SEFI) and Bloomberg New Energy Finance, Paris.

United Nations Environmental Program (UNEP). (2011). Global Trends in Renewable Energy' Analysis of Trends and Issues in the Financing Renewable Energy. Investment Report 2011, Nairobi: UNEP.

World Bank. (2009). *Global Economic Prospects 2009: Commodities at the Crossroads*. World Bank. Available online at: http://siteresources.worldbank.org/INTGEP2009/ Resources/10363_WebPDF-w47.pdf

Yadoo, A., & Cruickshank, H. (2012). The role for low carbon electrification technologies in poverty reduction and climate change strategies: A focus on renewable energy mini-grids with case studies in Nepal, Peru and Kenya. *Energy Policy*, *42*, 591–602. doi:10.1016/j.enpol.2011.12.029

Chapter 10
Ergonomic Analysis of Air Conditioned Bus Stops:
An Example of Gümüşhane

Ömer Faruk Efe
https://orcid.org/0000-0001-8170-5114
Afyon Kocatepe University, Turkey

Burak Efe
Necmettin Erbakan University, Turkey

ABSTRACT

Today, public spaces have changed depending on the differentiating human needs. The purpose of the study is to bring together a variety of users and to discuss a life in which various actions such as gathering and meeting take place in cities by taking an example based on the concept of ergonomics based on the human factor. In the study, the relationship between human behavior and physical environment, air-conditioned bus stops in Gümüşhane city, which is one of the places offered for different user groups, was examined. Arrangements for the use of the physical environment for people in many cities have not yet reached the desired level. Ergonomic evaluation of the effectiveness of all human-oriented studies will increase the importance and value of the work done. The more psychological and emotional needs of people provide, the more people will have access to a society with increased satisfaction.

INTRODUCTION

Ergonomics, which has an important place in human life, has spread over a wide area. In product design for human use with ergonomic and anthropometric data, it is very important to think and produce the space used, taking into account the physiological limits of the human being. The word "ergonomics" is produced from the words "Ergo" meaning ancient Greek work and "Nomos" meaning natural order.

DOI: 10.4018/978-1-7998-4915-5.ch010

Spaces are defined as common use areas where a large number of people of different qualities spend most of their time during the day. Ergonomics is often used in the design of the spaces. Ergonomically designed work spaces are known to have positive effects on human (Öz, 1992).

Urban space is defined as a whole, which is formed by buildings, perceived by the urbanites and related to the city's events (Bağbaşı, 2010). In general terms, urban spaces are social public textures where people come together to communicate, interact and realize access as shared living spaces. The public space, which is defined as a concept in which common interactions take place in terms of society, is the common area where discourses, actions and thoughts that favor the benefit of society are produced and developed (Özbek, 2004). According to Arendt (1970), while determining the need for public spaces to be accessible and used by everyone; the necessity of designing the gathering and circulation areas in these areas not to be irregular but to provide comfort and safety (Çelikyay and Karayılmazlar, 2016). In this context, the quality of being public is related to the participation of all individuals in the society in the process of meeting, meeting and gathering in the urban area. Therefore, the arrangement of urban spaces for the use of everyone in the society appears to be an important factor in the context of publicity. It is a fact that in today's information and communication age, most of the time spent by human life while awake is spent in offices. Another point that cannot be ignored is that the quality of life is gradually increasing; The products and services produced in the modern era are more concentrated on the human factor and produced to create a happy and healthy environment for the human. In this case, "living spaces" where we spend our day and even our whole years should be designed in accordance with user needs and satisfaction, especially with the effect of rapidly developing and developing technology (Efe et al. 2018).

Depending on the increasing needs with the developing technology, it is important to provide a healthy and people-oriented living environment for the employees on a scale ranging from any product to the scale of urbanism. These requirements, combined with an increasing standard of living, are physically and psychologically confronted with us; directly affects the quality of life.

The purpose of this chapter is to show that the individual design and workspace design are intertwined; It is the study of ergonomics, which is not stable, has a structure that can vary according to human factor and needs, and is one of the effective factors in meeting these requirements.

After examining the place of ergonomics in urban life and all the conditions to be considered during the design phase, it is thought that it would be useful to examine this information obtained on an example. Thus, in an example of an air-conditioned bus stop, it will be measured how the right and wrong things are thought, and how these affect all people, how they can affect working life and health, and how the ergonomic designs of the places we live and use affect human comfort.

This chapter consists of the following parts. A literature research about city ergonomics is given. Later, the questionnaire application used in the ergonomic analysis of air-conditioned bus stops is mentioned. In addition, information was given about the place of study. The findings obtained in the questionnaire application are mentioned. Subsequently, the findings were discussed and suggestions were made to further improve the ergonomic bus stops.

MAIN FOCUS OF THE CHAPTER

In this study, air-conditioned bus stops, which have become widespread in the cities, have been examined ergonomically. There is limited information in the literature on this subject. At this stage, ergonomic factors are tried to be examined with the help of the survey method.

METHODOLOGY

When it comes to ergonomic design, the thought that comes to our minds only as a choice of furniture and tools does not yield sufficient results. Ergonomic criteria should be sought in every factor affecting human beings. Ergonomic factors are also important in the "space - human" relationship, which includes the "vehicle - human" and "furniture - human" relationship (Güney, 2005).

Ergonomic criteria are as follows (Güney, 2005).

1. Feasibility: (short term) Anthropometric, psycho-physical condition.
2. Durability: (long term) Refers to the physiology of work. It is endurance in working with muscle.
3. Expectability: It is the acceptance of the conditions within the limits of endurance by individuals.
4. Satisfaction: It is the psychological interpretation of the individual.

The science of ergonomics shows its effectiveness not only in the business environment, but in almost all areas of life. It also shows the effect of cities, cities and spaces in the design. Considering that the concepts of expectedness and satisfaction are the criteria that reveal the ergonomic satisfaction of individuals, their importance increases in their studies on urban ergonomics. In their study, Göllü and Türkyılmaz (2019) evaluated a city square in terms of ergonomic criteria. In the study, they stated that ergonomic criteria are not only anthropometric data, but comfort criteria of ergonomics should be taken into consideration in order to ensure the safety of individuals. The element that is primarily considered in the design of an environment or an object is human (Doğan & Altan, 2007). Physiologically, it is desirable to arrange the space in comfort conditions suitable for the anatomy of the human. These; visual, auditory and thermal comfort (Kuru & Türkyılmaz, 2019). While trying to make people more comfortable and peaceful in harmony with the environment, efforts are made to make the environment suitable for the human. Space for shaping a space from an ergonomic point of view; anthropometric, physiological, psychological, information, work organization and safety aspects (Ercan, 1988).

When we compare the spatial organization stages with the functions of the definition and purpose of ergonomics, we see that these issues are intertwined with each other. If we take a look at the stages of ergonomic workplace formation, these are; It is observed that it is provided in terms of anthropometic, physiological, psychological, information, organization and safety (Ercan 1988).

Figure 1. Ligthing

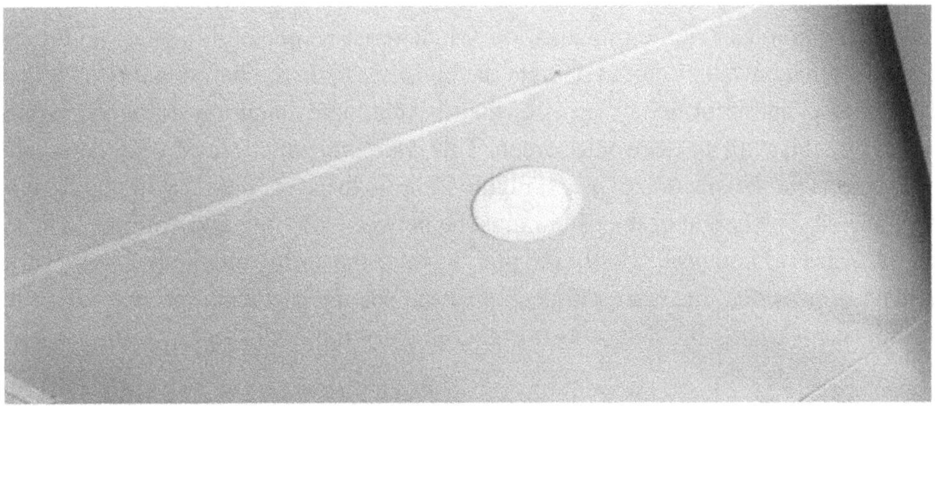

There are some factors that should be addressed in terms of ergonomics and comfort in the design and positioning of air-conditioned bus stops. These;

Noise: The noise defined as unwanted, disturbing sound harms the health and safety of the individual in the environment. Noise can affect psychologically different levels on individuals (Efe & Efe, 2015).

Lighting: Individuals have to look at different points in their area. In order to have a better view and to understand the necessary details, the place must be well illuminated naturally and artificially (Kariptaş, 2012). Figure 1 shows the lighting vehicle at the bus stops.

Ventilation (Thermal Comfort): Temperature levels in living spaces are one of the important factors affecting the quality of life. People are disturbed by sudden changes in body temperatures and they are not very resistant to these changes. In order to feel comfortable, they need to work at the level of temperature they are accustomed to and comfortable with.

The more suitable the weather conditions are, the more comfortable and healthy the employee will feel (Gülçubuk, 2005). The suitable ambient temperature for most people is 18-25 C. The effective parameters in the formation of thermal comfort in human-environment relations appear as activity level, clothing thermal resistance, air temperature, average radiation temperature and relative humidity (Fanger, 1970). Figure 2 shows the air conditioner in the bus stops.

Figure 2. Air conditioned in bus stops

Traffic flow and practices in the organization: Managerial responsibilities fall on the municipality and the companies that have the right to operate the bus lines from the municipality. If it is possible for individuals to be kept and informed in terms of their psychological and physiological satisfaction, it is important to ensure this with an electronic screen. This situation, which we can call cognitive ergonomics, positively affects the satisfaction of users. Figure 3 shows the schedule showing bus departure times.

Seating elements: The height of the chairs must be between 430 mm and 460 mm. These values are suitable standard sizes (Thompson, 1989). People; they use the sitting elements to rest, sit, chat, watch or observe. The dimensions, material and design of the seating elements directly affect the physical comfort of the users (Yücel, 2006) It is also ergonomically important to have lumbar support in the

Figure 3. Bus depature times

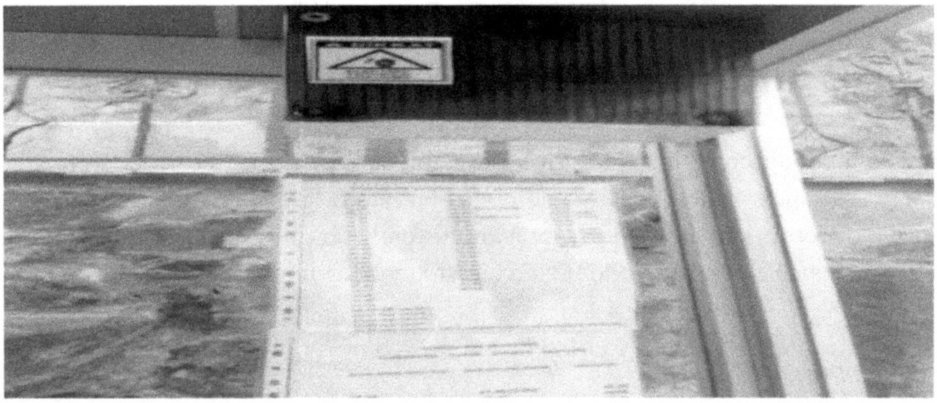

seats. It is also necessary to design suitable seats for individuals with disabilities. The seats in the bus stops are shown in Figure 4.

Figure 4. Seats

Storage areas: Creating appropriate storage areas that will not prevent the entrance and exit in the spaces used and will not restrict the movement, will increase the satisfaction for the users.

Also, when considered as a psycho-social factor, it should be taken into consideration in definitions and approaches related to privacy. It is seen that it is a concept that regulates the relations of people with their environment rather than their relations with the physical world. Therefore, the phenomenon of privacy is a unifying state between personal space and space determination processes. When considering ergonomic factors, psychosocial factors should be considered.

Method

One of the first air-conditioned bus stops Gumushane province started the practice in Turkey. For this reason, it was chosen as the study area. In recent years, the field of application has been increasing in other provinces.

In the research, the air-conditioned bus stops available in Gümüşhane province were ergonomically handled. There are approximately 15 air-conditioned bus stops throughout the city. The stops that are made for the use of both in summer and in winter are currently serving only in winter. 45 responses came to the survey. An example of an air-conditioned bus stop is shown in figure 5.

Figure 5. Air conditioned bus stop

In the study, the ergonomic properties of the air-conditioned bus stops in the sample area were examined by using different methods such as on-site examination, observation, and user survey. Firstly, the existing elements at the air-conditioned bus stops (ABS) were photographed and their positive and negative aspects were examined on the spot. Then, by making use of ergonomic studies, it was evaluated in terms of ergonomic criteria according to the purpose of use of air-conditioned bus stops. In the survey prepared for this purpose, a questionnaire consisting of personal questions, questions about traffic organization, environmental factors, equipment and, ultimately, general evaluation questions was applied to users using bus stops.

The purpose of this study conducted in Gümüşhane province is to determine whether the air-conditioned bus stops are ergonomically efficient, user satisfaction and usefulness, and firstly, how the sample area can be reached in an ergonomically better level and how ergonomics can contribute to the design of the air-conditioned bus stops.

Gümüşhane province, in the Black Sea region of Turkey, located in the eastern Black Sea section. Its population is around 160,000. There is no alternative road route in the city center, so it is an example and an example for the evaluation of air-conditioned bus stops.

APPLICATION

The air-conditioned bus stops were examined for spatial, visual, auditory and thermal comfort. Lighting, air conditioning, seats, book shelves, trash cans, direction and signboards at bus stops are discussed.

A questionnaire consisting of questions about air-conditioned bus stops was conducted for 45 users who actively use air-conditioned bus stops. The average duration of the respondents' living in Gümüşhane is 3 years. The majority of these people are students with 82%. 91% of the respondents use the bus at least 2 times a day. 75% of the respondents stated that they did not use any other means of transportation other than buses.

The data obtained from the questions are analyzed and presented in Table 1.

Table 1. Air conditioned bus stops survey application results

Question	Yes (%)	No (%)
Is the weekly journey time 10 hours or less?	84,4	15,6
When there are route changes, is it possible to adapt without any problems?	66,7	33,3
Can passengers easily follow the bus movement processes?	44,4	55,6
Is there storage space inside the bus stop or elsewhere?	17,8	82,2
Are bus stops located at the right angle according to the light sources?	66,7	33,3
Is lighting sufficient?	66,7	33,3
Is it positioned correctly for any disaster risk?	44,4	55,6
Is there fresh air circulation at bus stops?	40	60
Is there a temperature difference between the air conditioner and other places?	73,3	26,7
Are the voices of other waiting passengers or the voices of other cars unobtrusive?	26,7	73,3
Is the noise at the bus stops quiet enough that it does not interfere with communication?	51,1	48,9
Are the areas of the seats at a distance that will not hinder the entrance and exit?	75,6	24,4
Are the seats not to hinder the movement of the hands and arms?	71,1	28,9
Can you easily use the bus stop locations without encountering any problems?	82,2	17,8
Are passengers exposed to sudden environmental changes?	60	40
Are any disinfectants made against viruses, bacteria, microbes?	66,7	33,3

Then, users were asked to make an overall assessment of the ergonomic satisfaction levels of air-conditioned bus stops. Bus stops, air conditioned by users, have been rated as 1-5 scale (very bad-bad-neutral-good-very good). This analysis made through the survey application is presented in Table 2.

Table 2. General evaluation points

Ergonomic Criterion	1	2	3	4	5	Total Point
Noise in the environment air conditioned stops	3	5	24	10	3	140
Lighting in air-conditioned stops	3	6	16	17	3	146
Ventilation of the environment (thermal comfort)	8	9	18	9	1	121
Traffic flow and applications in the organization	6	13	15	7	4	125
Seating surfaces comfort levet	5	10	19	8	3	129
Availability of seats (chairs)	4	8	21	7	5	131
Use of storage areas	15	13	8	6	3	90

For the general evaluation, the users were asked to make a scoring under the titles of these criteria evaluated at the end of the questionnaire. Thus, although the answers given to the previous questions allow some information to be accessed, the scoring enables the users to evaluate the environment they use from their own perspective, thus helping the survey to evaluate the ergonomic and comfort level. The accuracy of the survey will be tested by comparing whether it supports the results.

Considering that it can get a maximum of 5 points in the survey applied to 45 people, the total evaluation is over 225 and the best scoring is given to the lighting level (146 P). As a matter of fact, the answers to the questions about lighting are mostly positive in the examination. This means that the passengers (customers) are not psychologically aware of the ergonomic factors, although they are not intended to create an ergonomic environment individually.

According to the evaluation points, the lowest criterion was "usage of storage areas" (90 P). Book shelves have been put in place for the users to benefit from while waiting for the bus and they have a useful function. In addition to bookshelves, users can also find suitcases, bags, items, etc. It is thought that the construction of the shelves where the objects can be placed will increase the satisfaction.

FUTURE RESEARCH DIRECTIONS

Ergonomic analysis will continue to gain even more importance in the coming years. For this reason, it will be important to identify the persons / institutions that will carry out this service. As it is apparent, all mentioned subjects involve considering ergonomic analysis to determine and solve city problems. For this, many different methods can be proposed but ergonomic analysis might be applied as one of the best improvement techniques. Moreover, ergonomic with hybrid fuzzy MCDM methods and other techniques such as data envelopment analysis or goal programming will severely be recommended in term of applications and precedent studies.

CONCLUSION

Undoubtedly, the effects of living in a unique and comfortable area are very important in ensuring an efficient quality of life. To achieve this, both the tools used, ergonomic designs, furniture designs and organization and space designs are very important.

Ergonomic factors always affect the person, as can be seen in both the work life and the non-work daily life of the employees. It is known that the job performance of the employees is not only due to the factors related to the working environment. By examining these factors, making specific improvements will contribute to further improving the quality of life of people.

In Turkey, the air-conditioned bus stops, was first applied in gümüşhane province. In the following years, it started to be applied in other cities. Necessary precautions and studies should be done in order not to pose any risk for human health and safety. For this reason, air-conditioned bus stops are examined ergonomically. This study was made and it was thought to make a contribution for the air conditioned bus stops planned to be done after that. When the criteria are evaluated in terms of ergonomics in air-conditioned stops, it is seen that they are sufficient in terms of noise (140) and lighting (146) criteria. However, it should not be overlooked that there is a storage rack space (90) where passengers can put their belongings, luggage in the station, and the environment is kept warm all the time, so the need for ventilation (121) also occurs.

One of the most important ways to increase the satisfaction and learn something new is to learn by seeing. Proper use of tools such as monitors and displays provides some convenience and benefits. The municipality has been proposed to design the vehicles to show the traffic flow in air-conditioned stops or to place them at appropriate points on the screens introducing the services and services, and to support the lower part of the seats. The air-conditioned bus stops should be disinfected periodically against all kinds of chemicals, viruses, bacteria and microbes.

In this chapter, a ergonomic analysis was examined by considering an air conditioned bus stop. Lighting in air-conditioned stops (146) were the highest criterion in the ergonomic analysis of air conditioned bus stop and "Use of storage areas" (90) was the least score criterion.

In order to make the air-conditioned bus stops into a healthier and safer place, the ratings made must be taken into consideration. These criteria should be taken into consideration especially in the design of any place where ergonomic improvement will be made in the city.

Other municipalities should make use of such works to make the air-conditioned bus stops more comfortable for the user.

REFERENCES

Arendt, H. (1970). *The Human Condition. The University of Chicago Press.*

Bağbaşı, G. (2010). *İstanbul Kent Meydanlarının Peyzaj Mimarlığı İlkeleri Açısından İrdelenmesi: Sultanahmet, Beyazıt, Taksim, Beşiktaş, Ortaköy Meydanı Örneği. Yüksek Lisans Tezi.* Bartın Üniversitesi, Bartın.

Çelikyay, S., & Karayılmazlar, A. S. (2016). Bartın Kent Merkezindeki Kamusal Alanların Kentsel Ergonomi ve Kent Kimliği Açısından İncelenmesi. *Bartin Orman Fakültesi Dergisi*, *18*(2), 224–238. doi:10.24011/barofd.278907

Doğan, C., & Altan, O. (2007). Kamusal Alanda Oturma Eylemi ve Ergonomik İlkeler. *Megaron Dergisi*, *2*(3), 159–166.

Efe, B., Efe, Ö. F., & Kurt, M. (2018). An integrated approach for ergonomic product design. *Sakarya University Journal of Science*, *22*(2), 192–202. doi:10.16984aufenbilder.306684

Efe, Ö. F., & Efe, B. (2015). Tekstil Sektöründe İş Kazalarının Oluşumuna Ait Ergonomik Risklerin Değerlendirilmesi. *Süleyman Demirel Üniversitesi Mühendislik Bilimleri ve Tasarım Dergisi*, *3*(3), 623–629.

Ercan, M. N. (1988). *Çalışma Yerlerinin ve Yaşam Ortamlarının Ergonomik Şekillendirilmelerinde Genel Prensipler. İTC Milli Prodüktivite Merkezi 1.* Ulusal Ergonomi Kongresi.

Fanger, P. O. (1970). *Thermal Comfort- Analysis and Applications in Environmental Engineering*. McGraw-Hill.

Göllü, S. K., & Türkyılmaz, Ç. C. (2019). Kent Meydanlarının Ergonomik Ölçütler Açısından Değerlendirilmesi: Kadıköy Rıhtım Meydanı Örneği. *Ergonomi Dergisi*, *2*(1), 32–48. doi:10.33439/ergonomi.479502

Gülçubuk, A. (2005). İnsana Uygun İş Tasarımında İklimlendirmeye İlişkin Ergonomik İlkeler ve Optimum Hava Koşulları. *Celal Bayar Üniversitesi Soma Meslek Yüksekokulu Teknik Bilimler Dergisi*, *3*(1), 95–100.

Güney, Ş. (2005). *Bürolardaki Mekan-Mobilya Organizasyonundaki Ergonomi Faktörü ve Verimliliğe Etkisi: Bir Banka Örneği. Yüksek Lisans Tezi*. Yıldız Teknik Üniversitesi Fen Bilimleri Enstitüsü.

Kariptaş, F. S., Yarael, B., & Ünver, B. (2012). *Ofis Mekanlarında Ergonominin Çalışanlar Üzerindeki Psikolojik Etkileri. 18*. Ulusal Ergonomi Kongresi.

Kuru, R., & Türkyılmaz, Ç. C. (2019). Kütüphane Yapılarının Mekansal Organizasyonunun Ergonomik Açıdan Değerlendirilmesi: Bahçeşehir Üniversitesi Kütüphane Binası Örneği. *Ergonomi Dergisi*, *2*(3), 153–166.

Öz, O. Z. (1992). Kütüphane Binalarının Mekansal Organizasyonlarında Ergonominin Önemi ve Standardizasyon. *Türk Kütüphaneciliği*, *6*(3), 159–171.

Özbek, M. (2004). *Kamusal Alan*. Hil Yayınları.

Thomson, G. (1989). *Planning and Design of Library Buildings*. Butterworth Architecture.

Yücel, G. F. (2006). Kamusal Açık Mekanlarda Donatı Elemanlarının Kullanımı. *Ege Mimarlık Dergisi*, *4*(59), 26–29.

ADDITIONAL READING

Bube, J. L. (1985). The ergonomics/human factors approach to health sciences libraries. *Bulletin of the Medical Library Association*, *73*(3), 254–258. PMID:3161572

Efe, B., Boran, F. E., & Kurt, M. (2015). Sezgisel Bulanık Topsis Yöntemi Kullanılarak Ergonomik Ürün Konsept Seçimi. *SDÜ Mühendislik Bilimleri ve Tasarım Dergisi*, *3*(3), 433–440.

Efe, Ö. F., & Efe, B. (2015). Tekstil Sektöründe İş Kazalarının Oluşumuna Ait Ergonomik Risklerin Değerlendirilmesi. *Süleyman Demirel Üniversitesi Mühendislik Bilimleri ve Tasarım Dergisi*, *3*(3), 623–629.

Rapoport, A. (1977). *Human Aspects of Urban Form: Towards a Man-Environment Approach to Urban Form and Design* (1st ed.). Pergamon Press.

KEY TERMS AND DEFINITIONS

Air Conditioned Bus Stops (ACBS): They are transportation stops that are designed according to the climatic conditions.

Ergonomic (E): It can affect people; Lighting, thermal comfort, heavy lifting, repetitive movements, reaching, pulling, turning, etc. movements, long periods of intense or standing work, improper working postures.

Thermal Comfort (TC): It expresses a certain state of comfort while carrying out both physical and mental activities in terms of climatic conditions such as temperature, humidity and air flow.

Chapter 11
Contemporary Developments in Waste Water Treatment Technologies

Mohit Dev Awasthi
ITM University, Gwalior, India

Mukesh Kumar Pandey
ITM University, Gwalior, India

Trilok Chauhan
ITM University, Gwalior, India

Mir Sayed Shah Danish
https://orcid.org/0000-0002-5196-2280
University of the Ryukyus, Japan

Dheeraj Kumar
National Institute of Technology, Durgapur, India

Nadeem Faisal
ITM University, Gwalior, India

ABSTRACT

Water is the sole reason for the existence of life and prime nutrient to sustain it. So, it becomes extremely important to preserve wastewater to utilize it efficiently for various purposes. The methods and techniques that have the least investment capital, qualitative output, and environment-friendly content are desired. Such unaccustomed methods are our hope for a cleaner, safer, and brighter future. This chapter revolves around the methods employed to brawl the threats posed to the viability of water at earth from emerging contaminants and the efforts made by mankind to prevent it. Nanotechnology has potent attributes

DOI: 10.4018/978-1-7998-4915-5.ch011

for water decontamination. Several nanotech-based devices, methods, and treatments have been used in recent times for wastewater treatment such as nanomembranes, nano photocatalysts, nano-cellulose, nano-absorbent, and nanocomposite materials. The chapter discusses these contemporary developments in nano-based technologies and the various advancements, which have been recently made in conventional techniques used for wastewater treatment technologies.

INTRODUCTION

Water – Irreplaceable, Reliant, Vital

WATER, the entity which allowed our planet to carry life; emboldening Earth as witnessed today, savoring it; is not merely a resource for satisfying the thirst for our human-centric pursuits but an element that roots the planet with approximately 8.7 million distinct life forms. Much exciting are the disputed study results claiming, that the water isn't native to our planet; affirming it as a boon. Apart from being the very foundation of our planet and essence of life, water architects the lifestyle of each organism on Earth.

In context to the human race, a clean supply of safe and sustainable water fuels our daily lives; From keeping our societies healthy and hydrated to running our cities and growing our economies it is a deep-rooted heritage of our daily lives. Water, the wonder liquid is the main ingredient for our food production and it very acts as a solvent, digesting that same food in our bodies. It also removes the toxins from human bodies as well as cleanses our vicinities. It allows the movement of nutrients in human bodies as well as transports nutrients and rich soil from mountains to the surface. As the human civilizations evolved, they steadily realized the potential of water and many crucial and vital methods for the sustenance of civilizations were invented; chiefly farming and transportation. Earlier human civilizations were set-up near the banks of various water bodies. The discovery of water as a means of transportation gave Homo sapiens the aid to explore the habitats and wield the opportunities offered by this planet. Not soon enough after the discovery of waterways, Humans turned towards water bodies for food, the sea life, and also utilized water as transport means.

With the market expansion, and due to population growth, the manufacturing industries, food processing plants and power production units gravitate to increase their production rate which contaminates the water like never before. Some general industries producing gallons of wastewater are textile, dyeing, petrochemical industry, organic chemicals, iron & steel industries, mines& quarries, food industry, nuclear industry, and electric power plants (Sonune & Ghate, 2004). According to the United Nations, several unprivileged communities across the globe are experiencing a scarcity of a clean supply of water, consequently leading to this - 4.5 billion people don't have safe access to adequate and equitable sanitation services. Approximately, 2.1 billion people are deprived of containment free water sources, available at all (UNICEF, 2017). Figure 1 represents the worldwide utilization of freshwater.

Dreadful Results Due to Water Contamination:

Figure 1. Representation of freshwater utilization across the world

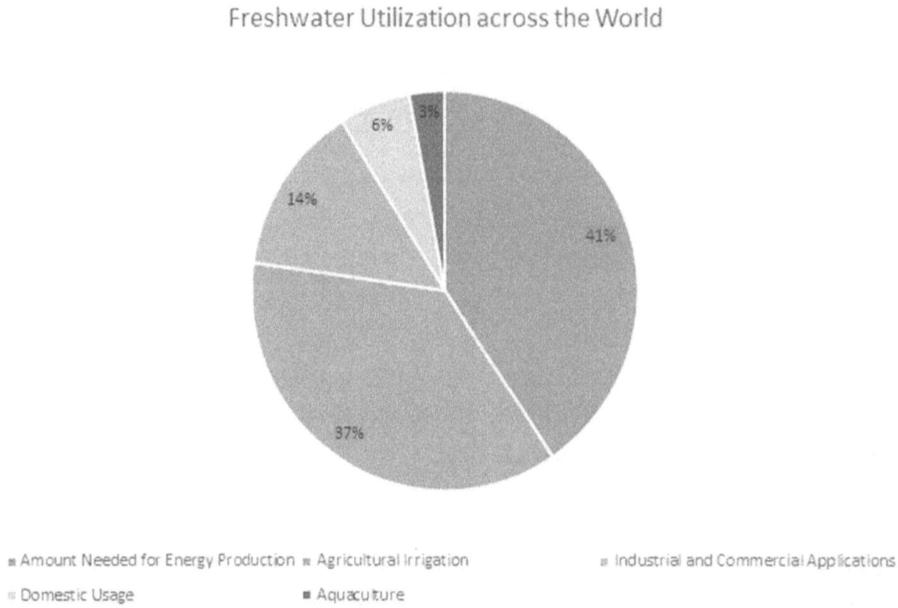

- Consequently, about 5,25,000 cases of child deaths (below the age of five) are annually reported face due to trivial diseases like diarrhea. (UNICEF, 2015)
- Nearly 40% of the world population is facing water scarcity (WHO)
- 90% of the epidemics, diseases, and inevitable natural accidents are due to water (UNISDR)

Threat to the Ecological Balance

- The ecosystem receives back a whopping 80% of untreated wastewater (UNESCO, 2017).
- The two-third of the planet's transboundary water bodies lacks a cooperative management framework. (SIWI)
- Energy production exhausts approximately 75% of all the industrial water apportion (UNESCO, 2014).

HISTORICAL PERSPECTIVE OF WASTEWATER TREATMENT

Processing, Re-Use and Management of Wastewater by Ancient Civilizations

Indus Valley Civilizations

Most of the prehistoric notable civilizations had somewhat an ideal structure for utilizing the water sources and treatment of wastewater. It cannot be compared with modern-day processes but they were sufficient enough for the pollutants and containments generated at that time. Due to the lack of technological resources managing water both wisely and efficiently was a necessity. Clay was the primary universally adopted element for water purification; it also had several applications in storing and transporting water through crafted vessel and conduits; as evident in the historical and archeological studies. Paleolithic cultures discovered effective applications but never realized the property of clay to kill waterborne pathogens (Smith, 2017).

At the beginning of the Bronze Age (at 3000 BC), several novel water-purification materials and methods were adopted. Bronze and copper-based utensils, vessels were used for water purification during religious practices; the Indus valley civilization along with the Sumerians placed the utmost importance on the usage of clean, filtered water for ayurvedic and medical needs; respectively. The reports of Sumerians using copper spikes into the storage vessels for cleaning the water are recorded in their tabs (Smith, 2017).

Beginning from the earliest known civilized habitat, the Indus valley civilization of the Harrapan period had an ideal understructure for its infrastructure as validated by its ruins today. It was cleverly established along the now dried up tributaries of Indus and ghaggar-hakra river bed (Warnes & Keevil, 2011). The infrastructure developed and fittings installed for utilizing freshwater and channelizing the wastewater into drains is still considered to be ideal; although many later societies had a similar structure, not one was this absolute. It was by the virtue of water channels made up of clay bricks and terracotta pipes, that the habitants efficiently used water for chief purposes, i.e. sanitation and agricultural needs. The destroyed structures of public and private baths, dedicated underground drainage and sewage systems, public water supply, water-storing tanks along with the maiden private drinking wells and step wells and the canalized irrigation system exhibits the brilliant management employed for managing the natural sources of water (Khan, 2014). Ruins of the structure of water reservoirs were discovered at the sites of dholavira (Warnes & Keevil, 2011). The only major techniques adopted by them could have been sewage water treatment and filtering of drinking water by the processes of segregation, dilution, and filtration. Despite being limited by the technology of their time, they learned to treat the contaminated water, and then discharge it to the open by utilizing sumps through the plumbing system entirely made up of earthenware pipes (Lofrano & Brown, 2010).

The used water from households and public places was channelized through the plumbing of terracotta pipes which collected the prime solid waste such as human excreta, food leftovers, and other biodegradable wastes into cistern sumps while the waste liquid was discharged into drainage channels, which further lead to soaking pits and water bodies depending upon the municipal structure of the respective provinces (Lofrano & Brown, 2010). The inhabitants of Indus valley not only wisely processed and managed their water resources but also reused the water wisely. In certain regions the sewage wastewater from households was discharged in agricultural fields, both acting as manure and dampening the land (Khan, 2014).

Mesopotamian Empire

The Mesopotamian Empire known for its pre-eminent irrigation systems and waterways is an epitome example of water management due to its construction of maiden dams and canal systems. The entire civilization flourished around the valleys of the Tigris and Euphrates river beds. To control the natural flow of rivers for desired needs in agricultural fields, and to prevent the mighty rivers from flooding the cities, huge hydraulic structures such as inlet dams also known as barrages and Inundation canals; such as Engabara-canal were constructed (Rost, 2019).

Mesopotamian civilization was the first to standardize the effective irrigation systems and also built efficient drainage and sewage systems, evident from the ruins discovered at the sites of Ur and Babylon.

The water supply system and sewage understructure of the Mesopotamians were very similar to the ones discovered at the sites of Indus and Minoan civilization. The drainage plumbing pipes were made up of baked clay bricks (Tamburrino, 2010). With the beginning of scientific discoveries, the desire and need for clean water resulted in the adoption of certain techniques and the invention of various methods.

Amongst all the early urban civilizations, Egyptians were the first to embrace scientific innovations and inventions sophisticatedly. Certain records evident the adoption of principle techniques of flocculation, filtration, and coagulation majorly based upon plant fragments for filtration of drinking water by ancient Egyptian civilization (Saleem & Bachmann, 2019). The Egyptian civilization was the first to use copper, for making vessels to store drinking water and plumbing fixtures (Tzanakakis et al., 2014). Initially, both the pipes and fittings were quite crude; similarly, like Mesopotamians, the pipes were constructed from clay and straw (Gohary, 2014). Ancient Egyptians discovered the property of Copper to naturally purify water as it ionizes, making water consumable; but never realized how. To obtain clean and disinfectant drinking water certain procedures involving the addition of botanical fragments such as herbs, plant leaf, and roots, fermented fruits which resulted in flavored beverages such as wine and beer was a common practice amongst the masses. Discovery of copper was not the only discovery they made for making consumable water safe, the employment of Moringa oleifera seeds for purifying drinking water was a method adopted by certain Egyptian communities (Smith, 2017).

There are also records of water clarification by using almonds in coarse silt ware vessels. Apart from these methods, Egyptians had also invented a water purification device, made from copper alloys based upon the principle of sedimentation (Baker, 1981). There is recorded evidence of the existence of clayed pipes that were employed for channelizing water for domestic and agricultural needs; before the discovery of smelting copper and its alloys were adopted. During 5700–3200 BC (the Neolithic age), dams and irrigation systems were employed for controlling and diverting the river flow. This was done primarily in Egypt and Mesopotamia's irrigational needs. The Egyptians termed this management system as the 'basin irrigation', which worked upon the natural principle of elevation and fall of the river. A water device known as 'the shaduf' was used at both Egypt and Mesopotamia, around 1500 BC. It was employed as a water-lifting device (Postel, 1999). Ancient Egyptians have developed the fabrication methods for copper alloys, along with the concept of drainage pipes (Gohary, 2014).

Greek Civilization

A forerunner of urban developments, the Greek civilization was an epitome for water management and reuse technologies. Practices for land applications of wastewater, since the origin of Mesopotamian and Indus regions, had been universally developed with time; with the evolution of technology, knowledge, and the method regulations of the particular regions. The Archeological sites in Greece, are evidence of the employment of sanitation technologies and principals since greek and roman civilizations (Tzanakakis, 2014).

The employment of urban methods for water management in ancient Greece has been established by archaeological studies, under the ruins of urban sewages and stormwater drainages. The toilets and sewage systems of proximity with the Egyptian sewage structures were discovered at Phaistos; adequately functioning (Angelakis et al., 2005). The facilities of public toilets were accessible to the masses in the region of 300 BC -500 AD, the latrines had a connection for the drains utilizing pipes, which also transferred the stormwater from streets into the collection basins situated at remote places. This collected wastewater was used to irrigate agricultural lands and fertilize crops and orchards, which was done through the channels, known as conduits made up of the brick layout. This whole process was based upon the structured channelization of the piping system from the streets to the large sewer. Most of these structures were found in major urban settlements such as Athens, Pompeii as these were not required nor practical at rural settlements (Lofrano & Brown, 2010).

In terms of water purification, several innovations and inventions were made in that era, "Hippocratic sleeve," is one of them. It was a filter cloth bag which functioned upon the principle of sedimentation, also accounts of a machine, which dispensed purified water at prayer halls were discovered. Most of the methods and techniques adopted for water purification were based upon the principals of sedimentation and filtration, with the phenomenon believe that water being colorless and odorless makes it consumable, which was later realized to be falsified (Smith, 2017). With the discoveries of several water management devices at major urban settlements, like the tunnel of Eupalinos for water supply at Samos such as aqueducts, which acted as canals are the proof of modern advancements in the progression of urban water supply. Major structures and devices were built for flood protection, irrigation, and drainage process. At the archeological sites of Aegean civilization (ca. 3400–1200 B.C) sewer systems of the urban model were discovered, these sewer systems were connected through Ceramic clay pipes, these clay pipes were rough and primitive. Also, a similar section of pipes, made up of clayed pipes was used in aqueducts. The application of ceramic pipes for both water supply and drainage systems was a remarkable feat. The era of Greek civilization witnessed many inventions and innovations in technology, which made the routine process much efficient. The invention and implication of Archimedes's water screw pump was a revolutionary advancement in water management technology. The Minoan society was the first to adopt the practice of irrigating their agricultural fields through reusing the wastewater on considerably extensive planes, evident by the connecting sewer structures. Historical records of ancient cities such as Sparta, Athens, and katerini in 2600 BC also evident the employment of similar practices, i.e. using wastewater for irrigating agricultural fields. The implication of this method twice benefited the citizens, as both the functions, wastewater management, and irrigation of agricultural fields were achieved. The wastewater also performed as a fertilizer, hence increasing the crop yielding. The structures constructed for sanitation and water supply needs were of moderate scale, provided abundant water and their construction material and operation were in harmony to nature (Golfinopoulos et al., 2016).

Ancient Roman Period

For the modern-day technologies, Romans brilliantly engineered and managed their water resources. Most of Rome's water system derived inspirations from the eastern and Assyrian civilizations and effectively improved them. The concept of the integrated water supply was invented and implemented by Romans, which involved the completion of the water cycle; from supply up until collection and disposal. Romans collected and used spring water for consumption while the groundwater was either directly disposed of or used for other activities. The wastewater from baths was recycled and was used to flush the toilets, discharging excreta waste into Tiber river, channelized through complicated city sewers. Each major Roman city had a complex channel of sewages which demonstrated their engineering and construction skills (Lofrano & Brown, 2010). The employment of coal for water purification dates back to 32,000-30,000 years ago, all the ancient cultures i.e. Sumerians, Egyptians, Indus Valley cultures, Israelites, and Greeks had applications of coal for water purification.

During the end of the Roman Empire and the beginning of dark ages, which continued for over a millennium(476–1800), the human civilizations didn't progress much, and consequently, the adopted sanitation and water treatment practices were rather abanded instead of evolving. The culture of treating water as a tool for wellbeing and progression was abused and so were the infrastructure developed for the conveyance of water during roman civilizations, and many more civilizations before it. Water was usually acquired from rivers, streams, and wells; which was then returned to the ground and the stream sources. The release of contaminated water, without any treatment, became a pollution source. At the beginning of the 18th century, during the Industrial Revolution, the wastewater treatment was introduced. It took half of the nineteenth century for the water treatment methods to be widely adopted and used.

Emergence of Various Methods, Treatments and Technologies in the "Enlightenment-Era"

The beginning of the 18th century embarked on many revolutions, sanitation, and healthcare being the significant fronts, lead to the emphasis upon purification of consumable water and the treatment of wastewater. It is noteworthy that in some cities, drainage systems existed and were often used to discharge wastewater (Burian et al., 2000). With the majority of industrialization and urbanization focused at Britain, the wastewater disposal practice was adopted immediately, due to its consequences. Amongst the very first nation, it was Britain who started to work towards degrading environmental conditions, during the same time Germany came up with the first planned sewer system, designed by an English man Lindley in 1840. A similar method was quickly adopted by the United States and the rest of Europe. The primary aim of these systems was to eliminate water contamination and stop the widespread of water-borne diseases. After the 1850s the centralized sewer systems were viewed as necessary long term investment for managing wastewater. With the generalization, the piping system also evolved and various other materials were considered such as cement, cast iron, and lead. During the late nineties, between 1870–1915 in Italy the great public works started, and most of the work was built upon existing roman infrastructure (Lofrano & Brown, 2010).

Mechanically and Pneumatic based vacuum operating systems were introduced in the second half of the 19th century. This was done for sewage collection and cleaning purposes. In 1866, Engineer Liernur patented his vacuum operating sewage in England and the Netherlands. The Liernur systems were installed and functional at several European cities such as Leiden, Amsterdam, Hanau, Dordrecht, Prague, Stansted, and Trouville. These were operational until 1980 in France. Deriving the inspiration, several pneumatic systems were developed, and the Excreta discharged through these vacuum sewer system were used as a fertilizer (Baker, 1981).

By 1870 the principle of soil filtration was proposed by Edward Frankland, which platformed the concept of tricking filter and other successive developments in the late 19th century. The tricking filter worked upon the principle of microbial degradation. It was first adopted in the United Kingdom in 1893, at Salford near Manchester and since then up to the year 1920, it was widely used throughout the United Kingdom for treating contaminated water from cities and towns. Table 1 represents the methods and wastewater treatment adopted during the 19th century. The tricking filter worked upon the principle of microbial degradation. It was first adopted in the United Kingdom in 1893, at Salford near Manchester and since then up to the year 1920, it was widely used throughout the United Kingdom for treating contaminated water from cities and towns.

In 1890 the biological treatment experiments were initiated to reduce water contamination, stream pollution, and ensure safe drinking water. The first operation was conducted at Massachusetts in Lawrence Experiment Station. It was a significant step towards the beginning of biological treatment for treating contaminated water by mineralizing organic matter through degradation by microorganisms. Concurrently in the same year Col. George E. Waring developed a filtration technology and the first large scale was built in berlin (Tzanakakis, 2014).

Table 1. Major advancements in wastewater decontamination throughout the 19th Century

Period of Origin	Method of Wastewater Treatment
1804-1827	Sand and charcoal-based filtration processes were adopted across Europe, majorly in the United Kingdom and France.
1835	It was in the year 1835 that the chlorination of wastewater begun, for both removing the bad odor and decontamination of foul water.
1835- 1855	During these two decades, the efficiency of chlorination was recognized, resulting in further scientific developments, focused around Great Britain.
1869	In the United Kingdom, the first patent for a desalination process was filed and a distillation plant was built in Yemen.
1887	Introduction of safe drinking water and construction of the first sewage treatment plant in the United States.
1879-1889	The chlorination process for sewage treatment and its scientific development progressed in Great Britain.

By the end of the 19th century, 2 major systems were developed for collecting wastewater: static systems and sewer systems. The static systems can also be termed as the first step towards the collection of contaminated water for treatment; sewerage. By canalization and plumbing of pipes, the second generation of centralized sewerage, the sewer systems put an end to the decentralized and chaotic systems like cesspools and privy vaults. At the end of the 19th century, in the wake of industrial revolution and urbanization; the increase of the population and the use of plumbing fixtures (e.g., water closet), producing more wastewater, lead to the increment of concerns of public health and hygiene. Indeed, both lined and unlined decentralized systems were unable to manage urban wastewater effectively because the lined ones required too frequent cleaning to be cost-effective over the long term, and the unlined ones contaminated groundwater and the surrounding soil (Burian et al., 2000). The evolution of Septic tank treatment was a revolution in the industry as it dealt with the sludge formation due to chemical precipitation. A primary sedimentation tank was based upon the principle of horizontal-flow, resulting in a longer retention period. Sewage was treated with anaerobic microbes, below the surface level of the tank, but due to accumulation the sludge was never entirely removed. By the end of the century, due to benefits such as natural decomposition of solid organic waste, cost efficiency, and no need for chemical treatment made them widely adopted. With time, the design of Septic tanks was improved and patented by Donald Cameron in 1895 (Beder, 1996).

The industrialization in the 18th century accompanied immigration, which invited epidemics such as yellow fever, typhoid fever, dysentery, influenza, cholera, TB, and malaria; due to the unsafe public water supply and inefficient wastewater treatment. Embarkment of the 20th century brought a reduction in the diseases with the aid of clean water supply and organized wastewater treatment, under the concept of collective "public health" action (Centres for Disease Control and Prevention, 2012). Adaption of sewer systems, a network of pipes for transporting organic waste before fermentation from urban settlements was a major advance as it reduced the chances of a disease outbreak (Roccaro, Santamaria & Vagliasindi, 2014). Since the 1930s through the 1950s, many states in the U.S. adopted substantial programs for sewage disposal, water treatment, and organic solid waste disposal. The Chlorination treatment for drinking water was established in the early 1900s; drastically reducing the rate of typhoid and cholera victims (CDCP, 2012). Designed by a German engineer, Karl Imhoff in 1906 the Imhoff tank was an advanced upgrade that still has applications worldwide.

Crisis and epidemic breakout during the 20th Century, which influenced the water decontamination processes:

- In 1905, typhoid fever breaks out in England. It was controlled through water Chlorination, suggested by Dr. Alexander Cruikshank Houston. Since then, the permanent chlorination of consumable water began worldwide.
- In 1980 Pennsylvania US, 'hepatitis A' broke out because of a water well, being feces contaminated.
- The Contaminated filtration system at western Georgia was responsible for Carroll County Cryptosporidiosis outbreak in 1987
- Several people were poisoned at Camelford in 1988 because of aluminum sulfate, being mixed in the wrong tank at a water supply facility (Lofrano & Brown, 2010).

By the end of the 19th century, Industrialized countries have overcome the threat possessed by the water pollution and health problems associated with it. However many developing countries still face the same threats due to lack of proper sanitation, toilets, and clean drinking water. At the end 20th Century, the water purification processes were limited to Primary treatment, which changed in the U.S. after the introduction of the Clean Water Act (1972), which implemented the secondary treatment of wastewater for public use. The primary treatment is based upon the principle of gravity sedimentation (Lofrano & Brown, 2010).

Based upon a secondary treatment, procedures for purifying wastewater were developed in 1900 and 1916 using a moving wooden cylinder with slats and agglomerated slag as a cylinder, which rotated in the horizontal direction; respectively. These were the parent designs for another biotic attached based processes known as Rotating Biological Contactors (RBCs). In this process instead of wooden slats the metal disks were used as it reduced the threat of clogging, which was later replaced by expanded polystyrene disks around the 1950s and succeeded by polyethylene disks as they were cost-effective. The RBC plants were first installed in Germany in the year 1960, followed by the US and UK. In 1913, the aeration method was adopted for decontaminating wastewater, the activated sludge process was patented by Edward Arden and W.T. Lockett. Their studies resulted in high-quality effluent due to the concentrated micro-organisms, these microorganisms transformed the carbonaceous based contaminants such as CO_2, H_2O, and regeneration of energy. After the implementation of the Clean Water Act 1972, the activated sludge plants became mandatory secondary treatment plants in the US (Lofrano & Brown, 2010).

In the 1960s a filtration based technology such as membranes was developed by Dorr-Oliver Inc, however, Ultrafiltration was a contempt replacement for activated sludge process and membrane sewage treatment begun by late 1960s. Within a decade of its industrial application, the MBR (membrane bioreactors) processes were introduced, and by 1980s full scale commercial aerobic sludge removal and biomass separation through ultrafiltration of membrane bioreactor systems had applications across Europe. Table 2 represents the invention, introduction, and adoption of various treatments and methods adopted throughout the 20th century, across the world. By the year 1990, two successful patents were registered based upon the research studies of Dorr Oliver; for wastewater treatment technologies (Lofrano & Brown, 2010).

Table 2. Major advancements in wastewater treatment methods in 20th century

Origin Period	Process Adopted for Decontamination of Water
1902-1910	Adoption of sand filtration for drinking water, Employment of chlorination technique through compressed liquefied chlorine gas to safeguard public health in the United States.
1903	Liquefied chlorine gas was used in India for decontaminating wastewater.
1913	Discovered and Employed in the United Kingdom, an Activated Sludge process for decontaminating wastewater from sewages and industries.
1914	In the United States, The Public Health Service Act is passed which conducts periodic surveys for checking the contamination level of water.
1919	In the United States, Abel Wolman and Linn H. Enslow published research of chlorine absorption for municipal water supplies, initiating reform in the water treatment industry.
1938	The first water treatment plant is built in Saudi Arabia.
1951-1953	Introduction of Water fluoridation in Brazil and the United States.
1959	Establishment of a Dual-media, multiple-tray water treatment plant in the United States.
1960	In Kuwait MSF (multi-stage flash) evaporation technology was developed which provided dual benefits, i.e. freshwater and electricity.
1968	The largest capacity wastewater Treatment plant is constructed in the United States.
1996	The introduction of UV-Light operated, low maintenance, and energy-efficient water purifiers. These were invented in the United States.

CONTEMPORARY DEVELOPMENTS IN WASTEWATER TREATMENT INDUSTRY

Due to a better understanding of wastewater treatment and the threats posed due to its negligence, advanced treatment with an impact of wastewater on the environment, and scarcity of water resources; sophisticated analytical methods are being developed. Table 3 represents the type of contaminants found in urban and industrial discharge and their treatment method. With progression in the scientific endeavors and consequent advancement in the industrial practices, new contaminates have developed in the 21st century and the disinfection of wastewater has become much challenging, but concurrently many new methods and treatment processes for treating wastewater have emerged, Figure 2. depicts the condition of wastewater recollection and treatment.

Nanotechnology for Water Decontamination

The comprehensive approach for the potential and viable applications of materials at nanoscale, i.e. the range from 1-100 nanometers, is termed as nanotechnology. In general, it's the pursuit to miniaturize the bulk matter to achieve revolutionizing results which are only possible at the nanoscopic scale of the matter material. At this region, the standard property of the bulk material begins behaving differently which are both, beneficial and alarming. The phenomenon of nanotechnology takes place at mesoscale, the standard field of physics which domains the fundamental physical properties at both macroscopic and microscopic scales of matter. The mesoscale range is the ground for applications of both classical and quantum mechanics (Sia, 2017). The term nanotechnology was first coined in the year 1959 by Richard Feynman and since then there has been significant development at the scientific level, by early 90's the commercialization of nanotechnology has begun.

Figure 2. Representation of wastewater treatment

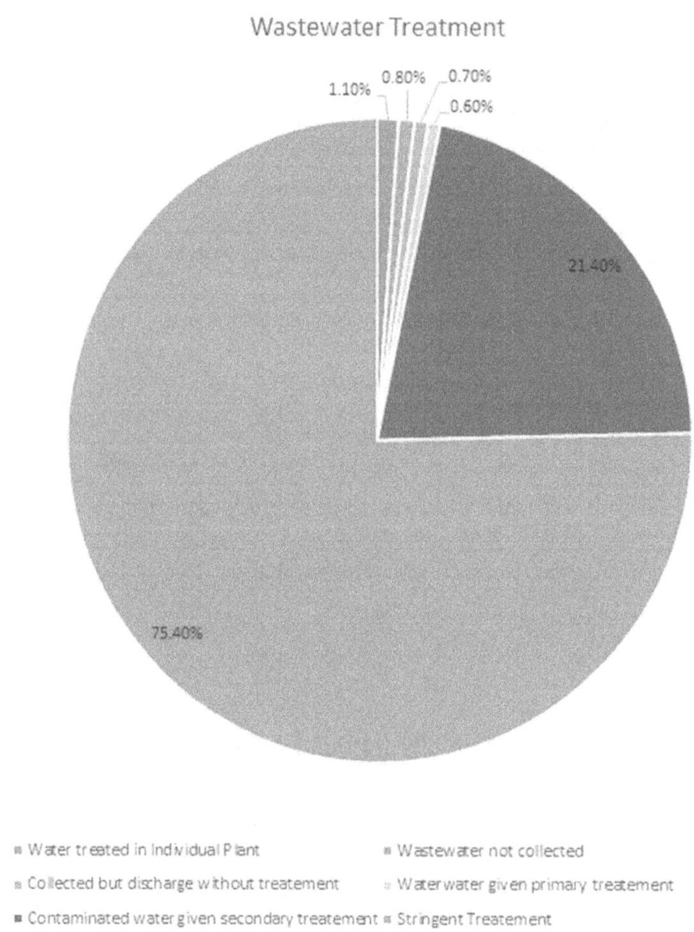

Certain characteristics exhibited by the materials at nanoscopic scale, specifically the manipulation/ mutation of biological, chemical, and physical properties of the matter material, which is independently different from the bulk matter at the macro scale makes nanotechnology favorable for novel technical applications. Significant alteration in the standard properties of the matter material, which results in obtaining new, unique properties and enhanced performance by virtue of decreasing the dimension size below 100 nm, paves the way for infinite discoveries and inventions (Nasrollahzadeh et al., 2019).

Viable Applications of Nano-Tech in Wastewater Treatment

The fabrication of the desired material with a specific set of properties is the salient feature of nanotechnology, and by its virtue, the fabrication of nanomaterial for treating wastewater is a potent solution. The nanomaterials for water treatment apparatus could be fabricated according to the required features of treatment processes, such as desired chemical reactivity of material, the exact ratio of volume to the surface area, enhanced mechanical properties, tunable pore volume hydrophilic and hydrophobic interchanges, and electrostatic. These properties influence the water treatment methods such as adsorption,

Table 3. Wastewater treatments introduced in 21st century

Contamination reduction	Methods and Treatment Process used
Reduction of contaminated suspended solids	Flocculation, UF (Ultrafiltration), Sedimentation, Coagulation, Media filtration, and MF (Microfiltration).
Neutralizing the chemical concentration from wastewater	NF (Nanofiltration), ED (Electrodialysis), GAC (Granular-activated carbon), BAF (Ion exchange Biologically Active Filtration), EDR (Electrodialysis reversal), and RO (Reverse osmosis).
Trace reduction of Organic Compounds and water disinfection.	Advanced oxidation processes such as (O_3/H_2O_2, UV/H_2O_2, UV/Cl_2) – AOP based), Pasteurization Ozone Chlorine dioxide, Chlorine/chloramines - PAA (Peracetic acid) and UV (Ultraviolet disinfection).
For Stabilizing the PH level of water	Sodium hydroxide (NaOH), Lime stabilization and Calcium chloride (CaCl) Blending
For influencing Aesthetics, i.e. color, odor and taste.	O_3/Biologicall3y Activated Carbon (BAC) MF/RO
Reduction of contaminated suspended solids	Flocculation, UF (Ultrafiltration), Sedimentation, Coagulation, Media filtration, and MF (Microfiltration).

catalysis, sensing, and optoelectronics catalysis (Kunduru et al., 2017). Nano-engineered materials are durable and possess high SBET, i.e. exact aspect ratio which regulates the interaction between containments and filtrates surface (Parvin et al., 2019).

Attributes like these make the wastewater treatment processes proficient, compatible, modular, and cost-effective. The integration of nanotechnology, i.e. nanostructured adsorbent, nanocatalysts as well as nanocomposite polymers along with other conventional water treatment techniques like ion exchange, reverse osmosis, and membrane-filtration is immensely beneficial as it conserves both monetary and infrastructural resources and much effectively separates heavy metals and organic containments. With the specific intent of high removal rate, cheap operation, and regeneration many nano-engineered materials are synthesized, which possess their own set of specified properties for unique operations (Gupta et al., 2011).

Due to these attributes, the potential of nanotechnology in advancing the processes of wastewater treatment is evident. The applications of nanomaterials are already commercialized and established as nano adsorbents, nano metal oxides, nanomembranes, and nano photocatalysts materials. Also, these materials are compatible and can be integrated with existing conventional wastewater treatment methodologies in their existing module structures (Suthar & Gao, 2017). Nanotechnology is a sustainable resource for wastewater treatment because of its rational attributes, i.e. reprocessing, cost-effectiveness, and resource conservation of material synthesis. Also, at the nanoscopic scale, the molecules of nanomaterials, efficiently access and interact during treatment and remediation processes averting, sensing, following, and eliminating pollutants (Nnaji et al., 2018).

Owing to its vast range of advantageous properties, nanotechnology has immensely improved the current scenario of wastewater treatment technologies; clearly on a small scale yet it exhibits the potential to revolutionize the wastewater treatment methodologies under fabricating potent and effective nanomaterials in near future. Broadly nanomaterials are classified as fullerenes (0D structure), carbon nanotubes (1D structure), graphene sheets (2D structure), and nanoparticles (3D structure) according to their lattice arrangement in the three-dimensional space (Gopakumar et al., 2019).

Nanomaterials for Wastewater Treatment

Nano-Sorbents

The adsorption method consists of the chiefly employed methods at wastewater treatment plants, universally. The main reason for its wide implementation is the simple and convenient operation method, design arrangement, and the adaptability with hybrid technologies (Safarpour & Khatee, 2019). Nanoparticles employed for membrane operations are characterized via SEM (scanning electron microscopy), TEM (transmission electron microscopy), and XRD (X-ray diffraction) for the applications in the wastewater industry and are generally fabricated through laser pyrolysis, photo-thermal synthesis, and inert gas condensation. The fabrication technique depends upon the desired properties, application requirements, and material dimensions. The most sought out new generation of absorbent materials in the industry are TiO2, NiO2, Al, and Si which is fabricated by the sol-gel method (Basheer, 2018).

An adsorbent work upon the phenomenon of adhering to physical containments present in wastewater upon the material adsorbent which involves the separation of any physical contaminant from the wastewater by the principal of adhesion of the waste adsorbates upon the material adsorbent is termed as adsorption (Bhadra, 2019). The adsorbent material used for the procedure is the most crucial element which influences the whole operation, i.e. the cost and efficiency of the adsorption process. Also, certain parameters such as temperature, adsorbates and adsorbent material, particle size, contact time, and chemical environment moderately affect the adsorption process. The adsorption sites are selected based on adsorbents Surface area and pore size. Conventional adsorbent materials are less efficient due to the low ratio of surface area or active sites to pore size, reduction in selectivity, and adsorption kinetics (Suthar & Gao, 2017).

The chief requirements for an efficient adsorbent material are effective adsorption rate, high rate of selectivity, abundant supply, the longevity of operation, and considerate cost.

Conventionally employed adsorbent materials include activated carbon, organic and synthetic zeolites (based upon aluminum-silicate polymers), Natural clay minerals, silica gel, and aluminum-based adsorbent materials. These are generally used for their effective polar bonding sites and uniform pore distribution; effective adsorption rate of organic and inorganic matter ions & large polar adsorption capacity for water from agricultural and industrial waste (EMIS, 2010). Amongst all these adsorbent materials, activated carbon is the only rational candidate due to its standard properties i.e. high adsorption rate, immense surface area, high rate of surface reactivity, and exact microporous structure (Mallakpour & Rashidimoghadam, 2019). Relatively, the applications of carbon-based adsorbents are reducing due to its high production and regeneration cost. The low surface of the adsorbent is also a demeaning factor. Due to its limited capacity and selectivity, the adsorption-regeneration cycle is also reduced, consequently decreasing the adsorption process (Hasbullah et al., 2019).

Certain properties of nanomaterials-based adsorbents such as their surface chemistry, quick adsorption rate, high SSA (specific surface area), adaptable pore dimension, and efficient sorption sites, upon a broad pH scale; along with very less intra-particle diffusion exposure makes them much better than their conventional counterparts. Very less volume of nanomaterials is employed for the treatment of dye removal from contaminated water. Also, they can be simply fabricated at a lower price and smaller amounts are needed for efficient elimination of dyes. The fabrication cost of the respective material is very cheap, bringing the total cost very affordable, in comparison to the activated carbon. Nanomaterials are further categorized into CNT's, carbon fullerenes, carbon nanofibres, and graphene. These materials are termed as carbonaceous nanomaterials, as these possess a strong covalent bond between carbon molecules. This class of material has demonstrated higher water affinity and adsorption capacity than activated carbon-based adsorbents (Mallakpour & Rashidimoghadam, 2019).

The regeneration of used nano adsorbents is a safe, economically, and ecologically viable process, as it allows the recovery of adsorbate arsenic, for its proper disposal. This also renews the adsorption capacity of the adsorbent and monitors the components trapped in the adsorbed phase. This phenomenon makes the process very cheap (Lata & Samadder, 2016).

Carbon-Based Nano-Adsorption Materials

CNT's are well established as nano adsorbent employed for the decontamination of heavy metal from wastewater. This process is carried out in their oxidized state, by virtue of its quick kinetics. CNTs are considered as an ideal substitute for activated carbon. CNTs are distributed into two categories, i.e. single-walled nanotubes and multi-walled nanotubes (MWCNTs). The nomenclature depends upon the fabrication method employed. along with large assessable adsorption sites, CNTs also possess large surface areas (Parvin et al., 2019).

The surface chemistry of the CNT's can be modified accordingly. Due to their hydrophobic surface, in an aqueous medium, the CNTs form loose aggregates, consequently reducing the surface area. These aggregates act as high-energy sites for the sorption of organic contaminants. Even though CNTs have significant advantages, but possess certain limitations as well (Kunduru et al., 2017).

CNTs are coupled along with the organic compounds and merged with the metal ions or oxides, depending upon the desired upshot to be altered by the oxidation process. Concerning the conventional non-oxidized carbon sites, the carbonyl activated CNTs had better sorption rates for heavy metals such as Zn. Also, the heavy metal ions such as Cu^{2+}, Pb^{2+}, Cd^{2+}, and Zn^{2+} were efficiently removed by 8-hydroxyquinoline modified CNTs (Nnaji et al., 2018). The industrial adaptation of CNTs for wastewater treatment is limited due to coagulation formation as a result of some organic contaminants and algae, its high production cost, toxicity which results in the deformation of its nanostructure (Kunduru et al., 2017).

Graphene-Based Nanocomposites as Nano Sorbents

Results derived from various research studies conducted, regarding the induction of graphene into the nanocomposites reported increased efficiency of its thermal stability, mechanical strength, and electrochemical activity of the composite material. Enhancement in the gas barrier capacity of the material was also recorded. Even the slightest addition of graphene filler, regardless of the presence of various polymeric matrices in the composite positively influences the properties of nanocomposites. The induction of Graphene-based nanocomposites along with active carbon, polymer, and metal is a promising material for the fabrication of Nanosorbents. A GO-polymer based composite was fabricated, without covalent bonding between their respective atoms; resulting in the sublime union of certain effective properties such as mechanical robustness, efficient water transport capacity, and anti-swelling tendency of the material. These acquired properties are very suitable at high-pressure conditions such as reverse osmosis (RO) membrane process and any other harsh environmental factors such as oxidation (Kim et al., 2018). Presently, extracting heavy metal ions from wastewater graphene is widely used; which is due to its stable benzene six-membered ring. Graphene has the most stable basic structural unit amongst every other available organic material. By this graphene does not react with heavy metal ions. This efficient adsorption performance is due to the high surface area of GO material, influencing its adsorption rate (Nnaji et al., 2018).

The graphene sheets are synthesized through the exfoliation of hydrogen induction; followed by functionalization. By employing the super capacitor-based water filters, these functionalized graphene sheets treated the inorganic arsenic and sodium-based species, of higher concentration. It was also used for treating the desalination of seawater. The functionalized GO sheet–based water filters were found to be efficient for the removal of heavy metal ion (Lee &Yang, 2012).

Graphene Oxide

This is the graphene oxidized form, which consists of several epoxy, hydroxyl and oxygen-containing groups, upon the basic planes and at the edge, carboxyl groups are placed. Due to its effective adsorption rate, it is considered as the ideal material candidate (Liu et al., 2019).

The obtained GO nanoparticles were fabricated via synthesis of modified hummer's technique. In this technique the graphite powder is transformed into graphite oxide (aqueous medium) via oxidation, followed by its exfoliation to graphene oxide. The fabrication of GO nano-sheets, from graphite via hummer's method results in the acquiring of carboxyl and hydroxyl groups, upon the surface of nano-sheets. This functional group enhances the high-sorption rate of heavy metals present in wastewater. The synthesis cost and probability of GO nanoparticles are certain limitations of this method. This can be treated by impregnating the GO mixed matrix membrane upon polymer-supported CeO2 and γ-Al2O3; as these nano-membranes have demonstrated effective adsorption capability for Cr (VI) (Parvin et al., 2019).

The carbon-based nanomaterials are toxic for human cells, which makes the treated water unfit for human consumptions. Also, Mg-Graphene based adsorbents greater than 10 nm of size highly binds the As3+ and As5+. The high hydrophilicity, stable dispersion, and high-speed centrifugation of Graphene oxide in the water further limit its industrial applications (Nnaji et al., 2018).

Metal and Metal Oxides Based Nano Adsorbents

Due to the possession of certain specific properties like effective para-magnetism, effective surface area to volume ratio, and the nano-size range offered by the FeO_2 based nanoparticles their synthesis and application in the industry aspire. Also, their chemical inactivity, fewer amounts of toxicity, and compatibility with other metal oxides and alumina for the formation of biotech for removing heavy metal ion during water decontamination prove out to be cost-efficient and effective materials (Gopakumar et al., 2019).

The mechanism process which takes place is due to the presence of O_2 based metal oxides, which reacts along with the heavy metals' ions dissolved in the industrial wastewater. With a reduction in the particle dimension, the adsorption rate increases drastically.

With an effective adsorption rate for material collection and removing the toxic elements from wastewater the magnetic nano adsorbents like Maghemite (γ- Fe_2O_3), and Hematite (α-Fe_2O_3) and spinel ferrites ($M^{2+}Fe_2O_4$, where M^{2+}: Mg^{2+}, Cu^{2+}, Fe^{2+}, Mn^{2+}, Cd^{2+}, Ni^{2+}, Zn^{2+}, Co^{2+}) have proven to be good adsorbing material due to their beneficial magnetic nature and ecological benefits. The adsorption film can be separated through an external magnetic field (Kunduru et al., 2017).

Characterized through the measure of qualitative and quantitative tools, Mg and Cu associated ternary nano adsorbents; of their metal oxides are fabricated for treating the arsenic metal ions contaminating the fresh groundwater reserves. To convert As(III) to As(0) through redox-active behavior for establishing synergistic adsorption as the adsorption process is not competently supported by As (III) whereas the As(V) and As (0) samples favorably carries out the adsorption process by various spectroscopy techniques. Considering the post adsorption leaching effects, the use of Cu-Al-Fe and Mn-Al-Fe based nano-adsorbents, to remove the arsenic metal ion from wastewater by converting As(0) insignificant proportion (~ 70 at. %) in the toxic conditions was eminent. Whereas the existing adsorbents for various focus systems (~90% transformation in 1 ppm aliquots) utterly converting, toxic As(III) to less toxic non As (III) species (Penke et al., 2018).

Copper and titanium oxide-based nano adsorbents are also used for Arsenic removal. Magnetite based nanorods are also used for removing heavy metal ions such as Zn^{2+}, Fe^{2}, Cd^{2+}, Pb^{2+}, Ni^{2+}, Cu^{2+} from contaminated water (Karami, 2013). In comparison to the nanotubes, nanorods demonstrated a high adsorption rate for Pb^{2+} and Zn^{2+} but the low rate for Cu^{2+} (Kunduru et al., 2017).

Nano-Metal Based Oxides

The recent effective results in the wastewater treatment by nanoparticles based upon TiO2, ZnO, CuO, IONPs, and noble metal nanoparticles have significantly increased the R&D for their large-scale industrial applications. With the advancement of nano-scale, the efficiency of ZVZn (zero-valent zinc) and ZVI (Zero valent iron) based nanoparticles have significantly increased. For adsorption-based nanomaterials, most crucial factors include oxidation, effective precipitation, and efficient adsorption rate. The NZVI (nanoscale zero-valent iron) possesses the exact combination of properties, such as efficient adsorption rate, strong reduction capacity, low operation cost, and effective precipitation and oxidation; because of its smaller dimensions, but the larger surface area (Parvin et al., 2019).

The higher adsorption rate was witnessed for Ce-Ti, Fe-Ce, GOMnFe, Fe-Cu, Mn-Al, Fe-Mn; the bimetal oxide magnetic nanomaterials concerning single metal oxide-based magnetic nanomaterials. The amalgam properties such as higher adsorption capacity is witnessed upon the metal composition of two, or greater to the individual metal oxide. 2050 times much arsenic's adsorption rate increased due to bimetal oxide, amongst Fe_3O_4, Mn_2O_4, and $MnFe_2O_4$ nano-adsorbent.

An aqueous solution having a magnetic field easily gets treated of its heavy metal ion concentration due to the Paramagnetic nature of magnetic nanomaterials by modified ligands combination electrostatic interaction, magnetic selective adsorption rate and site binding (Parvin et al., 2019).

Polymer-Based Nano-Material

As much as different nanomaterials are efficient in removing heavy metals from wastewater, their agglomeration hinders their employability. This issue is resolved by converting the nanomaterials into polymer-based nanocomposites. This is achieved through the impregnation of nanoparticles into the polymer skeleton. By chelation process, the metal ions interact with carboxylate anion, i.e. polymer-based nano adsorbents. The acid-catalyzed polymerization process, through the encapsulation of metal oxides in polymeric resin, helps achieve the synthesization of Polymer-based nanocomposite (Parvin et al., 2019).

Efficient thermal stability and effective adsorption capacity of the polymer-inorganic nano adsorbents, over a large scale of pH value, is their most influential advantage. Also, resistance offered to the linkage of acid and base hydrolysis, along with polymeric groups is a valuable advantage. Heavy metals ions such as Cd^{2+}, Zn^{2+}, Pb^{2+}, and Cu^{2+} were effectively separated from an aqueous solution of ph value 5.5, by virtue of Fe_3O_4 magnetic nanoparticles (Fe_3O_4 MNPs) modified with 3-aminopropyltriethoxysilane (APS) and copolymers of acrylic acid (AA) and crotonic acid (CA) as polymer shells (Fe_3O_4) were prepared. Reduction in Interparticle aggregation and increment in the dispersion stability of nanostructures was registered through the employment of Fe_3O_4 based polymer shells. This polymer-based nano adsorbent has a life expectancy of four cycles (Kunduru et al., 2017).

Nanocellulose

Because of its cost efficiency, Nanocellulose is an efficient decontaminant for dyes, heavy metals, and viruses. The easy availability, eco-inertness, and favorable ratio of surface area to the volume of the Nanocellulose make it an ideal candidate. The functionalize OH group on its surface facilitates the binding capacity of contaminants to the surface of Nanocellulose material. The CNF (Cellulose nanofibers) are another ideal candidate due to a similar class of cellulose nanomaterials. Due to their cost efficiency, easy availability, eco-inertness, and a favorable ratio of surface area to the volume, the CNFs are the ideal substitute adsorbents (Gopakumar et al., 2019).

Nanocatalysts

By virtue of its surface-to-volume ratio, nanomaterial tends to develop a high rate of adsorption along with an increment in its chemical reactivity, the property which turned out to be quite favorable for catalysis and wastewater treatment. By the acquisition of unique physicochemical properties, nano catalysts have proved to be efficient for wastewater treatment. By their ferromagnetic property, Iron nano catalysts can undergo a series of recycling and reuse, simplifying the post-treatment separation. The process involving the Synthesization of Silver fiber-based nano catalysts is based upon synchronization interactions and under tetrahydrofuran treatment; it can be reused in repeated cycles. To degrade organic dyes, these are the ideal candidates. When compared with their conventional catalyst counterparts, Nano catalysts requires less amount of input energy due to their much lower operational temperatures (Nnaji et al., 2018).

Nano-Structured Membranes

The main objective of wastewater treatment is to rectify the contaminated water and make it fit for human consumption. The membrane-based technology is employed in the water transportation systems, where the membrane filter acts as a physical barrier for decontaminating the water from waste matter. Due to the presence of microstructure openings, the water molecules are obtained at the other end, but the containment bigger than the openings are trapped. (Suthar, R. G., & Gao, B. 2017). The branes allow effective automation and modular configuration, minimizes the chemical consumption, requires less space; consequently, resulting in a flexible design (Li et al., 2013). In order to trap all the physical coagulation, the system is operated at a higher flux. Also, the membrane material chiefly influences the system's performance.

CNT and Biomimetic Membranes

These are based upon the principle of nanochannels, both biomimetic membranes and the CNT's provides passage for the water molecules, at an exceptional rate (Li et al., 2013). These channels act as nanotubes for CNT's and are termed as aquaporins in biomimetic membranes. The diameter of the nanotube and chemical structure of the nanotubes opening determines the water selectivity for CNT's and aquaporins, respectively. The frictionless passing of water through CNT'S, along with retaining a higher containment volume is another desirable requisite for water purification. Due to the atomic smoothness of aligned and nano-sized CNTs along with one-dimensional permeation of water molecules, the flow is incremented multiple times via nanotubes (Suthar & Gao, 2017).

The development and fabrication of high hydrophilicity, longevity, and permeability compatible nano-composite membranes are a major trend. The polymeric membranes are modified from inorganic nanomaterials, for the construction of nano-composite membranes. The fabricated material proved to be an efficient solution for water permeability and solute selectivity issues, along with their longer life span. The anti-microbial property, conductivity, hydrophilicity, mechanical & thermal strength along with solute selectivity of the polymeric membranes can be influenced via nanomaterials employed (Safarpour & Khataee, 2019).

The virtue of nanoparticles has attracted the attention of several membranologists, to study and employ them as nano-fillers. This is due to the hydrophobic nature of the polymeric membranes; and the membrane fouling is directly dependent upon hydrophobicity (Favvas et al., 2014). In order to directly influence the flux improvement and fouling mitigation, these hydrophilic nanoparticles are installed at the membrane pores. Apart from the wastewater treatment industry, the membrane applications are employed at several other ventures as well such as metal oxide-based nanoparticles for gas separation, nanofiltration, and ultra-filtration for desalination (Ng et al., 2013). Many studies demonstrate an increment in the mechanical strength and water permeability (Hasbullah et al., 2019).

Nano-Photocatalysts

Nano-photocatalysis has numerous applications in the wastewater treatment industry, based upon advanced oxidative processes using nanomaterials such as anti-microbial activity; deterioration of organic contaminates, and dyes; accompanied by the OER (oxidative elimination reaction) method, for treatment of micro-pollutants and microbial pathogens. Several inorganic semiconductor materials are employed for its applications in photocatalytic based materials, but due to the abundance of TiO_2, lesser toxicity, and cost efficiency; it is generally employed. It is also used for UV-based processes. The activity and degradation are greatly influenced by the reactor layout, its surface area, reactor design, and the UV irradiation's intensity (Laohaprapanon et al., 2015).

The application of UV/TiO based processes via the moxifloxacin method, are widely adopted for the treatment of contaminated water at several industrial fronts (Van Doorslaer et al., 2015). Due to the fine structure of its particles, the recollection, reuse, and separation of the TiO_2 particle is a challenging operation. To enhance the quantum yielding rate, Ti-nano-particle based nano-composites are chemically functionalized. These are effective in degrading environmental pollutants and treating contaminated water. Several research studies have reported the adoption of photo-catalytic, TiO_2 based nano-material mechanisms, whilst the decontamination process (Bhadra, 2019).

Ceria, CNT's and titanium are usually employed for the degeneration of the photo-catalytic process for nanomaterials. The cost efficiency, non-toxicity at ingestion, and stability of Ti-based nanoparticles are ideal for wastewater treatment. Also, the production of free radicals in water, oxygen, and UV irradiation which subsequently decomposes into the Carbon compound is an advantage of Ti-based nanoparticles. Titanium is often employed as slurry, a thin film coating, or membranes. This method is widely adopted, commercially at wastewater treatment industries.

ADVANCEMENTS IN THE CONVENTIONAL TECHNOLOGIES

Adsorption Based Technology

For tertiary removal of non-biodegradable micro-scale residue, which couldn't be separated at the up-stream treatment processes, the Blue CAT™ method is proposed. This process is an integration of an excellent oxidation process (based upon ferric with ozone), and the Blue PRO™ adsorption method for filtration. Several benefits of employing this oxidation process include providing efficient disinfectants, without any chlorine-based compounds.

An apparatus of the Blue PRO method, made up of $H_2Fe_2O_4$ coat-sand filter and patented BlueCAT treatment pre-reactor, is used for adsorption of phosphorus and other contaminants in an up-flow. As per the U.S. EPA (2013), an efficient increment in the filtration process was recorded, due to the reduction of estrogenic compounds, along with total organic compounds. Less percentage of color removal disinfection was discovered in the water, along with a decrement in the phosphorus level as well. These results were obtained during unpublished pilot experiments and were shared in the document.

The waste stream obtained through the Blue CAT system can be further recycled at the plant head to remove excessive containment and for secondary process treatments. The blue CAT treatment is contempt enough for organic destruction, enhanced containment removal rate, and disinfection from the wastewater; by combining the blue CAT technology (U.S. Environmental Protection Agency, 2013). When compared to the current counterpart technologies such as, oxidation processes the Blue CAT treatment performed much efficiently with the minimum usage of power due to its catalytic configuration which maximizes its oxidative capacity.

During the process, no polymer and a negligible amount of iron reagent were employed. It was also stated, that this process requires less amount of chemical dosing, comparatively; and also produces a lesser amount of solid remains. By virtue of Iron-based reactive agents, the odor emission is also minimized. The Blue CAT treatment can be considered eco-friendly, lesser than 10 MGD (U.S. EPA, 2013).

PHOTOCATALYSIS

By virtue of the photocatalytic process, major organic contaminants such as azo dyes, biocides can be potentially reduced to simpler compounds like water, CO_2, and mineral acids from the wastewater discharged from textile industries. TiO2based materials have the potential to be utilized as the photocatalyst semiconductor for the photocatalytic degradation of textile and the emerging contaminants, owing to their application favorable properties such as non-toxicity, low environmental impact, chemical stability, efficient performance at low concentration, absorption of UVA and low cost of operation. Owing to similar properties such as low cost of operation, high redox potential, and non-toxicity the ZnO also finds numerous applications in the photocatalyst degradation (Daoud Bouras et al., 2019).

The implication of photocatalytic processes in the wastewater, obtained from the textile industry is dually beneficial, as along with de-colorization of the wastewater it also mineralization. Yet, the practical applications of the TiO_2 compound are limited due to the wide band-gap, very less lifespan of photo-induced electron-hole pairs, and inefficient treatment rate of high-strength solutions. Therefore, to increment the adsorption capacity, along with an efficient photocatalytic activity of the method, synthesizes other conventional treatment methods is essential.

Moderate Biotic Applications

The decontamination of pollutants and biomass Production is a very efficient and appealing wastewater treatment method, although it is not feasible at the industrial level; but at the municipal extent, it is very attractive because of its cost-effective operation method (Rajani et al., 2016).

The Purple non-sulfur bacteria (PNSB) is an efficient biochemical substance for treating wastewater as the production of biomass also yields high-value substances and in this manner, polluted water is both treated and utilized. The PNSB treatment can be employed for treating several organic wastewaters, producing high-value biomass resources (Kim et al., 2000).

The PNSB process method is based upon the phenomenon of orbital completion; it requires an external valence electron for development, which is easily gained by the organic and inorganic contaminant compounds, acting as electron donors. Abundant of required organic and inorganic as a cheap but effective resource, required for the development and bioremediation are found in the polluted water bodies. This is evident from the study results, demonstrating the coupling of PNSB strains for bioremediation of wastewaters, effective tolerance of various pH levels, and efficient utilization of organic and inorganic contaminant compounds, acting as electron donors, along with their development and yielding of high-value biomass (Rajani, 2016).

Upon using organic wastewaters as substrates, the generation of biomass was very less. Due to the Enhancement of conversion efficiency of organic compounds, striving in contaminated water.

The generation of a bacterial cell is a promising alternative for promoting PNSB. Addition of Mg2+ in the polluted water, increases the PNSB biomass accumulation, as observed in the PNSB metabolic activity, Mg2+ directly affects PNSB growth and metabolism (Van Doorslaer et al., 2015).

CONCLUSION

The whole chapter has focused upon an ever-existing threat to the hormonal balance of the planet, the contamination of water. Since the evolution of mankind, their development and hence after their advancement the water pollution has grown simultaneously. With the discovery and invention of many elements and materials respectively, the level of threat possessed by water contamination has also grown manifold. As the efforts are made to make human life more comfortable so is the abuse to the water resources of the planet. With the beginning of our civilizations, our ancestors recognized the brawling threat and did everything in their power to neutralize it.

But with the development of the cities and lifestyle of people, their focus shifted much upon the materialistic elements than the natural ones upon which they survived. The results of this ignorance were witnessed in the dark ages where people died due to natural calamities. But with the age of industrialization and the ongoing persecutions of several centuries people came to awake and once again the efforts for neutralizing this threat begun in the late 18th century. As the applications of our scientific endeavors grew so did the application for treating contaminated water.

Due to the discovery of nano-science and the beneficial virtues it offered, the research upon nanotechnology was soon initiated and as a result, there are various nano-based applications in the industry available today. There are certain limitations because of recent technology, but with dedicated researches more potent and permanent purification systems based upon nanotechnology would be available shortly. Cost efficiency and the requirement of industry scale platform is one of the major challenges faced by nanotechnology-based applications that can be compensated through advancement in the conventional technologies, which are easily employable at municipal or small urban level platforms.

The advancement of our scientific applications has possessed a threat to our natural resources, affecting us but the very same applications allowed us to neutralize the effects and maintain an ecological balance.

FUTURE SCOPE

Globally the trend of using nanotech for decontaminating wastewater is increasing, due to their favorable features and properties. The integration of nanotech with conventional treatment methods is also increasing rapidly. However, most of the nanotech-based processes are at the initial laboratory research and experimental pilot studies; which can be quickly established commercially. Nano adsorbents, along with nano photocatalyst and nanomembranes are some of the most promising fields which can revolutionize the wastewater treatment technologies if proper research is promoted in these fields. The aim of these researches in future should be to make these technologies more cost-efficient, easily available and ease of implementation at rural level.

REFERENCES

Angelakis, A. N., Koutsoyiannis, D., & Tchobanoglous, G. (2005). Urban wastewater and stormwater technologies in ancient Greece. *Water Research*, *39*(1), 210–220. doi:10.1016/j.watres.2004.08.033 PMID:15607179

Baker, M. N. (1981). The quest for pure water: the history of water purification from the earliest records to the twentieth century. American Water Works Association.

Basheer, A. A. (2018). New generation nano-adsorbents for the removal of emerging contaminants in water. *Journal of Molecular Liquids*, *261*, 583–593. doi:10.1016/j.molliq.2018.04.021

Beder, S. (1996). The nature of sustainable development. *Metascience*, *10*, 304.

Bhadra, J., Abdulkareem, A., & Al-Thani, N. (2019). Nanotechnology in Decontamination. In *Composite Nanoadsorbents* (pp. 119–137). Elsevier., doi:10.1016/B978-0-12-814132-8.00006-X

Burian, S. J., Nix, S. J., Pitt, R. E., & Durrans, S. R. (2000). Urban Wastewater Management in the United States: Past, Present, and Future. *Journal of Urban Technology*, *7*(3), 33–62. doi:10.1080/713684134

Gohary, A. (2014). Evolution of sanitation and wastewater technologies in Egypt through centuries. *Water Intelligence Online*, *13*. doi:10.2166/9781780404851

Golfinopoulos, A., Kalavrouziotis, I. K., & Aga, V. (2016). Prehistoric and historic hydraulic technologies in stormwater and wastewater management in Greece: A brief review. *Desalination and Water Treatment*, *57*(58), 28015–28024. doi:10.1080/19443994.2016.1179224

Gopakumar, D. A., Pai, A. R., & Pasquini, D. (2019). Nanomaterials—State of Art, New Challenges, and Opportunities. In Nanoscale Materials in Water Purification (pp. 1–24). Elsevier. doi:10.1016/B978-0-12-813926-4.00001-X

Gupta, V. K., Agarwal, S., & Saleh, T. A. (2011). Synthesis and characterization of alumina-coated carbon nanotubes and their application for lead removal. *Journal of Hazardous Materials*, *185*(1), 17–23. doi:10.1016/j.jhazmat.2010.08.053 PMID:20888691

Hasbullah, H., Sabri, N. S. M., Said, N., Rosid, S. M., Roslan, M. I., Ismail, A. F., Jye, L. W., & Yusof, N. (2019). Nanoengineered Materials for Water and Wastewater Treatments. In *Nanotechnology in Water and Wastewater Treatment* (pp. 303–335). Elsevier. doi:10.1016/B978-0-12-813902-8.00016-2

Karami, H. (2013). Heavy metal removal from water by magnetite nanorods. *Chemical Engineering Journal*, *219*, 209–216. doi:10.1016/j.cej.2013.01.022

Khan, S. (2014). Sanitation and wastewater technologies in Harappa/Indus valley civilization (ca. 2600–1900 BC). *Water Intelligence Online*, *13*. doi:10.2166/9781780404851

Kim, M. K., Choi, K.-M., Yin, C.-R., Lee, K.-Y., Im, W.-T., Lim, J. H., & Lee, S.-T. (2004). Odorous swine wastewater treatment by purple non-sulfur bacteria, Rhodopseudomonas palustris, isolated from eutrophicated ponds. *Biotechnology Letters*, *26*(10), 819–822. doi:10.1023/B:BILE.0000025884.50198.67 PMID:15269554

Kim, S., Park, C. M., Jang, M., Son, A., Her, N., Yu, M., Snyder, S., Kim, D.-H., & Yoon, Y. (2018). Aqueous removal of inorganic and organic contaminants by graphene-based nanoadsorbents: A review. *Chemosphere*, *212*, 1104–1124. doi:10.1016/j.chemosphere.2018.09.033 PMID:30286540

Kunduru, K. R., Nazarkovsky, M., Farah, S., Pawar, R. P., Basu, A., & Domb, A. J. (2017). Nanotechnology for water purification: applications of nanotechnology methods in wastewater treatment. In *Water Purification* (pp. 33–74). Elsevier., doi:10.1016/B978-0-12-804300-4.00002-2

Laohaprapanon, S., Matahum, J., Tayo, L., & You, S.-J. (2015). Photodegradation of Reactive Black 5 in a ZnO/UV slurry membrane reactor. *Journal of the Taiwan Institute of Chemical Engineers*, *49*, 136–141. doi:10.1016/j.jtice.2014.11.017

Lata, S., & Samadder, S. R. (2016). Removal of arsenic from water using nano adsorbents and challenges: A review. *Journal of Environmental Management*, *166*, 387–406. doi:10.1016/j.jenvman.2015.10.039 PMID:26546885

Lee, Y.-C., & Yang, J.-W. (2012). Self-assembled flower-like TiO2 on exfoliated graphite oxide for heavy metal removal. *Journal of Industrial and Engineering Chemistry*, *18*(3), 1178–1185. doi:10.1016/j.jiec.2012.01.005

Liu, L., Luo, X.-B., Ding, L., & Luo, S.-L. (2019). Application of Nanotechnology in the Removal of Heavy Metal From Water. In *Nanomaterials for the Removal of Pollutants and Resource Reutilization* (pp. 83–147). Elsevier. doi:10.1016/B978-0-12-814837-2.00004-4

Lofrano, G., & Brown, J. (2010). Wastewater management through the ages: A history of mankind. *The Science of the Total Environment*, *408*(22), 5254–5264. doi:10.1016/j.scitotenv.2010.07.062 PMID:20817263

Mallakpour, S., & Rashidimoghadam, S. (2019). Carbon Nanotubes for Dyes Removal. In *Composite Nanoadsorbents* (pp. 211–243). Elsevier. doi:10.1016/B978-0-12-814132-8.00010-1

Nasrollahzadeh, M., Sajadi, S. M., Sajjadi, M., & Issaabadi, Z. (2019). Applications of Nanotechnology in Daily Life. In Interface Science and Technology (pp. 113–143). doi:10.1016/B978-0-12-813586-0.00004-3

Ng, L. Y., Mohammad, A. W., Leo, C. P., & Hilal, N. (2013). Polymeric membranes incorporated with metal/metal oxide nanoparticles: A comprehensive review. *Desalination*, *308*, 15–33. doi:10.1016/j.desal.2010.11.033

Nnaji, C. O., Jeevanandam, J., Chan, Y. S., Danquah, M. K., Pan, S., & Barhoum, A. (2018). Engineered nanomaterials for wastewater treatment: current and future trends. In *Fundamentals of Nanoparticles* (pp. 129–168). Elsevier. doi:10.1016/B978-0-323-51255-8.00006-9

Parvin, F., Rikta, S. Y., & Tareq, S. M. (2018). 8 -Application of Nanomaterials for the Removal of Heavy Metal from Wastewater. In Nanotechnology in Water and Wastewater Treatment: Theory and Applications (pp. 137–157). Elsevier. doi:10.1016/B978-0-12-813902-8.00008-3

Penke, Y. K., Anantharaman, G., Ramkumar, J., & Kar, K. K. (2019). Redox synergistic Mn-Al-Fe and Cu-Al-Fe ternary metal oxide nano adsorbents for arsenic remediation with environmentally stable As(0) formation. *Journal of Hazardous Materials*, *364*, 519–530. doi:10.1016/j.jhazmat.2018.10.069 PMID:30388635

Postel, S. (1999). *Pillar of sand*. W. W. Norton & Company Ltd.

Rajani, B., Kumar, R. S., Devi, M. U., & Nayak, J. B. (2016). *Role of Purple Non-Sulfur Bacteria Rhodopseudomonas Palustris Rsou000 and Rhodopseudomonas Thermotolerance Rsou555 in Waste Water Treatment.* doi:10.20959/wjpps20168-7428

Riahi, K., Van Vuuren, D., Akimoto, K., Dellink, R., Andersson, M., Bazilian, M., De Cian, E., Cosgove, B., De Vries, B., Fischer, G., Gielen, D., le Blanc, D., Gruebler, A., Hallegatte, S., Heaps, C., Hellmuth, M., Howells, M., Jiang, K., Kok, M., … Loulou, R. (2013). *Sustainable Development Scenarios for Rio+20 A Component of the SD21 project.* https://sustainabledevelopment.un.org/content/documents/793SD21 scenario report.pdf

Rost, S. (2019). Navigating the ancient Tigris – insights into water management in an early state. *Journal of Anthropological Archaeology*, *54*, 31–47. doi:10.1016/j.jaa.2019.01.005

Safarpour, M., & Khataee, A. (2018). Graphene-Based Materials for Water Purification. In *Nanoscale Materials in Water Purification* (pp. 383–430). Elsevier., doi:10.1016/B978-0-12-813926-4.00021-5

Saleem, M., & Bachmann, R. T. (2019). A contemporary review on plant-based coagulants for applications in water treatment. *Journal of Industrial and Engineering Chemistry*, *72*, 281–297. doi:10.1016/j.jiec.2018.12.029

SIWI. *Cooperation Over Shared Waters - Stockholm International Water Institute*. (n.d.). SIWI. Retrieved May 18, 2020, from https://www.siwi.org/priority-area/transboundary-water-management/

Smith, L. (2017). Historical Perspectives on Water Purification. In *Chemistry and Water* (pp. 421–468). Elsevier. doi:10.1016/B978-0-12-809330-6.00012-X

Sonune, A., & Ghate, R. (2004). Developments in Wastewater Treatment Methods. *Desalination*, *167*, 55–63. doi:10.1016/j.desal.2004.06.113

Suthar, R. G., & Gao, B. (2017). Nanotechnology for drinking water purification. In *Water Purification* (pp. 75–118). Elsevier. doi:10.1016/B978-0-12-804300-4.00003-4

Tamburrino, A. (2010). Water Technology in Ancient Mesopotamia. In *Ancient Water Technologies* (pp. 29–51). Springer Netherlands., doi:10.1007/978-90-481-8632-7_2

Tzanakakis, V. A., Koo-Oshima, S., Haddad, M., Apostolidis, N., & Angelakis, A. N. (2014). The history of land application and hydroponic systems for wastewater treatment and reuse. *Water Intelligence Online, 13*. doi:10.2166/9781780404851

UNESCO. (2014). Water and Energy Sustainability. *UNESCO*. http://www.unesco.org/open-access/terms-use-ccbysa-en%0Ahttp://www.un.org/waterforlifedecade/pdf/01_2014_sustainability_eng.pdf

UNESCO. (2017). *World Water Scenarios*. Author.

UNICEF. (2015). *Diarrhoeal Disease - UNICEF DATA*. https://data.unicef.org/topic/child-health/diarrhoeal-disease/

UNISDR. (2015). The human cost of weather-related disasters 1995-2015. *UNISDR Publications, 1*, 30. doi:10.1017/CBO9781107415324.004

United Nations Educational, Scientific and Cultural Organization. (n.d.). Retrieved May 19, 2020, from http://www.unesco.org/new/en/unesco/events/prizes-and-celebrations/celebrations/international-days/world-water-day-2017/

U.S. Environmental Protection Agency. (2013). *Emerging Technologies for Wastewater Treatment and In-Plant Wet Weather Management EPA 832-R-12-011 March 2013*. Office of Wastewater Management.

Van Doorslaer, X., Haylamicheal, I. D., Dewulf, J., Van Langenhove, H., Janssen, C. R., & Demeestere, K. (2015). Heterogeneous photocatalysis of moxifloxacin in water: Chemical transformation and ecotoxicity. *Chemosphere, 119*, S75–S80. doi:10.1016/j.chemosphere.2014.03.048 PMID:24735961

Warnes, S. L., & Keevil, C. W. (2011). Mechanism of Copper Surface Toxicity in Vancomycin-Resistant Enterococci following Wet or Dry Surface Contact. *Applied and Environmental Microbiology, 77*(17), 6049–6059. doi:10.1128/AEM.00597-11 PMID:21742916

WHO. (n.d.). *Drinking-water*. Retrieved May 19, 2020, from https://www.who.int/news-room/fact-sheets/detail/drinking-water

Chapter 12
Urban Socio-Ecosystems Green Resilience

José G. Vargas-Hernández
https://orcid.org/0000-0003-0938-4197
University Center for Economic and Managerial Sciences, University of Guadalajara, Mexico

ABSTRACT

This chapter is aimed to analyze the system of green resilience eco-urban land uses oriented in urban social-ecological systems. It reviews and analyses the relevant literature in green social-ecosystem resilience concept and presents a discussion in relation to the sustainable development and ecological sustainability. It further discusses and gives an in-depth overview of the urban social ecosystems as a working structural and functional unit, describes decision support tools that could be applied to sustainable green land uses and development, and offers some strategies for engaging in urban ecosystems, ecological sustainability and adaptive development. It is concluded that the urban land use that through the innovative pro-environmental solutions can, in a natural way, support the system of green resilience eco-oriented urban land uses in urban eco-systems and serve to improve the quality of life in the city.

INTRODUCTION

Human development have had a profound imprint on nature and coevolving ecosystems which had resulted in complex, economic-socio-ecological challenges for sustainability and future development. Sustainability is a set of goals combining social equity, economic viability and ecological integrity (Curwell, Deakin, & Symes, 2005; Jenks & Jones, 2010). Humankind alter the dynamics of ecosystem through anthropologic activities that change the atmosphere, climate, land surface, forest, sea, and waters. Humankind another organism's survival is dependent on healthy and resilient social-ecological systems and sustainable environments.

Human wellbeing, economic growth and social development are dependent on the interrelationships between and within the regions and environmental sustainability (Arrow et al., 1995; Folke et al., 1998). Uncertainty, diversity and variability of social-ecosystem increases their inevitability reducing

DOI: 10.4018/978-1-7998-4915-5.ch012

their capacity to cope with disturbance and change within functional groups in the adaptive capacity of ecosystems (Folke et al. 2002; Jackson et al. 2001; Scheffer et al. 2001).

Cities have been portrayed as predominantly monumental static architectural structures of increasing ecological complexity that change over the long term. Contextual and conceptual factors of urban change can be assembled into a framework of ecological urban design. Disturbances change the resilient capacity of nature to supply ecosystem services, degrading socioecological systems and leading to social and economic vulnerability. Inhabitants of urban areas pose a high impact on the ecosystems services with trade and consumption, claiming support in waste absorption, carbon emissions, residential water use, wood for industrial purposes (Folke et al., 1997, Grimm et al., 2008). Adverse human impacts on the social-ecosystem and the biosphere can be minimized through the use of resilient and sustainability environmental approaches such as environmental resources management, conservation biology, environmental and ecological economics, bio economics, green technology, etc.

According to Levin (1999), fragility is inherent to the ecosystems services on which human's dependent and not nature. A global health control of the ecosystems, published in 2005 by The United Nations Millennium Ecosystem Assessment (MA) observed that the technological advancement is the main cause of the Earth's ecosystem services degradation and are used unsustainably. The growing eco-deterioration is an impediment to combating poverty (Millennium Ecosystem Assessment (MA), 2005). If the ecosystem and natural resources are controlled by few people, they do for short-term economic gain.

Sustainability is the capacity of a system or process to be preserved, enhanced, upheld, or maintained. Sustainability is the capacity of a biological systems and processes to endure disturbances and remain vigorously diverse. Sustainability is the systematic combination of environmental science and sustainable development (Lynn et al., 2014).

Sustainability has the core value the sustainable development comprised by interconnection of ecology, culture, politics, and economics domains (James, et al., 2015). Ecology refers to the paradigm, knowledge, methods and procedures of contemporary ecological science (Kolasa & Pickett, 2005). Ecology is a multidimensional and complex concept that requires an interdisciplinary framework of analysis for its application in urban spaces (Pickett & Cadenasso, 2002). The dimensions of ecology are the notion, the models and the metaphors used to communicate assumptions, values and experiences (Cadenasso et al., 2006a)

Sustainability is a social challenge that has an impact on economic efficiency, production, distribution, urbanization, transportation, lifestyles and ethical consumerism. Promoting sustainable and resilient lifestyles in sustainable urban social-ecosystems to conserve ecological resources, range from among other activities from green spaces, sustainable architecture, green housing and building, ecovillages, green business and green economic sectors, sustainable urban agriculture and farming, green technologies, renewable energies and designing reversible and flexible systems (Zhang and Babovic, 2012). The specific types of sustainability are sustainable architecture or ecological economics, sustainable technology, and agriculture (Costanza and Patten, 1995).

Sustainability includes concerns on global climate change, ozone depletion, biodiversity, wild life conservation, deforestation, genetically engineering, urbanizations, etc. Sustainable environmental and ecological system is being identified in the resilient social ecosystems with the capacity that have to continue functioning despite facing natural and anthropogenic disturbances, but anticipating and preparing ahead for them.

GREEN SOCIAL-ECOSYSTEM RESILIENCE

Human beings have had a profound imprint on nature and on dealing with the resulting challenges. One of the most challenging events is the global warming phenomenon which is increasing the average temperature of organisms while speeding up their metabolism. An ecosystem is a structural and functional unit of the biosphere and the complex interrelationships with human activities that are threatening its sustainability. Resilience has its source of thinking in ecology (Holling, 1973; Wu & Wu, 2013) a concept that que be operationalized to be applied in urban environments and cities (Musacchio, 2008). Resilience deals with the opportunities arising from ecosystemic changes in its structures, processes, emergence of great developments and coping with disturbances.

The resilience approach provides a conceptual and theoretical framework for interdisciplinary collaboration with ecological economics, sustainable development and governance (Lambin, 2005). Resilience is a concept and model framework used to operationalizing normative sustainability (Childers, Pickett, Grove, Ogden, & Whitmer, 2014). Resilience is a non-normative conceptual model for the achievement of economic, social and environmental goals disaggregating process (Walker, Holling, Carpenter, & Kinzig, 2004) into capacities for adaptation (Yohe & Tol, 2002).

Resilience is a dynamic process interfering symbiotic relationships within and between the social ecosystem and its environment. Resilience is a fundamental capacity that individuals, groups, organizations, and ecosystems have that respond to significant disruptive changes. Resilience is essential to sustainable development in dynamic equilibrium with the natural disturbances of the social – ecosystem functioning. Resilience refers to the ecosystem ability to respond productively to significant disruptive change and adapt for successfully to external variables that threaten its existence.

Resilience is the capacity of the social-ecological system to reorganize itself after any disturbance. Social-ecosystem resilience lies in the continuous cycling movement of adaptation and transformation while maintaining the system integrity and viability. Resilience involves the reaction of the socio-ecological system under disruption and destruction with the capacity to recovering and developing in a state of uncertainty, discontinuity, and emergency.

Human-induced erosion of resilience has consequences on shifts of the social ecosystems. Response diversity in sustaining ecosystem states, functional groups and ecosystem services are linked when facing disturbances and environmental change. Response diversity among species within a functional group is a component of resilience. Response diversity interacts operating across spatial and temporal scales. High response diversity increase the insuring buffer for resilience and the effectiveness of ecosystem management policies and actions.

Moderate perturbations and disturbances events on the stability of the social ecosystems either may be absorbed by the resilient ecosystems through reorganization it may bring small changes in resilience supported by response diversity (Deutsch et al. 2003). For example, response diversity maintains resilience in lakes stressed by toxic chemicals and acidification (Carpenter and Cottingham 1997).

The social system structure equilibrium in its interactions with the ecosystems by accepting a specific level of stability. Sustainable development involves local and global efforts to meet core human needs without distorting the ecosystem (Kates et al., 2005; IISD, 2009; EurActiv, 2004). Sustainable development implies proactive and responsible decision making innovation processes to maintain social ecological systems equilibrium between resilience, economic efficiency, social inclusion and equality, political justice and vibrant cultural values aimed to ensure a livelily and desirable ecosystem on the long term for all the species (Liam et al., 2013).

Resilience is an approach for cogitating and critiquing of social–ecological systems, with policy implications for sustainable development (Folke et al., 2002). Resilience of social-ecological systems integrates self-organization, learning, adaptation and persistence. Social-ecological resilience measures the degree of the system self-organization and coping with disturbance and still maintaining its inertia, attraction and capacities for learning and adaptation (Carpenter et al. 2001).

Sustainability is a normative social goal promoted by using the mechanisms of ecological resilience. The ecological resilience includes the patch ecology or landscape, the notion of metacity and experimental modeling and design.

The ecological resilience is the capacity of a site in urban systems to adjust and control of interactions with external vulnerabilities, disruptions and shocks making more sustainable the urban component of the ecosystem, given its nature. It is different from engineering resilience which has the capacity of returning to the previous existing state before perturbation. The resilience sustains the urban ecosystem states under uncertain, variable and complex environments. Response diversity is critical to resilience during ecosystem reorganization.

The terms urban and city are used as the totality of densely settled regions, and urban systems have spatial and functional contexts combine social, biological, built and geophysical components (Graham & Marvin, 2001; Seto et al., 2010; Naveh, 2000; Pickett & Grove, 2009). Some social dimensions of urban systems are economics of production and consumption, the economic and political power, social inclusion, identity, equality and change, social justice and vulnerability, the nature of livelihood and lifestyle (Dow, 2000; Grove et al., 2006; Machlis, Force, & Burch, 1997).

Response diversity does not necessarily support that high species biodiversity entails high ecosystem resilience and less vulnerable to environmental change. Migration of people to the large cities have results into a reduction in the ecosystems biodiversity (MA, 2005; Sala et al., 2000). Biodiversity is critical factor of social-ecosystems while cultural diversity and common property systems are related with the resilience building of urban socio-ecological-systems. Urbanization processes have led to higher levels of cultural diversity in cities (Zanoni and Janssens, 2009).

Urban space influence environmental values of urban populations (Miller, 2005; Tidball et al., 2010). The urban space and territory are changing rapidly and becoming globally interconnected across contrasting types of land and facing new environmental, demographic and social threats and pressures. Urban green commons (UGCs) have potentials to facilitate civic participation and cultural integration, to manage urban land and biological diversity in urban spaces and promoting urban resilience building (Colding and Barthel, 2012). Urban resilience in the context of an urban system is an ecological approach to analyze the changes, disturbances, vulnerabilities and mutability of city development and their relationships with the climate change.

Urban transformation is benefiting from the shift of urban ecology framework from earlier approaches such as the metabolic urban energy budgers advance to the city resilience approach with components of these hybrid systems such as the biophysical - social structures and processes (Cadenasso et al., 2006b; Cadenasso & Pickett, 2013; Cadenasso & Pickett, 2008; Pickett, Cadenasso, & Grove, 2004).

The vulnerability of an ecosystem states is related with other organisms in a functional group. Contraction of spatial resilience increase the disturbance to catastrophic level (Nyström and Folke 2001). The ecosystem resilience has a cross-scale response diversity, so, biodiversity has an important role in formulating policy on ecosystem resilience for sustained economic and socioecological development. Managing for resilience copes with uncertainty in the biosphere shaped by human action (Folke et al. 2002).

The preservation of congenital features leads to robustness of the ecosystem despite behavior alterations of its component parts and environment. The resilience of ecosystems provides backups to replace failures in the species through the diversification and redundancy of biodiversity with overlapping ecological functions in natural ecosystems. Ecological functions across a range of scales can have diverse disturbances. Ecological disruption occurs across a limited range of spatial and time scales allowing undisturbed scales to persist and ecological functions to operate.

Response diversity is relevant to resilience because the diversity of species contributing to ecosystem functions of renewal and reorganizations in environmental changes. Response diversity contributes to resilience when planning, management and restoration social-ecosystem. Response diversity provides adaptive capacity in uncertain and complex systems and human-dominated environments.

The property of response diversity is a variability in responses of species to environmental change is critical to social ecosystem resilience. Response diversity is the set of reactions among species contributing to the same ecosystem function to environmental change. Species and populations diversity within functional groups maintain ecological redundancy in ecosystem services (Luck et al. 2003). Resilience equated with species should consider that species loss is non-random in relation to response diversity and ecosystem functioning. However, response diversity in ecosystem resilience is related to ecosystem disturbances and environmental changes. Biological diversity is essential in social ecosystem resilience and in sustainable ecosystem change (Peterson et al. 1998). Biological diversity enhances the social ecosystem resilience required to secure the production of ecosystem services.

Resilience can interfere and be in conflict with other beneficial social objectives, such as economic efficiency because the costs of redundancy. So, economic efficiency reduces resilience. Resilience of social ecosystems require more adaptability to stress and still maintain the stability domain from extrinsic disturbances, that is, to find a solution of the conflict between stability and resilience for sustainable development in terms of complex system cycles.

Resilience is concerned with the management of sustainable interactions between human-developed systems and natural ecosystems. Social resilience linked to social change is the communities' capacity to cope with extrinsic disturbances to their social infrastructure, such as political turbulence, socio-economic reforms, and environmental variability (Adger 2000; Anderies et al., 2004).

The ecosystem resilience is the capacity to accommodate and adapt to disturbance, and the buffer capacity of persistence. Ecosystem resilience is the amount of disturbance a system can absorb and still may remain within the same state and domain of attraction (Holling 1973, 1996). Ecosystem resilience is the capacity degree of self-organization, learning and adaptation (Carpenter et al. 2001), encompassing its ability subject to disturbance and change to reorganize and renew itself. The dynamic changes of the ecosystem resilience are related to the diversity of species and the functional groups (Walker 1992, 1997; Norberg et al. 2001).

Social-ecosystem resilience is an essential factor to uncertain and complex systems for sustainable natural resources and ecosystem services (Gunderson and Holling 2002). Any resilience disequilibrium between sustainability and development change leads to collapse of the ecosystems, thus, it is extremely difficult to transform a resilient ecosystem to a more congenial one (Scheffer et al., 2001; Gunderson and Holling, 2002; Walker et al., 2004). Congeniality with nature to prevent degradation is doing things on the nature's way.

Sustainable planning urban ecosystems framework draws from urban ecology, green resilience, green infrastructure, multifunctional and sustainable landscapes planning, etc., with the full purpose of creating and developing healthy and sustainable economic, social–ecological urban systems. Urban ecology is rooted in landscape ecology and has been integrating some principles from physical and atmospheric sciences, soil, hydrology, social sciences, etc. (Sukopp, 1990; Collins et al., 2000; McDonnell & Hahs, 2009; Pickett, Burch, Dalton, & Foresman, 1997; Redman, Grove, & Kuby, 2000). Green resilience deals with the coping of the planet earth with anthropogenic disturbances and still remaining viable for future generations. Ecological resilience is the ecosystem´s capacity to cope with disturbance and retaining the its structure and functions.

Resilience concerns the capacities to absorb the uncertainty resulting of shocks and conflicts to maintain functions and for renewal, growth, development and re-organization (Gunderson and Holling, 2002; Berkes et al., 2003). The concept of green resilience may be from a narrow to more broad interpretation of the socio-economic and ecological context. An ecological unit is the functioning of components and their relationships and interactions with each other forming a complex and dynamic whole.

The accountability of natural resources in an ecosystem occurs through the adaptation process after an externa disturbance. Adaptive processes contribute to the adaptive cycle in urban socio-ecological systems. Sustainability measurement requires transparent accountability between the resources depleted and those used to replace them in the ecosystem. Sustainable and green resiliency of an ecosystem can be measured at events' junctures with naturally occurring regenerative forces interacting with energy released into disturbances. These regenerative forces are the solar energy, water, soil, atmosphere, vegetation, and biomass (Ben, 2013). Green resilience is an institutional capacity to cope and deal with stresses and conflicts arising from climate change, unforeseen contingencies, unsustainable development to live and other emerging environmental issues within the ecosphere.

The old perspective of resilience assumed stable and static equilibrium to build socio-economic and ecological adaptation to nature (Berkes et al., 2003, Smit and Wandel, 2006). Resilience has some tools based on structured scenarios and adaptive management to build in complex and uncertain systems. These tools increase capacity to build ecological resilience. Natural resilience sources of complex adaptive systems are more than preservation of ecosystems and resistance to change. Traditional and dominant theoretical perspectives have been assuming a stable and resilient environment where nature and its flows of natural resources are in static equilibrium after stressors are controlled or removed (Holling, 1973).

Green resilience sources are in intertwining relationships in complex adaptive systems of dynamic changes in the ecosystem. Dynamic adaptive capacity of the ecosystem provided by resilience interplays between development change and sustainability (Smit and Wandel 2006). The institutional capacity of the social system to cope with change in the ecosystem by anticipating and combating disturbances leads to adaptive development integrating resilience in the interactions in ecological sustainable development. Adaptive development in the social ecosystems has as elements the ability of ecological assessment of actual events and taking corrective action.

Resilience is the system´s ability to adapt to any kind of disturbance and self-re-organize while undergoing transformation processes retaining its initial forms, roles, identity, and feedbacks characteristics (Walker et al., 2004). Any disturbance of a strong resilient socio-economic and ecological system, may have the potential capacities to create new opportunities for innovation and advancement. However, when the system is weak, any disturbance may cause great disasters (Adger, 2006). A sustainable and resilient social-ecosystem is more performed by nature in interaction with small dependence on human inputs, because the anthropogenic interferences may interrupt the provision of inputs.

Sustainable development pretends to reduce social and ecological damage on both a local and a global scale through the operation of fairness and social inclusion, protection of the environment, and economic efficiency. Green resilience-building in complex, uncertain and unpredictable urban ecosystems is supported by structured scenarios and active adaptive management for sustainable development. Green resilience management enhances sustainable development in changing complex environments where the future is uncertain and unpredictable (Walker et al., 2004; Adger et al., 2005). Strategies in any system of urban green resilience focus on green infrastructure transform and adapt resources to face future challenges such as climate change and food insecurity.

The social ecosystem is currently threatened by anthropogenic activities at the brink of collapse and requires prompt action. In this context of social-ecological systems, it is required management and flexible collaboration aimed to develop policy frameworks as a basis to build adaptive capacity. Nature should be strengthened to stimulate development through the interaction and interdependence with humankind and to enhance resilience in social-ecosystems.

The phenomenon of resilience in sustainable development may lead to some policy recommendations to improve the interrelationships between the economic efficiency and the social-ecosystems and biosphere, to develop some flexibility and innovative relationships of collaboration to achieve sustainability and its operationalization in the context of socio-ecological resilience. Other policy recommendation should focus on the development of indicators to measure any change of the level of resilience and to signal and monitor uncertainties of social – ecosystem variables and to manage diversity.

URBAN SOCIAL-ECOSYSTEMS

In 1930, the term ecosystem was coined by Roy Clapham describe the biological and physical environmental components as a unite where all are related to each other. The ecosystem was described as the interaction between the living creatures or bioconenosis and the environment where they live, or biotope (Tansley, 1935). For the Convention on the Biological Diversity (1992), an ecosystem is a dynamic complex formed by plants, animals and micro-organism communities and their non-living environment interacting as a functional unit.

The biosphere is the largest ecosystem. The biosphere interacts with and exchanges matter and energy with the lithosphere, hydrosphere and atmosphere. The level of the ecosystem is above the ecological community corresponding to the biomes and the biosphere, also considered as the regional ecosystems. Ecosystems are complex and adaptive systems characterized by historical dependency and non-linear dynamics, besides they have multiple basins of attraction (Levin 1999). The ecosystem refers to the level of biological organizations formed by living organisms interacting with each other and in a symbiotic relationship with their environment. An ecosystem is capable of harnessing solar energy during photosynthesis and converted into carbon dioxide and other inorganic chemicals functional to organic life.

Ecosystem processes are seasonally dynamic responding to the solar input driving the largest biogeochemical carbon cycle of the global cycles (Odum, 1971). An ecosystem-level process can be represented by a biogeochemical cycle of exchange of matter between organisms and their environment (Golley, 1993). Energy transfer and matter cycling processes are essential in determining ecosystem structures and functions of a wide diversity of species, as well as in defining the interactions between organisms and their environment (Golley, 1993). The living things are continuously competing to reproduce and survive with one another. When dominant species decline, other subdominant species in the same functional group survive (Elmqvist et al. 2001)

Sustain social ecosystem capacity copes with and adapts to environmental changes in the context of multiple-equilibrium systems and human-dominated environments (Folke et al. 1996; Norberg et al. 2001; Luck et al. 2003). Environmental change with low response diversity may result on extinct or ecologically insignificant functional groups contributing to ecosystem services. Sustaining a social ecosystem from compounded perturbations requires functional groups of species available for renewal and reorganization (Lundberg and Moberg 2003).

Response diversity management sustains and enhances the flow of ecosystem services subject to disturbances and operating within and across special and temporal scales. The impact of disturbances on species loss have implications for social ecosystem and the sustained flow of ecosystem services (Zimov et al. 1995). The persistence of functional groups of species prevents shifts to ecosystem states to sustain the ecosystem services. Species loss may entail low rates of ecosystems processes. Low or absent response diversity redirects ecosystem development into a different pathway. Stress-sensitive species populations decline and response diversity deteriorates while less sensitive species have minimal changes in ecological processes (Schindler 1990; Frost et al. 1995, Rudd et al. 1988).

An ecosystem is an open system requiring a flow of energy and matter between diverse organisms and the environment driving biogeochemical processes. A functioning system has living and non-living components interacting with each other. Any ecosystem has as components the abiotic or living organisms and abiotic which refers to the non-living matter of the environment. The biotic components are the living forms that inhabit the ecosystem and have a biogeochemical energy cycle. The abiotic are the nonorganic components of the environment that determine the life forms that can thrive. Thus, the living organisms are in continuous interactions with the environmental biotic and abiotic components (Golley, 1993). An ecosystem consists of all natural living things interacting each other and with all non-living factors of the environment (Christopherson 1997).

An ecosystem is made up of a set of components including the relationships between living and non-living factors, a biotic element or some source of a single living plant with some energy and an abiotic environment with elements of the Earth system. Energetic processes in ecosystems are formed by thropic levels defined by the role of organisms in its flow of energy. Ecosystems fall between the extremes of complexity (Odum, 1971). The term ecosystem refers to the bionic community of living organisms and biocoenosis always in continuous interaction with their environment or biotope functioning in a space as a unit.

Urbanization is rapidly expanding in cities, which are becoming the frontier for landscape ecology in urban ecosystem science. Urban design and planning take into account the coordination of resources, capacities and efforts to be implemented in comprehensive agendas across the city. Landscape ecology comprises the concepts of landscape sustainability and ecosystem services in changing landscapes (Wu & Qu, 2013). Landscape sustainability is the capacity to consistently provide long-term, landscape-specific ecosystem services that are essential for maintaining and improving human well-being (Wu 2013).

Committed cities to urban planning and implementing of ecosystem services supporting greening projects for transforming urban green infrastructure are likely to benefit through the value creation for all the population. A community organization focused on the urban environment attains social ecologically sustainable urban development. Modernist model of urban development is portrayed as a life cycle of cities moving from commodity exchange and industrialization, through an ecological version that mitigates all the ills of the urban ecosystem. Urban development emerges in high levels of biodiversity areas with high ecosystem productivity (Ricketts and Imhoff, 2003; Hansen et al., 2004; Ljungqvist et al., 2010).

The movement of people into more densely urban settlement can lessen pressure on urban and sub-urban sprawl and more remotely located ecosystems (Colding, 2011). Biodiversity at suburbs hold more natural and semi-natural land with an increase in the semi-rural urban fringe (Blair, 2001; McKinney, 2002; Sukopp et al., 1979 Colding et al., 2006).

Urban green infrastructure is a strategically planned and managed urban network of natural lands, working landscapes, and urban open green spaces that provide a range of diverse benefits. Ecosystem services derive benefits to humans (Millennium Ecosystem Assessment 2005). The social-ecosystems capacity altered to buffer disturbances has an impact on the flow of the services requires for the well-being. Landscape services derive some benefits to humans working out as a structure–function-value chain to create landscape development (Termorshuizen Opdam 2009).

Urban ecosystem is a potential solution to future ecological and environmental issues created for the loss of natural landscapes. When the natural resources and the ecosystem is controlled by few privileged people, the capacity of adaptive development is reduced, the resilience diminish and there is more disequilibrium. The balance between natural and urban ecosystems and the surrounding regional landscape is limited. Urban ecosystems services are relevant for urban green spaces such as biodiversity, microclimate control carbon sequestration, soil infiltration, food production, recreation, etc.

Multifunctional landscape frameworks are applied to agroecosystems and for planning urban ecosystems. Multifunctional landscape provides beneficial production, ecological, and cultural functions, considering the needs and preferences of users and owners (Otte et al. 2007; Lovell et al. 2010). A commitment to multifunctional urban green infrastructure pays off through the value of ecosystem services for urban population. Optimization of urban green space functions might be challenged by using a multifunctional landscape framework for sustainable urban design and planning of green infrastructure. Multifunctional green infrastructure provides ecosystem services and biodiversity benefit the environment and society.

The relationship between biological diversity and the ecosystems functions is that biodiversity provides renewal and reorganization capacities for social-ecosystems change (Loreau et al. 2001; Kinzig et al. 2002). Organization and functioning of a social ecosystem can be conservative in relation to the patterns of species loss despite the diversity compositions and considerable species turnover (Schindler 1990; Kalin Arroyo et al. 1995; Forys and Allen 2002; Brown et al. 2001, Havlicek and Carpenter 2001). Otherwise, social ecosystems may change when some species are lost or when there is an invasion of other new species (Estes and Duggins 1995; Terborgh et al. 2001; Vitousek and Walker 1989).

Multifunctional urban planning involves participation of diverse stakeholders in decisions on green infrastructure planning and implementation of solutions. Urban ecosystem services are supplied through participatory planning processes focused on multifunctional green resilience infrastructure developed to contribute to sustainable ecological, social and economic welfare of the city. Ecosystem services transform urban green infrastructure through the integration of empowerment and creative process into small-scale greening projects.

Community organizations on nature conservation focus on the protection and conservation of urban social ecosystems from dysfunctional anthropogenic activities in order to adapt the development and achieve economic efficiency. Community greening is a community-based effort to transform underutilized sites and areas into valuable green spaces such as community gardens (Tidball and Krasny 2009).

Urban green space is an under developed piece of land located within the territorial context of a city and open to the public. Urban green spaces are under pressure to provide the production, economic, social, cultural and ecological functions. The functions provided by the ecosystems services benefit humans while simultaneously meeting the needs and preferences of local population. Urban green spaces play a critical role in conserving biodiversity, mitigating the impacts of climate change, sequestering carbon, improving microclimate, protecting water resources conserving biodiversity, protecting water resources, suppling fresh food, etc.

ECO-ORIENTED URBAN LAND USES

Sustainable urbanization has received an increasing emphasis since metropolitan areas are engaged on initiatives aimed at upgrading land use regulations, infrastructure, urban form and ecosystem services, etc., in an effort to protect urban biodiversity and create a better urban resilient socio ecosystem, to develop a sustainable urban environment with more social justice and economic growth while improving competitiveness and attractiveness of the city (de Jong et al. 2015).

The land use must have a priority linked to nature and biodiversity to support the urban sustainable urban development in a territorial context and networks of urban open spaces unite nature and gardening to become a green resilient city under a concept of nature in the city and biodiversity (Quincerot & Weil, 2009, p. 175; Daune & Mongé, 2011). Land uses can be appropriate to develop a network of green open spaces for gardening penetrating urban core areas (Quincerot & Weil, 2009). A multifunctional role of land uses can provide urban corridors of gardening spaces to preserve nature, agriculture, farming, to perform educational and recreational activities helping to protect biodiversity and the natural environment while improving resilient socio ecosystem services in the context of increasing housing, building and inner-city densification.

Urban land is a natural resource-constrain threatening the prospect to meet the needs of increasing urban population, the economic prosperity and socio-political stability and influencing policy decisions. Resource depletion and pollution are two related global issues to land-use change.

The accelerated conversion and change of land use from arable or forested lands to urban land use is threatening the biodiversity, species and natural resources, which historically had provided the sustenance, valuable goods and services, climate regulation, water recharger, nutrient recycler, agricultural products, timber, seafood, waste assimilations, etc. All of these benefits are in danger to be lost with the increasing urban land use. To respond to the growing challenges posed by the conversion of land uses from agricultural to urban, it is required for the local government to formulate and implement urban development strategies aimed to ensure a more eco-efficient benefits in terms of economic growth and productivity, more socially equitable and inclusive and environmental sustainability.

Pressures are arising from climate change is related, to a certain extent on the increasing scarcity of land for urban use and other natural resources, such as water, natural nutrients, etc., at a time when population is continuing to rise and becoming an important concern and relevant issue for the policy agenda. Resilient ecological land uses orientation is only one of the governance challenges, structures and processes of the urban social eco-systems converging with energy, fuel, climate, water, food, etc. These challenges incorporate and integrate the economic, social and environmental perspectives and initiatives of diverse stakeholders to invest in natural resources and capital projects in the generation and development of the green resilience of the city.

Urban land use is inextricably an interrelated function of the resilient socio-ecosystem influenced by the behavior of individuals, groups, organizations and communities targeted at specific and limited functions at a time, such as biomass and water passage and accumulation. Urban land use has some strong relationships with the urban land market, local government regulations, management practices, technology innovations, etc., and influenced by the complexity of societal processes but ignoring the systemic consequences and effects on the resilient socio-ecosystem services. A sustainable urban transport system using renewable resources and energy, operating efficiently and affordable, minimizes the use of land and limits emissions, waste and noise within the ability of the city to absorb them.

A contemporary ecological movement advocates the principles of sustainable land use involved in urban planning although this movement is not always engaged in specific project to demonstrate the principles (Seana, Johnson & Peters, 1999). besides, the integration of land use intentions and initiatives is formulated in urban planning, not always are implemented and realized, becoming negotiation spaces for less formalized land uses initiatives and practices. Land use issues requires bringing a multi-disciplined and cross-departmental approaches that have an impact on decisions with effects on economic efficiency and urban growth, employment, social inclusion and equality, health, sustainable environment, etc. Urban zoning, neighborhood plans, the spatial-temporal patterns of inequality are some other factors that may have some implications and impacts of land use.

Urban land uses can be supported by public actions to promote the social dimensions of urban public spaces as part of collective re-appropriated spaces for diversified uses of unused urban vacant land around housing areas. Recovered unused land is often used for agriculture, farming and gardening projects on a temporary basis created on constructible land and adapting to the demands of residents. These public actions for more integral city development are aimed to foster conviviality, proximity and social cohesion between residents (Canton de Genève, 2013) enhancing more urbanite social interaction and cohesion while improving quality of life.

Urban planners and landscape designers can coordinate activities across different fields and urban functions, allowing to live up to their full potential as a multifunctional sustainable land use. Multi-scale analysis of land use compositions is a useful element in the land use planning and policy design for urban sustainability. Multi-functional analysis of land uses may be conducted by a participatory design process of visioning, defining, relating the elements, elaborating the site plan and the implementation process. Multiple land use scenarios simulate the future changes in composition of their functions under consideration such as self-sufficiency in vegetable production. Changes in land use may lead to inaccurate counts of land abandonment and vacant land (Bowman and Pagano 2004).

Population trends and urban land use and vacancy patterns have to consider their impacts on vacant land and structures. In densely populates areas, urban farming, agriculture and gardening offers an alternative land use for integrating multiple functions. However, current urban land use practices in urban sustainable planning prohibit agricultural and farming activities in urban spaces while there is a significant loss of farmland nearby (Pothukuchi & Kaufman, 1999; 2000).

Urban land use for agriculture, farming and gardening have become an attractive because of their potential to meet multiple needs, supplying fresh and healthy food in neighborhoods with limited access, articulating the linkages between the built environment, food systems and health while offering opportunities to residents for employment, recreation and education (Hodgson et al., 2011; Redwood, 2011).

However, agriculture and farming as a forms of land use are not well accepted local and city governments because of perception of health risks. Urban agriculture and farming are practices socially divisive when the land used can be dedicated to other scarce economic activities. Urban land use planning and zoning is in inertia and conflict with urban agriculture and farming supported from a thinking that there is a better use. Land use requires enhanced security on physical limitations that makes the urban space unbuildable and the urban parcels underutilized (Pagano and Bowman 2000) although may have a potential productive quality.

Urban agriculture, farming and gardening increasingly have become a necessary urban land use functions supported by sustainable urban planning and designing to have an influence on city life and forms and publicly accepted benefits in terms of multicultural traditions, recreational trends and poverty alleviation (Hough, 1995, p. 230). Gardens in Motreal, the city that has the largest municipal urban agriculture program in Canada, have their land use designation (Cosgrove, 2001).

Sustainable urban planning seeks to create attractive land-use combinations to meet the needs of residents and environmental challenges, including urban farming agriculture and gardening. Urban planning seeks to create and incorporate to sustainable city development more attractive land use combinations to provide recreational, educational, tourism, cultural, agricultural and farming, gardening, urban forestry, aquaculture, health services, commerce, (van de Berg and van Veenhuizen, 2005; Deelstra et all, 2006).

The urban land-use planning system contributes to the economic growth, social development and environmental sustainability. Vancouver's Food Action Plan (City of Vancouver, 2003) supported land use decision to serve as a public resource to support the city's sustainability commitments. State regulations, like in Oregon, states that land use planning system requires that each city have to set an urban growth boundary to show the limits and control "sagebrush subdivisions, coastal condo-mania, and the ravenous rampage of suburbia" but protecting the farm and forest lands (McCall, 1973).

The awareness of sustainable urban land-use planning reinforces the development and maintenance of a data base urban development on issues of urban growth or decline, rundown and waste resources (Carley, 1995; Curwell and Cooper, 1998; Department of the Environment, Transport and the Regions (DETR), 1998). Decision making about land use urban development and parallel policy processes support urban sustainable planning and policy making based on an inventory of natural resources. Urban land-use planning for commercial farming in most of the cities requires permission and to follow zoning and building regulations, water sources, fire and energy codes, waste disposal, etc.

High land prices and exposure to pollution and contamination of water, air and soil by industrial and commercial activities and traffic affect the land use for housing, education, recreation, agriculture, farming, gardening, etc. hampering investments and leading to poor land use security (Mougeot 2000).

Urban land use planning and design authorities for a sustainable environment does not necessarily considers relevant the natural resources and land provision to support urban food production, except for urban planning policies supporting gardening. Urban land use policies and rules may negatively affect local food production and distribution (City of Portland, 2007). Innovation land use types includes vulnerable groups to benefit (Clement 2013, p. 76). The innovation productive land use as landscape typology is designated for local food production and with innovation ecological resilience, these open spaces require decommissioning of urban services to encourage residents to seek employment in agriculture.

Urban planning and food production policies developed by local authorities may connect the issues of land-use planning and sustainable development. Urban land-use and comprehensive planning of some cities have already included urban farming and agriculture providing a food policy planning and adjusting zoning and permit processes with infrastructure, materials, knowledge, and other resources (Mukherji and Morales 2010). The land-use planning system fills the middle layer vacuum. Although land availability and use are not a statutory requirement for sustainable urban planning, they are at the center of the development of urban food production.

Buildings and infrastructure or built environment put pressure on urban land use. Urban vacant, residential yard and industrial rooftop spaces are considered different land uses. Some land use maps do not distinguish between parking lots and other vacant spaces. A good example has been use of vacant abandon land used to plant flowers, ornamental plants, herbs and vegetables, next to government building, as reported by Hui (2011).

Land use with the purpose of reuse of wastes and disposal may offer another benefit in reducing transportation and city management in the long-term. Land use management calls for a more spontaneous dynamics of the soil as a consequence that cities struggle to manage the opposing external and internal frontiers of land (Berger 2007). That is a struggle between the city expansion on lands of the urban periphery and the unused spaces within the internal frontier and urban core as land uses move in more elastic cities (Rusk 1993).

Land-use planning has an important role in urban policy making functions to deliver sustainable development, as for example in urban gardening that contributes to accrue economic benefits by preventing and enhancing health-care (Hynes, 1996, p. 161). Urban land use policies and regulations are deemed as methods for increasing and decreasing urban land vacancies. Management and design of vacant land uses contributes ecologically to sustainable cities by creating webs of sustainable urban spaces to provide ecosystem services. Long-term vacancy requires more oriented regeneration policies (National Land Use Database, 2000). Vacant buildings require recording reliable and comprehensive information.

Innovative and integrated sustainable initiatives and strategies in urban land-use planning can contribute to green resilient economic growth, social development issues such as inclusion and equality and environmental sustainability. For the implementation of the strategies is relevant to create a typology of community land use types based on vacancy characteristics, incomes of residents in the area and vacancy market rates in order to convert these vacant spaces into urban green resilience spaces with their various inherent functions (Akers, 2013; Detroit Works Project, 2014, p. 51). Vacant land restoration should be interdisciplinary approached combining economic, socio ecosystem concerns from a holistic urban land use perspective.

CONCLUSIONS

Human beings are living in a time of qualitative changes in many areas of urban life. These changes affect many levels of city management. It seems that the time has come to redefine the city's functional structure. The functional and spatial division in place in many countries seems to be fossilized and ineffective. In the face of enormous civilization changes, if only related to the possibilities of introducing innovative solutions to cities, it turns out that many existing regulations constitute a barrier and block their rational development. Particularly noteworthy are the urban land use that through the innovative pro-environmental solutions can, in a natural way, support the system of green resilience eco-oriented urban land uses in urban eco-systems and serve to improve the quality of life in the city.

Urban land is a scarce natural and real-estate resource which can be used in a more sustainable manner. Ultimately, at the policy level, new forms of value are ascribed to public and private urban land property in order to join other urban land use priorities such as housing, commercial areas, urban green areas and open spaces, roads and other infrastructure. Efficiency of urban land use can be achieved through different means and variables, among others: energy consumption, water and other natural resources, waste management, etc.

Urban land use has different applications for multiple scales, from the entire city, a neighborhood and communities to the small vacant plots, houses, buildings, etc. Urban land use should be evaluated based on a landscape multifunctionality framework, accounting for the contributions and functions, services and benefits that can be provided to the urban socio ecosystem functions, biodiversity, ecological services, cultural functions, economic efficiency, social inclusion and equality, the benefits to the communities and neighborhoods and the society at large.

To determine the current characteristics of land uses is necessary to conduct some base data analysis from primary and secondary sources. This approach allows to broaden the vision by considering the multifunctional forms of land use of a specific site and defining the various components by comparing data with other cities. Then, land uses are identified for specific functions such as education, recreation, etc. Land use components and their dimensions and locations are proposed, this whole process serves to identify the needs and expectations of the urban community regarding land uses, spatial design and site plans.

City authorities and local governments should adopt innovative urban planning and sustainable development strategies based on the use of land, resource protection practices, environmental and ecological services, social inclusion and egalitarian issues, economic growth and efficiency, etc. A strategic orientations of initiatives can seriously obstruct zoning plans and sustainable urban planning to designate urban agriculture, farming and gardening as a land uses. Land use planning and landscape design should reconcile demands of socio economic activities and urban growth.

Urban sustainable planning system can encourage the use of urban land for more sustainable development context oriented such as the land use for urban food production. Sustainable urban planning does not consider that the food system is its turf, pointing the responsibilities to land use regulations and built environment. Temporal urban land-use in dense core with limited open green space may formalized some urban gardening and farming initiatives to provide more qualitative resilient ecosystem services that may contribute to develop a more resilient city.

Urban land uses sustainable planning must reconcile the demands posed by all the socioeconomic activities and the urban growth assured by an effective management to create value, health and environmental sustainability. However, the high value of competing land in relation to use needs in cities, some land uses, such as urban agriculture, farming and gardening may not seem to be alternatives for urban settings. Land use sustainable planning integrates urban agriculture, farming and gardening into planning and policy processes. Urban farming, agriculture and gardening offer a new frontier for sustainable land use planners and sustainable landscape designers to become involved in the urban sustainable development and transformation of urban spaces.

Agricultural and farm landscapes are considered by urban land use planners and landscape designers as areas for future sustainable urban development and environment instead to integrate new other production functions. The urban agriculture and farming land uses are inherently multifunctional, offering public benefits beyond the provision of sustainability and commodity outputs. Vacant urban land in cities is increasing in city centers where the land use densities are declining due to manufacturing activities declines after have been reached their limits on employment creation and population density.

Multiple urban land use scenarios are a tool that helps to allocate vacant urban spaces to vegetable production including residential green resilient areas, gardens, rooftops, etc. Land use control and building regulations constraint and limit farming and gardening in urban areas.

REFERENCES

Adger, W. N. (2006). Vulnerability. *Global Environmental Change, 16*(3), 268–281. doi:10.1016/j.gloenvcha.2006.02.006

Adger, W. N. (2000). Social and ecological resilience: Are they related? *Progress in Human Geography, 24*(3), 347–364. doi:10.1191/030913200701540465

Adger, W. N., Hughes, T. P., Folke, C., Carpenter, S. R., & Rockstrom, J. (2005). Social– ecological resilience to coastal disasters. *Science, 309*(5737), 1036–1039. doi:10.1126cience.1112122 PMID:16099974

Akers, J. (2013). Making markets: Think tank legislation and private property in Detroit. *Urban Geography, 34*(8), 1070–1095. doi:10.1080/02723638.2013.814272

Anderies, J. M., Janssen, M. A., & Ostrom, E. (2004). A framework to analyze the robustness of social-ecological systems from an institutional perspective. *Ecology and Society, 9*(1), 18. http://www.ecologyandsociety.org/vol9/iss1/art18/

Arrow, K., Bolin, B., Costanza, R., Dasgupta, P., Folke, C., Holling, C. S., Jansson, B. O., Levin, S., Mäler, K. G., Perrings, C., & Pimentel, D. (1995). Economic growth, carrying capacity and the environment. *Science, 268*(5210), 520–521. doi:10.1126cience.268.5210.520 PMID:17756719

Falk, B. (2013). The resilient farm and homestead. Chelsea Green Publishing.

Berger, A. (2007). *Drosscape: Wasting Land Urban America*. Princeton Architectural Press.

Berkes, F., Colding, J., & Folke, C. (Eds.). (2003). *Navigating Social–Ecological Systems: Building Resilience for Complexity and Change*. Cambridge University Press.

Blair, R. B. (2001). Birds and butterflies along urban gradients in two ecoregions of theU.S. In J. L. Lockwood & M. L. McKinney (Eds.), *Biotic Homogenization* (pp. 459–486). Kluwer. doi:10.1007/978-1-4615-1261-5_3

O'Bowman, A. M., & Pagano, M.A. (2004). Terra Incognita: Vacant Land and UrbanStrategies. Georgetown University Press.

Walker & Salt. (2012). *Resilience Practice: Building Capacity to Absorb Disturbance and Maintain Function.* Island Press.

Brown, J. H., Ernest, S. K. M., Parody, J. M., & Haskell, J. P. (2001). Regulation of diversity: Maintenance of species richness in changing environments. *Oecologia, 126*(3), 321–332. doi:10.1007004420000536 PMID:28547444

Cadenasso, M. L., & Pickett, S. T. A. (2008). Urbanprinciples for ecological landscape design and management: Scientific fundamentals. *Cities and the Environment, 1*(2), 4. doi:10.15365/cate.1242008

Cadenasso, M. L., & Pickett, S. T. A. (2013). Three tides: The development and state of the art of urban ecological studies. In S. T. A. Pickett, M. L. Cadenasso, & B. P. McGrath (Eds.), *Resilience in urban ecology and design: Linking theory and practice for sustainable cities* (pp. 29–46). Springer. doi:10.1007/978-94-007-5341-9_2

Cadenasso, M. L., Pickett, S. T. A., & Grove, J. M. (2006a). Dimensions of ecosystem complexity: Heterogeneity, connectivity, and history. *Ecological Complexity*, *3*(1), 1–12. doi:10.1016/j.ecocom.2005.07.002

Cadenasso, M. L., Pickett, S. T. A., & Grove, J. M. (2006b). Integrative approaches to investigating human-natural systems: The Baltimore EcosystemStudy. *Nature Sciences Sociétés*, *14*(1), 1–14. doi:10.1051/nss:2006002

Canton de Genève. (2013). Plan directeur cantonal 2030. A14: Promouvoir de nouvelles formes de jardins fa-miliaux et encourager la création de plantages. Geneva, Switzerland: République et Canton de Genève.

Carley, M. (1995). The bigger picture: Organising for sustainable urban regeneration. *Town and Country Planning*, *64*(9), 236–239.

Carpenter, S. R., & Cottingham, K. L. (1997). Resilience and restoration of lakes. *Conservation Ecology*, *1*(1), 2. doi:10.5751/ES-00020-010102

Carpenter, S. R., Walker, B., Anderies, J. M., & Abel, N. (2001). From metaphor to measurement: Resilience of what to what? *Ecosystems (New York, N.Y.)*, *4*(8), 765–781. doi:10.100710021-001-0045-9

Childers, D. L., Pickett, S. T. A., Grove, J. M., Ogden, L., & Whitmer, A. (2014). Advancing urban sustainability theory and action: Challenges and opportunities. *Landscape and Urban Planning*, *125*, 320–328. doi:10.1016/j.landurbplan.2014.01.022

City of Portland. (2007). *ENN-1.01: Food Policy Council (Binding city policy)*. Retrieved September 24, 2007, from http://www.portlandonline. com/auditor/index.cfm?a=ihci&c=checj

City of Vancouver. (2003). *Action plan for creating a just and sustainable food system for the City of Vancouver (RTS Number: 3755)*. Author.

Clement, D. (2013). *The spatial injustice of crisis-driven neoliberal urban restructuring in Detroit (Unpublished master's thesis)*. Miami, FL: University of Miami; Retrieved from http://scholarlyrepository.miami.edu/oa_theses/406/

Christopherson, R. W. (1997). Geosystems: An Introduction to Physical Geography (3rd ed.). Upper Saddle River, NJ: Prentice Hall Inc.

Colding, J. (2011). The role of ecosystem services in contemporary urban planning. In Urban Ecology. Patterns, Processes and Applications. Oxford University Press. doi:10.1093/acprof:oso/9780199563562.003.0028

Colding, J., & Barthel, S. (2012). *The potential of Urban Green Commons in the resilience building of cities*. Elsevier B.V.

Colding, J., Lundberg, J., & Folke, C. (2006). Incorporating green-area user groups in urbanecosystem management. *Ambio*, *35*(5), 237–244. doi:10.1579/05-A-098R.1 PMID:16989508

Collins, J. P., Kinzig, A., Grimm, N. B., Fagan, W. F., Hope, D., Wu, J., & Borer, E. T. (2000). A new urban ecology. *American Scientist*, *88*(5), 416–425. doi:10.1511/2000.5.416

Convention on the Biological Diversity. (1992). The Earth Summit. Held in Rio de Janeiro on 5 June 1992 and entered into force on 29 December 1993. United Nations, General Assembly.

Cosgrove, S. (2001). *Montréal's community gardening program*. Retrieved July 29, 2007, from http://www.cityfarmer.org/Montreal13.html

Costanza, R., & Patten, B. C. (1995). Defining and predicting sustainability. *Ecological Economics*, *15*(3), 193–196. doi:10.1016/0921-8009(95)00048-8

Curwell, S., & Cooper, I. (1998). The implications of urban sustainability. *Building Research and Information*, *26*(1), 17–28. doi:10.1080/096132198370074

Curwell, S., Deakin, M., & Symes, M. (Eds.). (2005). *Sustainable urbandevelopment, volume1: The frameworkand protocols for environmental assessment*. Routledge. doi:10.4324/9780203299913

de Jong, M., Joss, S., Schraven, D., Zhan, Ch., & Weijnen, M. (2015). Sustainable–Smart–Resilient–Low Carbon–Eco–Knowledge Cities; Making Sense of a Multitude of Concepts Promoting Sustainable Urbanization. *Journal of Cleaner Production*, *109*(1), 25–38. doi:10.1016/j.jclepro.2015.02.004

Daune, L., & Mongé, N. (2011). L'agriculture urbaine, un fondement dans le projet de territoire. *Urbia—Les Cahiers du développement urbain durable, 12*, 85-106.

Deelstra, T., Boyd, D., & van den Biggelaar, M. (2006). Multifunctional land use, promoting urban agriculture in Europe. In Sustainable Development: Opportunities for Change. Department of the Environment, Transport and the Regions. Consultation Paper on a Revised UK Strategy, DETR, London.

Detroit Works Project. (2014). *Detroit future city: Detroit strategic framework plan*. Retrieved from http://detroitfuturecity.com/wp-content/uploads/2014/12/DFC_Full_2nd.pdf

Deutsch, L., Folke, C., & Skanberg, K. (2003). The critical natural capital of ecosystem performance as insurance for human well-being. *Ecological Economics*, *44*(2-3), 205–217. doi:10.1016/S0921-8009(02)00274-4

Dow, K. (2000). Social dimensions of gradients in urban ecosystems. *Urban Ecosystems*, *4*(4), 255–275. doi:10.1023/A:1015767231137

Elmqvist, T., Wall, M., & Berggren, A. L. (2001). Tropical forest reorganization after cyclone and fire disturbance in Samoa: remnant trees as biological legacies. *Cons Ecol*, 5, 10. http://www.consecol.org/vol5/iss2/art10

Estes, J. A., & Duggins, D. O. (1995). Sea otters and kelp forests in Alaska: Generality and variation in a community ecological paradigm. *Ecological Monographs*, *65*(1), 75–100. doi:10.2307/2937159

EurActiv. (2004). *Sustainable Development: Introduction*. Author.

Folke, C., Carpenter, S. R., Elmqvist, T., Gunderson, L., Holling, C. S., & Walker, B. (2002). Resilience and sustainable development: Building adaptive capacity in a world of transformations. *Ambio*, *31*(5), 437–440. doi:10.1579/0044-7447-31.5.437 PMID:12374053

Folke, C., Pritchard, L., Berkes, F., Colding, J., & Svedin, U. (1998). The Problem of Fit Between Ecosystems and Institutions. International Human Dimensions Programme (IHDP). *IHDP Working Paper No 2*. www.uni-bonn.de/IHDP/public.htm

Folke, C., Jansson, A., Larsson, J., & Costanza, R. (1997). Ecosystem appropriation by cities. *Ambio, 26*, 167–172.

Folke, C., Holling, C. S., & Perrings, C. (1996). Biological diversity, ecosystems and the human scale. *Ecological Applications, 6*(4), 1018–1024. doi:10.2307/2269584

Forys, E. A., & Allen, C. R. (2002). Functional group change within and across scales following invasions and extinctions in the Everglades ecosystem. *Ecosystems (New York, N.Y.), 5*(4), 339–347. doi:10.100710021-001-0078-0

Frost, T. M., Carpenter, S. R., Ives, A. R., & Kratz, T. K. (1995). Species compensation and complementarity in ecosystem function. In C. G. Jones & J. H. Lawton (Eds.), *Linking species and ecosystems.* Chapman & Hall. doi:10.1007/978-1-4615-1773-3_22

Golley, F. B. (1993). *A History of the Ecosystem Concept in Ecology: More Than the Sum of the Parts.* Yale University Press.

Graham, S., & Marvin, S. (2001). *Splintering urbanism: Networked infrastructures, technological mobilities and the urban condition.* Routledge.

Grimm, N. B., Faeth, S. H., Golubiewski, N. E., Redman, C. L., Wu, J., Bai, X., & Briggs, J. M. (2008). Global change and the ecology of cities. *Science, 319*(5864), 756–760. doi:10.1126cience.1150195 PMID:18258902

Grove, J. M. (2009). Cities: Managing densely settled social–ecological systems. In F. S. Chapin III, G. P. Kofinas, & C. Folke (Eds.), *Principles of ecosystem stewardship: Resilience-based natural resource management in a changing world* (pp. 281–294). doi:10.1007/978-0-387-73033-2_13

Gunderson, L. H., & Holling, C. S. (Eds.). (2002). *Panarchy: Understanding Transformations in Human and Systems.* Island Press.

Hansen, A. J., Defries, R., & Turner, W. (2004). Land use change and biodiversity: a syn-thesis of rates and consequences during the period of satellite imagery. In G. Gutman & C. Justice (Eds.), *Land Change Science: Observing, Monitoring, andUnderstanding Trajectories of Change on the Earth's Surface* (pp. 277–299). Springer Verlag.

Havlicek, T. D., & Carpenter, S. R. (2001). Pelagic species size distributions in lakes: Are they discontinuous? *Limnology and Oceanography, 46*(5), 1021–1033. doi:10.4319/lo.2001.46.5.1021

Hodgson, K., Caton Campbell, M., & Bailkey, M. (2011). Urban agriculture: growing healthy, sustainable places. Washington, DC: American Planning Association.

Holling, C. S. (1996). Engineering resilience versus ecological resilience. In P. C. Schulze (Ed.), *Engineering within ecological constraints.* National Academy Press.

Holling, C. S. (1973). Resilience and stability of ecological systems. *Annual Review of Ecology and Systematics, 4*(1), 1–23. doi:10.1146/annurev.es.04.110173.000245

Hough, M. (1995). *Cities and Natural Process.* Routledge. doi:10.4324/9780203292433

Hui, S. C. M. (2011). *Green roof urban farming for buildings in high-density urban cities.* Invited paper for the Hainan China World Green Roof Conference 2011, Hainan, China.

Hynes, H. P. (1996). *A Patch of Eden: America's Inner-city Gardeners.* Chelsea Green Publishing Company.

IISD. (2009). *What is Sustainable Development?* International Institute for Sustainable Development.

Jackson, J. B. C., Kirby, M. X., Berger, W. H., Bjorndal, K. A., Botsford, L. W., Bourque, B. J., Bradbury, R. H., Cooke, R., Erlandson, J., Estes, J. A., Hughes, T. P., Kidwell, S., Lange, C. B., Lenihan, H. S., Pandolfi, J. M., Peterson, C. H., Steneck, R. S., Tegner, M. J., & Warner, R. R. (2001). Historical overfishing and the recent collapse of coastal ecosystems. *Science, 293*(5530), 629–638. doi:10.1126cience.1059199 PMID:11474098

James, P., Magee, L., Scerri, A., & Steger, M. B. (2015). *Urban Sustainability in Theory and Practice.* Routledge.

Jenks, M., & Jones, C. (Eds.). (2010). Dimensions of the sustainable city. New York: Springer.

Kalin Arroyo, M. T., Zedler, P. H., & Fox, M. D. (Eds.). (1995). *Ecology and biogeography of Mediterranean ecosystems in Chile, California, and Australia.* Springer-Verlag. doi:10.1007/978-1-4612-2490-7

Kates, R., Parris, T., & Leiserowitz, A. (2005). What is Sustainable Development? Goals, Indicators, Values, and practice. *Environment, 47*(3), 8–21. doi:10.1080/00139157.2005.10524444

Kinzig, A. P., Pacala, S. W., & Tilman, D. (Eds.). (2002). *The functional consequences of biodiversity.* Princeton University Press.

Kolasa, J., & Pickett, S. T. A. (2005). Changing academic perspectives of ecology: A view from within. In M. J. Mappin & E. A. Johnson (Eds.), *Environmental education and advocacy* (pp. 50–71). Cambridge University Press.

Lambin, E. F. (2005). Conditions for sustainability of human-environment systems: Information, motivation, and capacity. *Global Environmental Change, 15*(3), 177–180. doi:10.1016/j.gloenvcha.2005.06.002

Levin, S. (1999). *Fragile dominion: complexity and the commons.* Perseus Books.

Liam, M., Andy, S., Paul, J., Thom, J. A., Padgham, L., Hickmott, S., Deng, H., & Cahill, F. (2013). Reframing social sustainability reporting: Towards an engaged approach. *Environment, Development and Sustainability, 15*(1), 225–243. doi:10.100710668-012-9384-2

Ljungqvist, J., Barthel, S., Finnveden, G., & Sörlin, S. (2010). The Urban Anthropocene: Lessons for Sustainability from the Environmental History of Constantinople. In P. Sinclair, F. Herschend, C. Isendahl, & G. Nordquist (Eds.), *The Urban Mind, cultural and environmental dynamics Studies in Global Archaeol-ogy* (Vol. 15, pp. 368–394). University Press.

Loreau, M., Naeem, S., & Inchausti, P. (2001). Biodiversity and ecosystem functioning: Current knowledge and future challenges. *Science, 294*(5543), 804–808. doi:10.1126cience.1064088 PMID:11679658

Lovell, S. T., DeSantis, S., Nathan, C. A., Olson, M. B., Mendez, V. E., Kominami, H. C., Erickson, D. L., Morris, K. S., & Morris, W. B. (2010). Integrating agroecology and landscape multifunc-tionality in Vermont: An evolving framework to evaluate the design of agroecosystems. *Agricultural Systems*, *103*(5), 327–341. doi:10.1016/j.agsy.2010.03.003

Luck, G. W., Daily, G. C., & Ehrlich, P. R. (2003). Population diversity and ecosystem services. *Trends in Ecology & Evolution*, *18*(7), 331–336. doi:10.1016/S0169-5347(03)00100-9

Lundberg, J., & Moberg, F. (2003). Mobile link organisms and ecosystem functioning: Implications for ecosystem resilience and management. *Ecosystems (New York, N.Y.)*, *6*(1), 87–98. doi:10.100710021-002-0150-4

Kahle, L. R., & Gurel-Atay, E. (Eds.). (2014). *Communicating Sustainability for the Green Economy*. M.E. Sharpe.

Machlis, G. E., Force, J. E., & Burch, W. R. Jr. (1997). The human ecosystem. 1. The human ecosystem as an organizing concept in ecosystem management. *Society & Natural Resources*, *10*(4), 347–367. doi:10.1080/08941929709381034

McCall, T. (1973). *Legislative address, Governor Tom McCall, Oregon, (1973)*. Retrieved July 20, 2008, from http://arcweb.sos.state.or.us/governors/McCall/legis1973.html

McDonnell, M. J., & Hahs, A. (2009). Comparative ecology of cities and towns: Past, present and future. In M. J. McDonnell, A. Hahs, & J. Breuste (Eds.), *Ecology of cities and towns: A comparative approach* (pp. 71–89). Cambridge University Press. doi:10.1017/CBO9780511609763.006

McKinney, M. L. (2002). Urbanization, biodiversity, and conservation. *Bioscience*, *52*(10), 883–890. doi:10.1641/0006-3568(2002)052[0883:UBAC]2.0.CO;2

Millennium Ecosystem Assessment (MA). (2005). *Synthesis*. Island Press. Available on the Internet http://www.MAweb.org

Miller, J. R. (2005). Biodiversity conservation and the extinction of experience. *Trends in Ecology & Evolution*, *20*(8), 430–434. doi:10.1016/j.tree.2005.05.013 PMID:16701413

Mukherji, N., & Morales, A. (2010). *Zoning for urban Agriculture*. Zoning Practice 3.

Musacchio, L. R. (2008). Metropolitan landscape ecology: Using translational research to increase sustainability, resilience, and regeneration. *Landscape Journal*, *27*(1), 1–8. doi:10.3368/lj.27.1.1

National Land Use Database. (2000). *Previously Developed Land (PDL): Vacant Sites and Derelict Sites, and Other Previously Developed Land and Buildings that May be Available for Redevelopment, Data Specification*. NLUD – a partnership Rethinking urban capacity.

Naveh, Z. (2000). The total human ecosystem: Integrating ecology and economics. *Bioscience*, *50*, 357–361.

Nelson, D. R., Adger, W. N., & Brown, K. (2007). Adaptation to environmental change: Contributions of a resilience framework. *Annual Review of Environment and Resources*, *32*, 395–419.

Nyström, M., & Folke, C. (2001). Spatial resilience of coral reefs. *Ecosystems (New York, N.Y.)*, *4*(5), 406–417. doi:10.100710021-001-0019-y

Odum, E. P. (1971). *Fundamentals of Ecology* (3rd ed.). W. B. Saunders.

Otte, A., Simmering, D., & Wolters, V. (2007). Biodiversity at the landscape level: Recent concepts and perspectives for multifunctional land use. *Landscape Ecology*, *22*(5), 639–642. doi:10.100710980-007-9094-6

Pagano, M. A., & Bowman, A. O'M. (2000). Land in Cities: An Urban Resource. Washington, DC: Brookings Institution, Center on Urban and Metropolitan Policy.

Peterson, G., Allen, C.R., & Holling, C.S. (1998). Ecological resilience, biodiversity, and scale. *Ecosystems, 1*, 6–18.

Pickett, S. T. A., & Cadenasso, M. L. (2002). Ecosystem as a multidimensional concept: Meaning, model and metaphor. *Ecosystems (New York, N.Y.)*, *5*(1), 1–10. doi:10.100710021-001-0051-y

Pickett, S. T. A., Burch, W. R. Jr, Dalton, S., & Foresman, T. (1997). Integrated urban ecosystem research: Themes, needs, and applications. *Urban Ecosystems*, *1*(4), 183–184. doi:10.1023/A:1018579628818

Pickett, S. T. A., Cadenasso, M. L., & Grove, J. M. (2004). Resilient cities: Meaning, models, and metaphor for integrating the ecological, socio-economic, and planning realms. *Landscape and Urban Planning*, *69*(4), 369–384. doi:10.1016/j.landurbplan.2003.10.035

Pickett, S. T. A., Groffman, P. M., Brush, G. S., Galvin, M. F., Grove, J. M., ... Troy, A. R. (2008). Exchanges across land-water-scape boundaries in urban systems: Strategies for reducing nitrate pollution. *Annals of the New York Academy of Sciences*, *1134*(1), 213–232. doi:10.1196/annals.1439.012 PMID:18566096

Pothukuchi, K., & Kaufman, J. (2000). The food system: A stranger to the planning field. *Journal of the American Planning Association*, *66*(2), 112–124. doi:10.1080/01944360008976093

Pothukuchi, K., & Kaufman, J. (1999). Placing the food system on the urban agenda: The role of municipal institutions in food systems planning. *Agriculture and Human Values*, *16*(2), 213–244. doi:10.1023/A:1007558805953

Quincerot, R., & Weil, M. (2009). *Genève 2020: Plan di-recteur communal de la Ville de Genève*. Ville de Genève.

Redman, C., Grove, J. M., & Kuby, L. (2000). *Toward a unified understanding of human ecosystems: Integrating social sciences into long-term ecological research.* White Paper of the Long-Term Ecological Research. Albuquerque, NM: LTER Network.

Redwood, M., & Bahri, A. (2010). *Wastewater Irrigation and Health: Assessing and Mitigation Risks in Low-income Countries*. Earthscan/IDRC/IWMI.

Ricketts, T., & Imhoff, M. (2003). Biodiversity, urban areas, and agriculture: Locating prior-ity ecoregions for conservation. *Conservation Ecology*, *8*(2), 1. doi:10.5751/ES-00593-080201

Rudd, J. W. M., Kelly, C. A., Schindler, D. W., & Turner, M. A. (1988). Disruption of the nitrogen cycle in acidified lakes. *Science*, *240*(4858), 1515–1517. doi:10.1126cience.240.4858.1515 PMID:17798981

Rusk, D. (1993). *Cities without Suburbs*. The Woodrow.

Sala, O. E., Chapin, F. S. I., Armesto, J. J., Berlow, E., Bloomfield, J., Dirzo, R., Huber-Sanwald, E., Huenneke, L. F., Jackson, R. B., Kinzig, A., Leemans, R., Lodge, D. M., Mooney, H. A., Oesterheld, M., Poff, N. L., Sykes, M. T., Walker, B. H., Walker, M., & Wall, D. H. (2000). Global biodiversity scenarios for the year 2100. *Science*, *287*(5459), 1770–1774. doi:10.1126cience.287.5459.1770 PMID:10710299

Seana I., Johnson, L. & Peters, K. (1999). Community gardens and sustainable land use planning: A case-study of the Alex Wilson community garden. *Local Environment*, *4*(1), 33-46. DOI: doi:10.1080/13549839908725579

Seto, K. C., De Groot, R., Bringezu, S., Erb, K., Graedel, T. E., Ramankutty, N., & Skole, D. L. (2010). Stocks, flows, and prospects of land. In T. E. Graedel & E. van der Voet (Eds.), *Linkages of sustainability* (pp. 72–96). MIT Press.

Scheffer, M., Carpenter, S. R., Foley, J. A., Folke, C., & Walker, B. H. (2001). Catastrophic shifts in ecosystems. *Nature*, *413*(6856), 591–596. doi:10.1038/35098000 PMID:11595939

Schindler, D. W. (1990). Experimental perturbations of whole lakes as tests of hypotheses concerning ecosystem structure and function. *Oikos*, *57*(1), 25–41. doi:10.2307/3565733

Smit, B., & Wandel, J. (2006). Adaptation, adaptive capacity and vulnerability. *Global Environmental Change*, *16*(3), 282–292. doi:10.1016/j.gloenvcha.2006.03.008

Sukopp, H., Hejny, S., & Kowarik, I. (Eds.). (1990). *Urban ecology: Plants and plant communities in urban environments*. SPB Academic Publishing.

Sukopp, H., Trautmann, W., & Korneck, D. (1979). The soilflora and vegetation of Berlin'swastelands. In I. Laurie (Ed.), *Nature in Cities*. John Wiley.

Tansley, A. G. (1935). The use and abuse of vegetational concepts and terms. *Ecology*, *16*(3), 284–307. doi:10.2307/1930070

Terborgh, J., Lopez, L., & Nunez, P. (2001). Ecological meltdown in predator-free forest fragments. *Science*, *294*(5548), 1923–1926. doi:10.1126cience.1064397 PMID:11729317

Termorshuizen, J. W., & Opdam, P. (2009). Landscape services as a bridge between landscape ecology and sustainable development. *Landscape Ecology*, *24*(8), 1037–1052. doi:10.100710980-008-9314-8

Tidball, G. K., Krasny, E. M., Svendsen, E., Cambell, L., & Helphand, K. (2010). Stewardship, learning and memory in disaster resilience. *Environmental Learning Research*, *16*(5–6), 591–609. doi:10.1080/13504622.2010.505437

Tidball, K. G., & Krasny, M. E. (2009). *From risk to resilience: what role for community greening and civic ecology in cities?* Wageningen Academic Publishers.

van de Berg, L., & van Veenhuizen, R. (2005). Multiple functions of urban Agriculture *Urban Agriculture Magazine*, 14.

Walker, B. H., Holling, C. S., Carpenter, S. R., & Kinzig, A. P. (2004). Resilience, adaptability and transformability in social–ecological systems. *Ecology and Society, 9*(2), 5. http://www.ecologyandsociety. org/vol9/iss2/art5/

Wu, J., & Wu, T. (2013). Ecological resilience as a foundation for urban design and sustainability. In S. T. A. Pickett, M. L. Cadenasso, & B. McGrath (Eds.), *Resilience in ecology and urban design: Linking theory and practice for sustainable cities* (pp. 211–229). Springer. doi:10.1007/978-94-007-5341-9_10

Yohe, G., & Tol, R. S. J. (2002). Indicators for social and economic coping capacity – moving toward a working definition of adaptive capacity. *Global Environmental Change*, *12*(1), 25–40. doi:10.1016/S0959-3780(01)00026-7

Zanoni, P., & Janssens, M. (2009). Sustainable DiverCities. In M. Janssens, D. Pinelli, D. C. Reyman, & S. Wallman (Eds.), *Sustainable Cities. Diversity, Economic Growthand Social Cohesion* (pp. 3–25). Edward Elgar.

Zhang, S. X., & Babovic, V. (2012). A real options approach to the design and architecture of water supply systems using innovative water technologies under uncertainty (PDF). *Journal of Hydroinformatics*, *488*. www.frontiersinecology.org

Zimov, S. A., Chuprynin, V. I., Oreshko, A. P., Chapin, F. S. III, Reynolds, J. F., & Chapin, M. C. (1995). Steppe–tundra transition: An herbivore-driven biome shift at the end of the Pleistocene. *American Naturalist*, *146*(5), 765–794. doi:10.1086/285824

KEY TERMS AND DEFINITIONS

Eco-Urban: Those urban environments that are characterized by having a series of characteristics that define them as sustainable and ecological.

Green Resilience: Green resilience is based on the infrastructure of urban green areas applied to cities and is that they have the capacity to prepare, resist and recover from a crisis.

Land Use: Encompasses the management and modification of the natural environment to convert it into agricultural land: arable fields, grasslands; or human settlements. The term land use is also used to refer to the different land uses in zoning.

Resilience: The word resilience refers to the ability to overcome critical moments and adapt after experiencing some unusual and unexpected situation. It also indicates return to normal.

Urban: Makes reference to that belonging to or relating to the city.

Urban Ecology: Is a discipline whose object of study is the interrelationships between the inhabitants of an urban agglomeration and its multiple interactions with the environment.

Urban Socio-Ecological Systems: Are a complex urban structures that can be analyzed considering the social subsystem and the ecological subsystem.

Chapter 13
Feasibility Study of Visual Computing and Machine Learning Application for Textile Material Sorting

Siu Cheung Ho
The Hong Kong Polytechnic University, Hong Kong

Jiannong Cao
The Hong Kong Polytechnic University, Hong Kong

ABSTRACT

This project aims to study the feasibility of visual computing (VC) and machine learning (ML) method applied in the textile recycle industry for efficiently manages the post-consumer textile waste. It includes an image-based VC technology for supporting textile waste reuse and resale, and a material identification system for sorting textile materials by using near infrared (NIR)/hyperspectral spectroscopy technology to support efficiently recycling to reuse the textile fibre will be evaluated. The process involved collecting and validating reference samples and applying ML technique to auto recognize the garment type and features applying visual technology; afterward, the sorted garments would be measured and pre-treated by NIR/hyperspectral spectrum and building up the parameters for spectral patterns calculation for recycling process recover the fibre. The main part of the study is to proof of the concept for using VC and ML method for identifying the textile fibre in the recycling process.

DOI: 10.4018/978-1-7998-4915-5.ch013

INTRODUCTION

Nowadays, textile recycling in real world is a hot topic in these decades. Recycling raw material of textile consists of many types of materials with various blending fibre combinations. The most efficient way would be reuse and resale the post-consumer garment to best utilize the resources. For the garment that is not suitable for reuse or resale, the fibre fabrication process is needed. That needs a material identification for materials recycling process to keep the best quality of the recycle fibre. The current material sorting operation is handled by manual and haven't a system to sort the material automatically. The textile material sorting error rate will be increased. The labor cost of manual handling of textile material is very high and labor intensive. Automation for material sorting of textile recycling industry is necessary.

BACKGROUND OF RESEARCH

According to Helena et al (2017), it stated that "*the sorting of post-consumer garment and household textile waste for recycling has limited accuracy. Manual sorting relies to a large extent on label information attached to the garment and such information is often missing in collected goods. Erroneous labels and human errors in the manual sorting add to the difficulties of achieving the requested quality. A central issue is whether the quality requirements of the recycling processes can be met. This is especially critical for textile to textile recycling, which generally requires high-quality sorted textile waste fractions. For potential chemical fibre recycling processes, these are likely to need high quality-sorted fractions of post-consumer textile waste to produce new textiles. It provided that a material recognition technology can sort for textile to textile applications, this technology would be a promising solution for the textile sorting industry. The near-infrared (NIR) technology is still under development for textile sorting and a fully automated system has not yet been demonstrated and validated.*" The visual computing technology of NIR and hyperspectral are not mature in the market for textile materials identification and classification. High cost and recycle efficiency are the key issue in the textile recycles industry. Textile materials identification automation is important for improving the performance for recycle process.

AIMS AND OBJECTIVES

This project aims to study the feasibility of using VC and ML method applying in Textile Recycle Industry for efficiently manages the post-consumer textile waste. It includes an image-based VC technology for supporting textile waste reuse and resale, and a material identification system for sorting textile materials by using NIR / hyperspectral spectroscopy to support efficiently recycling to reuse the post-consumer textile fibre. The process involved collecting and validating reference samples, apply ML technique to auto recognize the garment type in visual technology, afterward, the sorted garments would be measured and pre-treated by NIR / hyperspectral spectrum, and building up the parameters for spectral patterns calculation for textile recycle process recover the fibre. The main part of the project is to proof of the concept for using VC and ML method for identifying the textile fibre in the recycle process. Figure 1 is the conceptual architecture of VC and ML for textile material sorting. The incoming bulk garment will be input to this vision system for training up the proposed model and database for textile material identification.

Figure 1. Conceptual architecture of visual computing and machine learning for textile material sorting

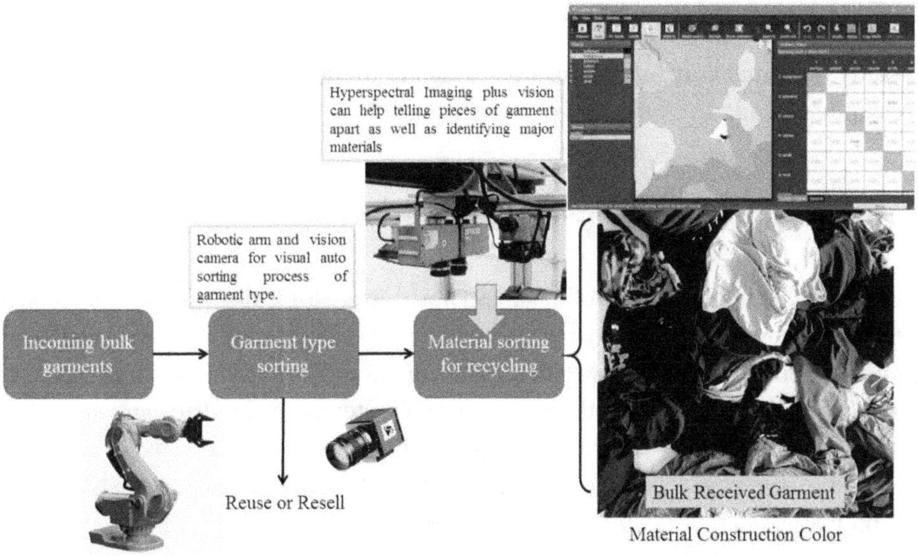

In order to facilitating the post-consumer garment for recycling, the garment type and material need to be classified first. The key feature of the collected recycle textile (See Figure 2) should be analysed by using visual computing. For the garment that couldn't be reused or resale, mechanical recycling is common way to recycle the fibre. The process involves mechanically deconstructing the fabrics, in order to end up with re-useable fibres and material that are ready to be used to make new yarn and fabric. It is an effective way to bring used materials back into the cycle.

Figure 2. Example of main feature and footprint of the bulk incoming garment

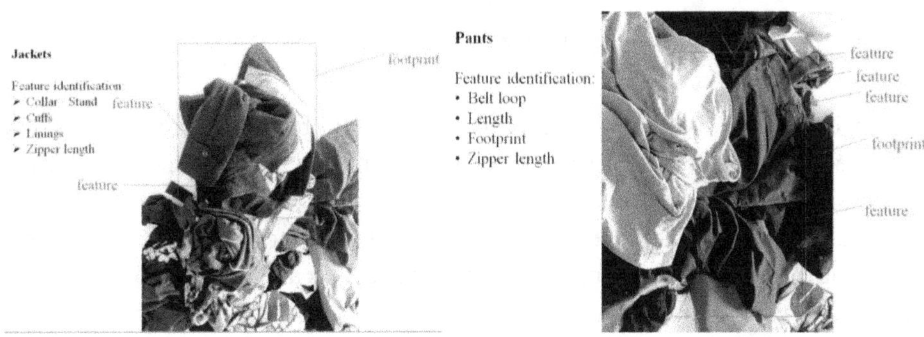

The proposed VC and ML material identification system using NIR / hyperspectral spectroscopy to identify the fibre and materials texture of the textile quickly and accurately followed the material characteristic spectrum. The identification method requires a standardized measurement distance, corrections for varying physical properties among the samples and identification tolerances that allow natural spectral variation but restrict impurities. It will cover a broad range of textile materials.

LITERATURE REVIEW

This study review the current issue and existing technology application in the textile recycling industry. The related study on visual computing, machine learning model, inspection and recognition for textile recycle were conducted and evaluated.

THE PROBLEM FACED OF THE INDUSTRY

Textile gathering, reuse, resale and recycling have a positive ecological and economic impact. Reuse and resale are regarded as the ways mostly keep the value of textile wastes. An auto identification garment type system could handle the textile waste in an occupational healthy way, meanwhile, efficient and environmentally friendly. This reduces the amount of all materials on landfills and hazardous emissions that are a consequence of natural materials rotting which produces methane and of a long-term decomposition of synthetic materials. Moreover, by reusing textile materials, current textile recycling sorting mainly relies on manual handling. Textile material recycling operation process involved a hug amount of manpower and the labour cost is very huge. The labour intensive textile materials recycle operation has many influence factors need to be considered as follow:

1. **Occupation Health and Safety**: Dusty in the recycling process to influence the mankind health
2. **Low Performance**: Slow textile material sorting speed and high rate of error such as materials color sorting, materials and textile type classification (e.g.: knitted high-twist vs woven (fabric opening steps)), and a small number of unambiguous elements can be processed in a sorting stage.
3. **Accuracy of Garment Type and Material Sorting**: Subject to human judgement and non-standard.
4. **Inadequate Information of Recycled Textile Materials**: Most recycled textile is scrapped and no content label in the cloth. Material identification and classification are difficult.

EXISTING SOLUTIONS AND DRAWBACKS

Textile waste consists of many types of garments and with various blending combinations. For the textile recycling process, all material sorting processes are relied on manual for handling the style and materials. The labor cost of manual handling of textile material is very high and labor intensive. Manual sorting is subjective to cause error rate increase. Therefore, automation process sorting for textile recycling is necessary. The garment type and textile material sorting process is a critical step in many types of textile recycling industry. Accuracy is important for greater value recovery through better separation of different garment type and range of textile material. Automation of sorting additionally helps in reducing worker issue. The existing automation material sorting process is commonly used NIR camera to capture the material spectrum for identification purpose. The drawbacks are the accuracy is very low. Since, the limitation of NIR, which could be influenced by many factors such as luminosity issue, fuzzy image, and effective data analysis, it is difficult to recognize the spectrum of textile materials or texture sorting in the factory. It needs a large and systematic database to analysis the spectrum.

RELATED STUDIES ON VISUAL COMPUTING AND MACHINE LEARNING APPLICATION FOR TEXTILE MATERIAL SORTING

Visual computing and machine leaning technology application are not new, but this is a hot topic in this decade in Hong Kong. The relevant research study for visual computing and machine leaning in textile recycling and materials sorting are related to three areas including:

1. Spectral Imaging and Convolutional Neural Network (CNN) Model Application (See Table 1)
2. Spectral Imaging Classification and Machine Learning (ML) Model (See Table 2)
3. Visual Computing and Machine Learning (ML) Model for Textile Inspection and Recognition (See Table 3)

DESCRIPTION OF THE PROPOSED R&D

ML is a part of Artificial Intelligence (AI). This means all ML counts as AI, but not all AI counts as ML. A VC and ML base garment type materials identification system with a comprehensive garment type and textile materials database is proposed. The preliminary system of Hyperspectral Camera with Halogen Lamp system and experiments results detail are shown in the part of Research & Methodology. The system development process involved collecting and validating reference samples, measuring and pre-treating NIR / hyperspectral spectra, and setting up parameters for matching spectral patterns. The goal is to achieve a robust and reliable method for sorting garment type such as cotton, polyester and wool. These are the fibre major textile waste from post-consumer textile materials collection. It is an applied research which can perform sustainable practice to recycle textile wastes. This research conceptual framework is divided into 4 stages:

1. **Stage 1**: Study the existing garment type imaging and material identification technologies, i.e. NIR imaging / hyperspectral spectroscopy.
2. **Stage 2:** Set up a garment type and fabric material database, build a garment features image and NIR / hyperspectral spectroscopy database for material identification
3. **Stage 3:** Develop a VC and ML material identification system for garment type and textile materials sorting. Figure 3 is the recycling material sorting process flow and proposed system modules.
4. **Stage 4**: For VC and ML material identification system will be developed to recognize garment type and fabric construction. It also can be compared with the content label information together for garment type identification.

Figure 4 is the referenced image process and NIR and hyperspectral spectroscopy module. The camera is planned to use the SPECIM FX Series hyperspectral cameras and high spectrum range NIR camera to enable the machine vision technology for materials screening and recording by providing more accurate information to detect whether the materials are matched with the database record. The spectrum identifies the reflection of the light of the materials for examining how light behaves in the target and recognizes the materials followed by their particular spectral signatures. These spectral signatures should

Table 1. Related Study on Spectral Imaging and CNN Model Application

Author	Methodology	Key Findings and Insights
Alam et al. (2016)	Hyperspectral Data and Convolutional Neural Network (CNN) Model	*"...a method that uses both spectral and spatial information to segment remote sensing hyperspectral images. After a hyperspectral image is over-segmented into superpixels, a deep Convolutional Neural Network (CNN) is used to perform superpixel-level labelling. This paper....attempts to combine the properties of CNN and Conditional Random Field (CRF)"*
Mei et al. (2017)	Hyperspectral Data and Convolutional Neural Network (CNN) Model	*"Inspired by recent achievements in convolutional neural network (CNN) based super resolution (SR), a novel CNN based framework is constructed for SR of hyperspectral images by considering both spatial context and spectral correlation."*
Paoletti et al (2019)	Hyperspectral Data and Convolutional Neural Network (CNN) Model	*"Convolutional neural networks (CNNs) exhibit good performance in image processing tasks, pointing themselves as the current state-of-the-art of deep learning methods."*
Paoletti et al (2019)	Capsule Networks (CapsNets), Convolutional Neural Networks (CNNs), Hyperspectral Imaging (HSI)	*"...we propose a new CNN architecture based on spectral–spatial capsule networks in order to achieve a highly accurate classification of HSIs while significantly reducing the network design complexity."*
Alipourfard et al. (2018)	Hyperspectral Image Classification, Convolutional Neural Network (CNN) Model	*"...combines the proposed CNN architecture and subspace reduction method to prepare informative features (from subspace method) and designing optimized CNN by considering limitation of training samples."*
Yu et al (2019)	Hyperspectral Image Classification (HSIC), Convolutional Neural Network (CNN) Model	*"The extensive experiments on real hyperspectral images results demonstrate that the proposed CNN network can effectively improve the classification accuracy after the embedding of the extracted semantic features."*
Li et al. (2018)	Hyperspectral Classification, Convolutional Neural Network (CNN) Model	*"...novel hyperspectral classification framework combining CNN and the orthogonal complement subspace projection (OCSP) balance solution... for further improving the classification accuracy. Experimental results demonstrate that the proposed framework can improve hyperspectral classification performance of CNN."*
Rao et al. (2019)	Hyperspectral Image Classification (HSIC), Convolutional Neural Network (CNN) Model	*"To augment the amount of input data, models operate on pixel-pairs have been proposed and Siamese convolutional neural network (S-CNN) is a typical one. S-CNN is used as a pixel-pair feature.... Taking advantage of the pairwise input and to make better use of spatial information, a voting strategy using neighbouring pixels is also employed to determine the final class label of the centre pixel."*
Mei et al. (2019)	Hyperspectral Image Processing, Convolutional Neural Network (CNN) Model	*"...sensor specific based transfer learning is proposed for hyperspectral images acquired from same sensors; in which knowledge learns from hyperspectral images..."*
Feng et al. (2019)	Hyperspectral Image Classification, Convolutional Neural Network (CNN) Model	*"...multilayer spatial-spectral features are fused to achieve an end-to-end classification, which learn complementary information among the shallow layers with basic information and the deep layers with abstract information...."*
Xu et al (2019)	Hyperspectral Image Classification, Convolutional Neural Network (CNN) Model	*"CNN based approaches may lead to over-smoothing effect in HSI classification task."*
Pooja et al (2019)	Hyperspectral Image Classification, Convolutional Neural Network (CNN) Model	*"This paper exploits CNN-based method along with multi-scale and dilated convolution with residual connection concepts for hyperspectral image classification on exclusive real time data set."*

be recognized the detected spectrum of the material for classification. The garment type image and key features can be identified (Reference to Figure 4).

Figure 5 is the preliminary NIR / hyperspectral spectroscopy module design and infrastructure proposed from Specim (2019). NIR and hyperspectral imaging system will be evaluated in this project to

Table 2. Related study on spectral imaging classification and ML model

Author	Methodology	Key Findings and Insights
Clark et al. (2010)	Hyperspectral, Feature Selection, Dimensionality Reduction	*"The Non-correlated Aided Simulated Annealing Feature Selection (NASAFS) method selects features in a pairwise manner, where the results of each pair are combined into a database of 'distinguishable features' for discriminating the reference class from all other classes. The database is then used by the detection algorithm to categorize unseen data samples. Although Simulated Annealing (SA) can be processing intensive, it is not a limiting factor in the processing of the feature set....Hyperspectral signal for 80% Polyester 20% Rayon blend with the respective feature set coefficients for the different feature selection methods..."*
Chen et al. (2015)	Hyperspectral Imagery (HSI) Classification, Spatial Feature Extraction	*"Hyperspectral imagery (HSI) captures a dense spectral sampling of reflectance values over a wide range of spectrum....From the results, we can see that the performance of classification with spatial features is much better than that with the spectral signatures only..."*
Rahaman et al. (2017)	Spectral Imaging, Band Selection	*"The extended CIE color matching functions based visualization method is most effective to render each multivariate data point by a single color. The defective areas of the surface in the resulting images appear strong to be detected readily"*
Zeng et al. (2017)	Hyperspectral Imaging Classification, Multiple Kernel Fuzzy Discriminant	*"...the results showed that the performance of fuzzy discriminant analysis with multiple kernels is better than that of fuzzy discriminant with single kernel for the Hyperspectral images' feature extraction and classification."*
Chu et al. (2019)	Hyperspectral, Texture, Non-uniformity	*"As validation, a texture classification scheme based on nearest neighbor classifier is applied on HyTexiLa dataset using rSDOM. The performance is close to Opponent Band Local Binary Pattern (OBLBP) with classification accuracy of 94.7%, but at a much-reduced feature size (0.24% of OBLBP's) and computational complexity."*

define which components should be suitable in the proposed system or combine both functions together. The basic functions of both cameras are 1) NIR systems perform spectroscopy on the entire scanning area and 2) Hyperspectral imaging system is able to perform per-pixel scanning.

NOVELTIES / INNOVATIONS OF THE PROPOSED R&D

In order to solve the material identification issue by using hyperspectral imaging technology, it has many literature and study in this decade for applying the method of hyperspectral imaging solutions in different recycle industry application. These include hyperspectral systems, hyperspectral cameras, a huge number of datasets and spectrographs. The proposed VC and ML garment type and materials identification system with a comprehensive garment type and textile materials database is used to detect the materials by using complex VC and ML garment identification and material sorting algorithm for building up

Table 3. Related study on VC and ML model for textile inspection and recognition

Author	Methodology	Key Findings and Insights
Karayiannis et al. (1999)	Feature Extraction and Classification, Neural Network	*"...defect detection and classification of web textile fabric in real-time is presented. The general hardware and software platform, developed for solving this problem, is presented while a powerful novel method for defect detection after multiresolution decomposition of the fabric images..."*
Fujiwara et al. (2004)	Visual Image Inspection	*"...visual inspection method of textile surfaces using the translation invariant Wavelet Shrinkage is presented...To deal with this problem, we use RI-Spline wavelets which are pseudo Complex wavelets consist of a pair of a symmetric bi-orthogonal spline wavelet and an anti-symmetric bi-orthogonal spline wavelet, for textile surface inspection..."*
Liu and Zhang (2008)	Mathematical model of quasi-regular patterns	*"It related quasi-regular pattern evolution mechanism with textile pattern design to translate formerly invisible information in athematic area into visible textile pattern of art design. So the visual field on design could be effectively expanded, and formed the new design concept and thinking"*
Liu et al. (2009)	Learning Vector Quantization (LVQ) neural network	*"In our experiments, we use 625 images of five grades and extract energy values from high frequency suabbands and take them as feature sets to train and test a LVQ neural network. Two criteria, including identification accuracy of each grade and average identification accuracy of five grades are used to evaluate the performance of the trained LVQ neural network"*
Wang et al. (2009)	Texture Features Extraction and Classification, Gray-Level Co-occurrence Matrix (GLCM)	*"To detect the interlacing area where weft yarn and warp yarn are crossed over each other, we apply a spatial domain integral projection approach"*
Sabeenian and Paramasivam (2010)	Gray- Level Co-occurrence Matrix (GLCM)	*"The proposed work succeeded in classifying the fabrics with repeated patterns as defective or non-defective based on the MRCSF that was performed."*
Su et al. (2011)	Visual Inspection, Back-propagation Neural Network	*"...use back-propagation neural network to classify type of textile texture. Firstly, wavelet transfer is applied to obtain vertical, horizontal and diagonal images of original image, and compute its wavelet energy to take them as texture features of this image. Finally, the backpropagation neural network is adopted to recognize texture feature of this image..."*
Lin et al. (2011)	Focus of Attention (FOA), Regions of Interest (ROI)	*"...image features were extracted by non-uniform sampling of image and computing "center-surround", which were used to simulate the human visual receptive field properties. Features became multi-scale conspicuity maps, which were combined into a saliency map..."*
Priya et al. (2011)	Fabric defects, Image Analysis, Bitplane Decomposition	*"...automated visual inspection systems play a great role in assessing the quality of textile fabrics. For the detection of fabric defects, we first decompose the image into its bit planes. The lower order bit planes are found to carry important information of the location and shape of defects..."*
Xue et al. (2012)	Fuzzy Inclusion Measure, Rank correlation measure (RC)	*"...the rough sets theory and the fuzzy sets theory was proposed in order to investigate from the evaluation data the extent to which the tactile properties of a fabric can be interpreted through specific visual representations of the corresponding apparel product..."*
Priya et al. (2012)	Bitplane Image Analysis, Weighted Morphology	*"...A method for defect detection and localization in woven fabric is proposed in this paper based on bitplane decomposition and weighted morphology. The algorithm is superior to the existing ones in terms of computational time and accuracy..."*
Liu and Lughofer (2015)	Visual Computing	*"...hierarchical feed-forward layer model represented by multiple linear regression to investigate the relationship between human aesthetic texture perception and computational low-level texture features..."*

Figure 3. VC and ML application in textile materials recycling process

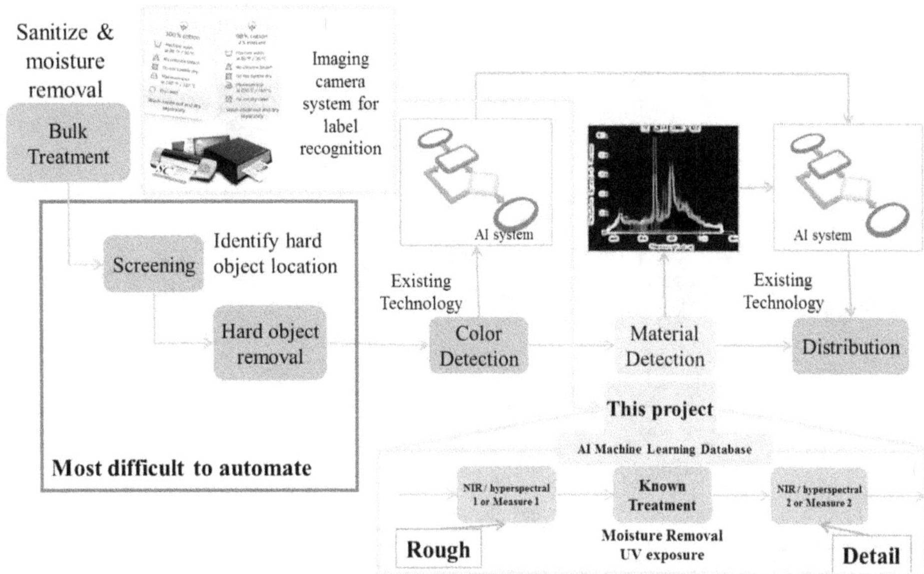

a machine learning module for textile recycling. The dataset is combined the data from garment type, fabric content label data and captured image of textile garment and material by using NIR / hyperspectral spectroscopy camera. The database will be built. Base on different garment type and material sorting methodologies and datasets including the data of 1) Optical Character Recognition (OCR) from garment image and fabric content label, 2) Artificial Neural Network (ANN) / CNN Algorithm, 3) garment image and textile material spectrum imaging from different spectral and light source, 4) Visual identification module by using NIR / hyperspectral camera with different image, wavelength, pixel, frame size. The detail is shown in section 3.5.

Figure 4. Proposed NIR and hyperspectral spectroscopy module (Specim, 2019)

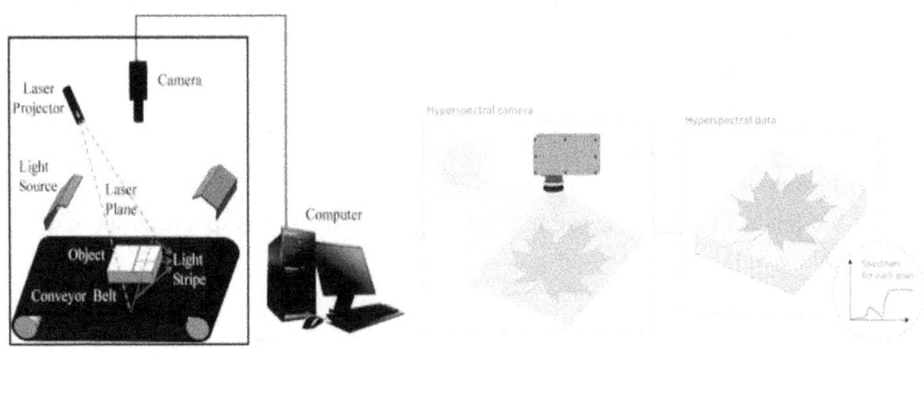

Figure 5. Preliminary NIR/Hyperspectral spectroscopy module and infrastructure (Specim, 2019)

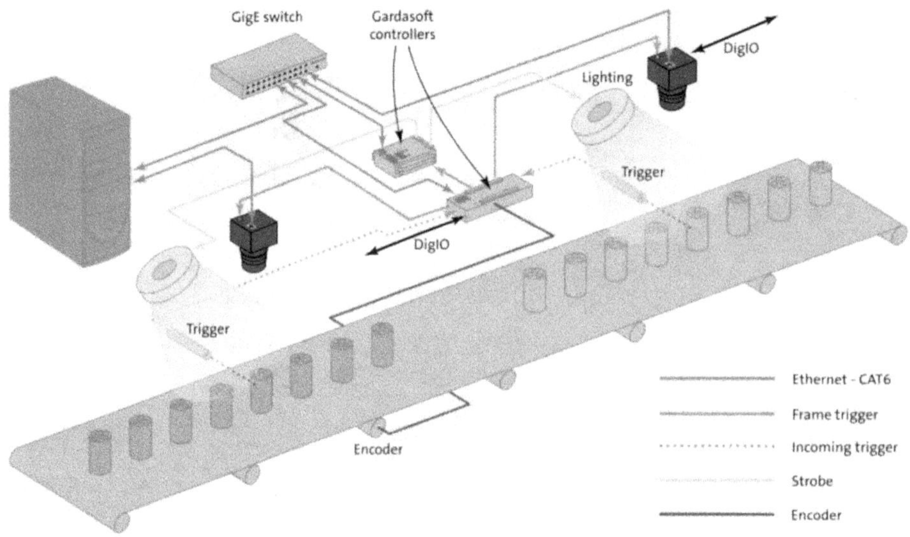

DESCRIPTION OF ANY EXISTING SIMILAR WORK

According to Hans Markowski (2012) report, it stated that "*Textiles for Textiles (T4T) project, a technology will be developed and demonstrated to sort post-consumer textile waste on chemical composition and / or colour. Discarded textiles are collected by many charity and commercial organizations. Wearable clothes are sorted and sold. The rest-fraction, often more than 50% of the collected textiles, end-up in low added value applications, which most often do not cover the costs of collecting, sorting and processing. In order to increase the value of this fraction, a lower cost sorting approach is needed. The materials sorting system is based on NIR-Identification, a method often used in plastics recycling, but new for the textile recycling industry. In the project much attention is paid to the further processing of sorted fractions into high added value textile and non-textile products in order to show the rationale behind the sorting process and to prove the economic sustainability of sorting these materials.*"

Nowadays, using AI, VC, and ML with hyperspectral imaging to identify the textile material is a limit. The normal operation is to classify the material texture (See Figure 6) by using imaging analytics without AI and different spectral and light source for material identification. Deep Learning is one of the proposed analytic methods to:

1. Study the supervised and unsupervised methodology such as material features, material characteristics and material particular spectrum identification
2. Analysis the hidden layer of the image captured from the NIR / hyperspectral for creating the dataset (Reference: Figure 3H)

Figure 6. Classify the material texture

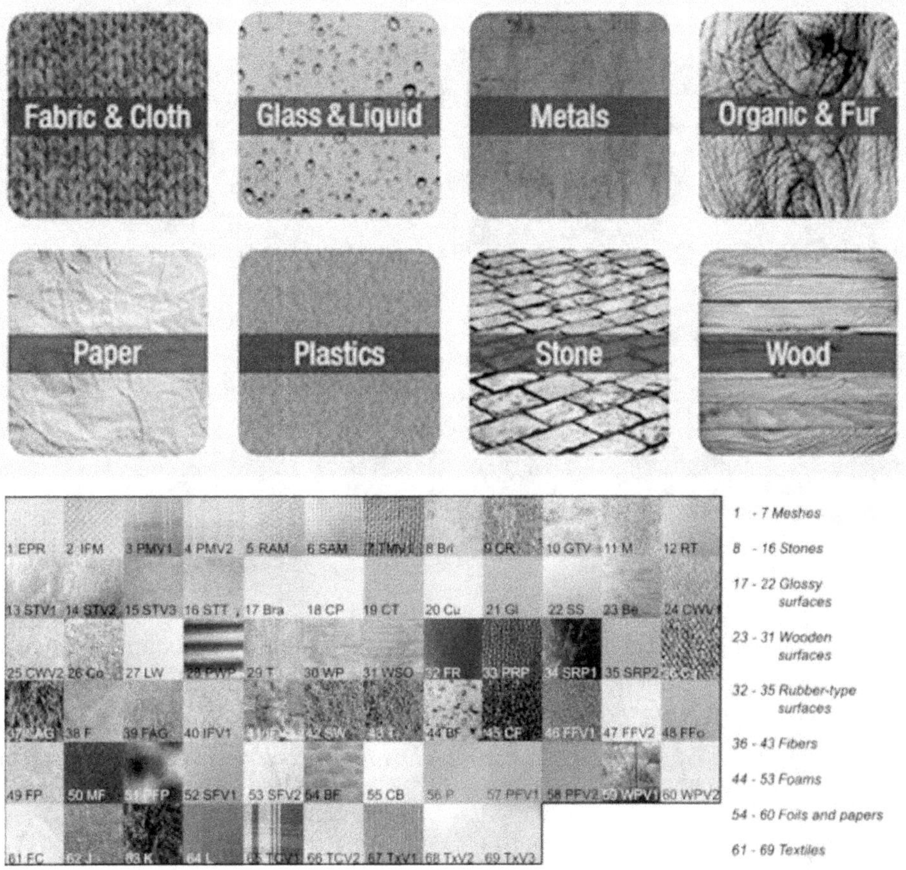

KEY DIFFERENCE OF THE PROPOSED R&D FROM THE EXISTING SIMILAR WORK

In this project, VC and ML garment and material identification system will be developed to detect the garment type and textile materials and identify the material texture for building up a comprehensive garment type and textile material database for textile material sorting. A visual textile identification system will be developed to identify garment type and fabric construction and combine the content label data for identification recycle textile material. For dataset identification of image, the VC and ML model need to be developed. A unique dataset for recycle textile, garment, and fibre by the database analysing the captured images that have been manually labelled and using the results to identify the material type of garment type, garment feature (See Figure 7) and textile material in other images. Base on the different garment feature image and spectral (See Figure 8), lighting, pixel, color and frame data (See Figure 10) to verify and identify the garment type and textile material and texture record in the database. By using CNN algorithm, hidden layer (See Figure 9) of neuron will be analysed and compared with upper and lower layer once by one. The system does this without any prior knowledge about garment type and textile material, for example, that the garment is suit, shirt, pants, and, material is cotton, nylon, polyester major

Figure 7. Garment Features for Identification

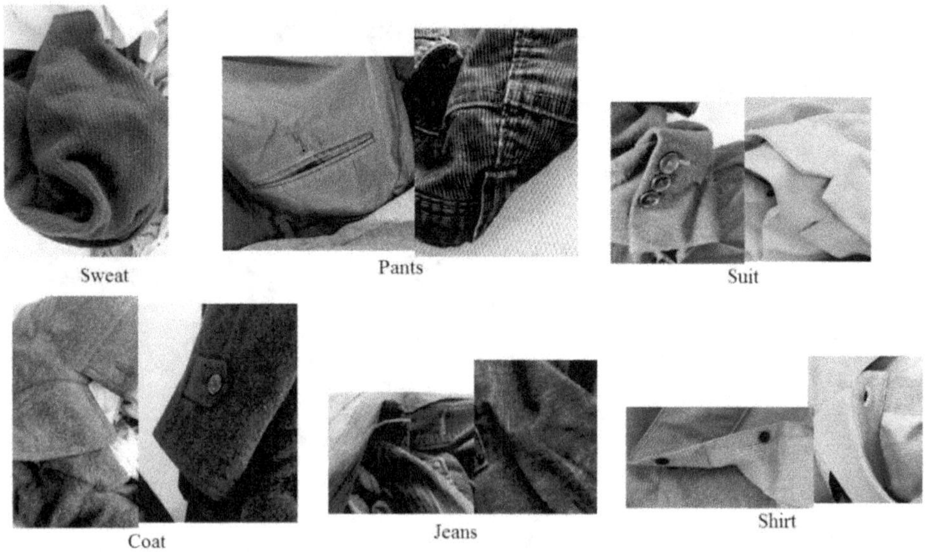

and so on. Instead, the system will be automatically generated identifying characteristics and garment features from the learning material that the database is created in the recycling process.

Refer to Middleton (2020), "*Machine vision applications span a wide range of research and manufacturing fields. Machine vision can be used to automate the identification of visible defects on the product,*

Figure 8. Difference between multispectral and hyperspectral data

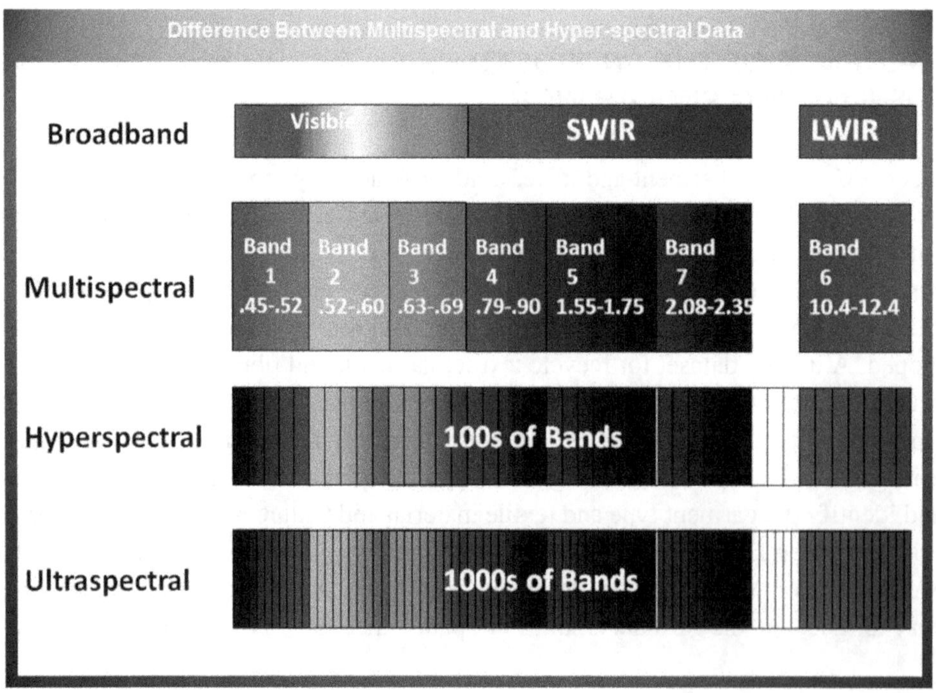

monitor a specific stage in the production process, or improve quality control of material recycle and other products. In machine vision systems, a camera is usually placed over the object or product line and acquires data via stationary or line-scan imaging. A computer controls the data collection, object or camera movement (if necessary) and analyses the data to report the specific results, such as material identification, a list of defect locations on the object or a warning of off-specification products." Figure 8 is the concept framework of CNN, the image color, feature, and characteristics are one of proposed method to use color bins, different layers, frame and pixels to build up a textile material data base to continuous analysis each hidden layer."

Figure 9. CNN – VC and ML material sorting algorithm for machine learning

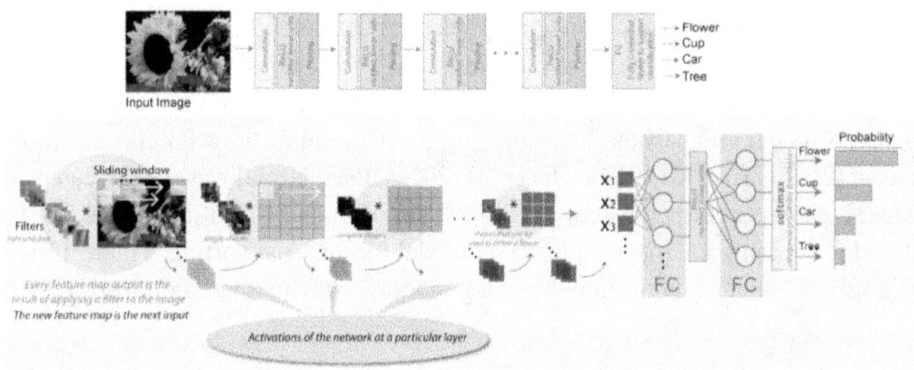

Figure 10 is the diagram of ANN / CNN. It has three basic layers including 1) Input Layer, 2) Hidden layer, and 3) Output layer (Priyanka and Nagaraj 2017). Chang-Yun (Chang-Yun et al., 2005) stated that "*Between software architecture conception and concrete realization, semantic gap exists, the structure problem spreads to inner parts of operating system and middleware, and software architecture cannot be observed and controlled by end user.*" Refer to Pat Research (Pat Research, 2018), "*AI platform was classified as either weak AI / narrow Ai, which was for a particular task or unfamiliar tasks to find out*

Figure 10. Data analysis for each hidden layer one by One and ML training

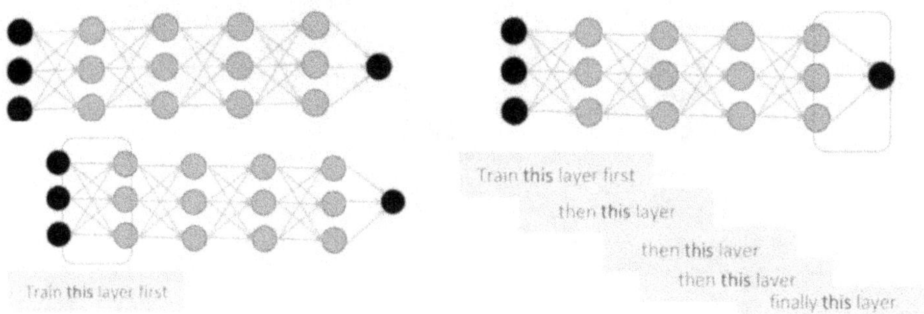

the solution. ML platform could be applied in Machine learning, automation, natural language processing and natural language understanding, and cloud infrastructure."

VISUAL COMPUTING AND MACHINE LEARNING MODEL

Refer to Jia and Ding (2004), the specific band of wavelength of different fibres has the different material spectrum characteristic. Regarding the potential material spectral bands for foreign fibres detection, it is important to find the most appropriate spectrum band for discriminating foreign fibres from cotton that is important in the sorting process. The absorption discrimination between cotton and foreign fibres at different wavelengths in NIR region was analysed. The problem of NIR wavelength selection for a wide range of foreign fibres detection techniques was evaluated. Refer to Alam et al (2016), Shan et al (2017), Mei et al (2019); hyperspectral imaging processing is a useful tool to capture the materials texture for identifying the material nature and characteristics. The material texture has a special wavelength through the hyperspectral imaging process. The captured image can be combined with different image dataset using CNN model for analysis and evaluating the material features for classification. According to Rahaman et al (2017), Yu et al (2019), Chua et al (2019), the material wavelength can be measured by the spectral and hyperspectral spectroscopy. Through the pixel analysis of the image, the hidden layers will be analysed for training the machine learning model for material sorting. Figure 11 is the research concept using spectral image and machine learning model for textile material identification.

Figure 11.Research concept for material feature and texture identification through VC and ML model

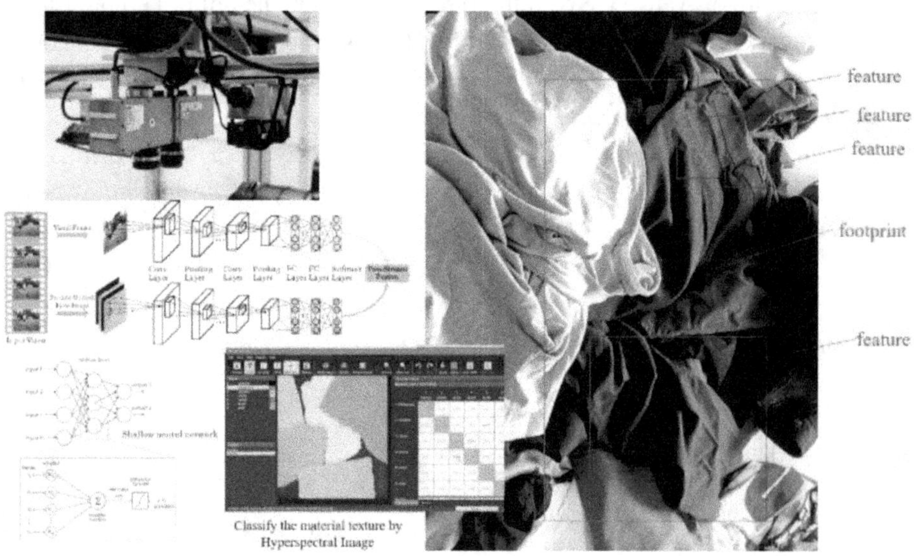

R&D METHODOLOGY

Task 1, Task 2, and Task 3 has been tested. The preliminary experiment was conducted. Task 4 and Task 5 are the proposed experiment method and it will be evaluated in the future study.

Task 1: I conducted literature review on possible VC and ML material identification system and requirements of developing machine learning algorithm. The finding is the CNN, NIR, hyperspectral camera and the potential tool for identification the recycle textile by using VC and ML methodology.

Task 2: I conducted literature review on NIR imaging / hyperspectral imaging and spectroscopy module for materials classification. The basic system infrastructure was developed.

Task 3: Design and sketch a systematic context and materials recognition architecture system that including hyperspectral imaging camera, sensors, lighting components, and imaging camera. A prototype of VC and ML material identification system is tested and evaluated.

Task 4: Define the detail specification of the VC and ML garment style and material identification system such as garment style, content label OCR annotation interface, spectral graph imaging materials composition identification model, content label & spectral graph data pipeline and infrastructure.

Task 5: Design and develop detail system specification of the VC and ML garment style and material identification system such as content label object character recognition annotation interface, spectral graph imaging materials composition identification model, content label & spectral graph data pipeline and infrastructure, API deployment & configuration interface.

Figure 12. Using NIR to identify different textile material

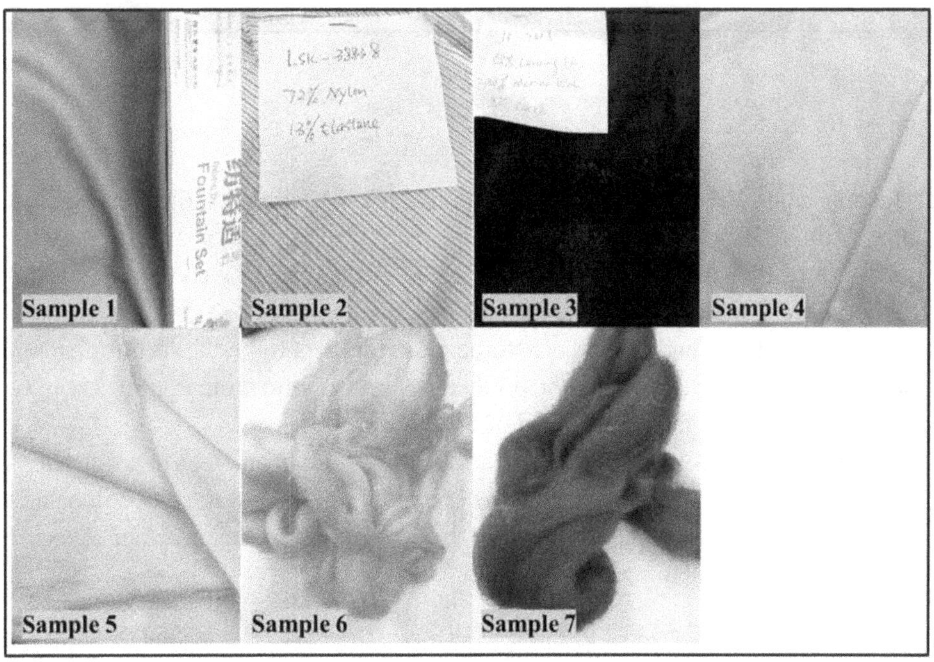

PRELIMINARY RESULTS AND SUPPORTING DATA

For the garment imaging and feature study, I worked with Hong Kong Applied Science and Technology Research Institute (ASTRI) Intelligent Sensing Technology team to use NIR camera to analysis different textile materials spectrum. The sample materials, test results, and textile and fibre materials spectrum are shown.

NIR TESTING RESULTS

In Table 1, sample 1, 2 and 3 has the content label to show the material information. Sample 4 to 7 haven't the material ingredient content information. After used the NIR camera, the results of sample 1, 2, 4, 5 and 7 are almost identified, but the disadvantage is the black dye material has strong absorption for the lighting, the material cannot be identified.

Table 3. Textile material result using NIR spectrum camera

Sample	Composition	Test Results
1	80% Nylon, 20% Spandex	Nylon dominant
2	72% Nylon, 13% Elastane	Nylon dominant
3	68% Lenzing FR, 28% Merino wool 4% Lycra	Unknown, the black dye has strong absorption
4	No label, unknown	Combination of cotton and polyester
5	No label, unknown	Cotton
6	No label, unknown	Polyester
7	No label, unknown	Wool

The summary for the testing results are as follow:

1. The reflection-mode NIR absorption method is capable to distinguish different textiles. To obtain the composition information quantitatively and correctly, a large database is required.
2. The dye in the textile may affect the testing result. A database for commonly used dyes is also needed in order to get a more accurate analysis result.
3. To obtain high sensitivity and avoid the disturbance of ambient light, the spectrometer probe should be as close as to the textile surface. In the production line, robotic arms equipped with spectrometer probes are recommended for highly efficient sorting.
4. There are two suggestions in solving the mentioned moisture problem. a) Design a drying process before the sorting section to reduce the effect of moisture; b) Take water as a composition of textile and collect a corresponding database, then use the developed machine learning algorithm to obtain the composition information.

Figure 13. Textile and fibre material wavelength

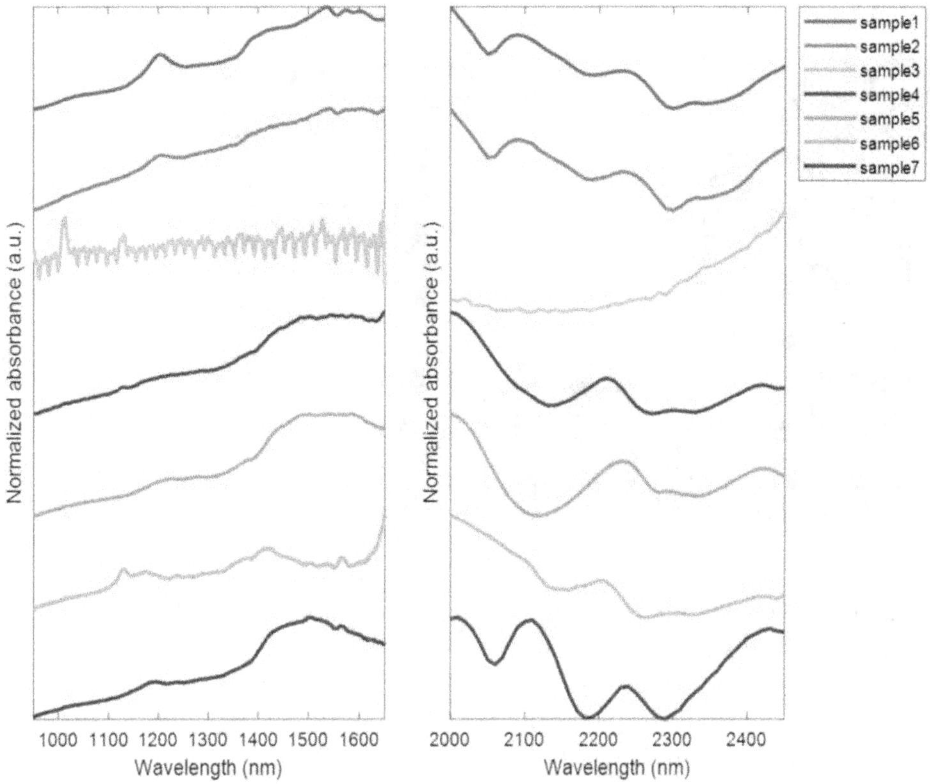

HYPERSPECTRAL CAMERA TESTING RESULTS

The preliminary results have been conducted by using the SPECIM FX10E series (Specim, 2019) camera to capture the material imaging shown in Figure 4C. Refer to Specim (2019), the FX series basic specification is as follow: "*Hyperspectral camera represents a breakthrough in hyperspectral imaging. It is a small size; wide field of view and excellent signal to noise open this camera up to a wide array of applications.*

1. **High Speed:** The FX-10e outputs 330 FPS with the Full Image mo de. With a MROI mode the speed can be up to several thousand frames per second depending on the number of selected wavebands. For examples with 20 bands the speed is about 2800 FPS. Fast Optics: FX10e has high quality optics with an F-number 1.7, which enables good signal with short integration times. High Signal to Noise Ratio (600:1) enables better detection accuracy on high speeds.
2. **Flexibility:** Free wavelength selection from 220 bands within the camera coverage. Plug 'n' Play: Each FX10e unit has the same spectral wavelength calibration, which means that all the units give identical results. Automatic image enhancement provides correction for keystone and reduced distortion.
3. **Size**: FX10 has much smaller footprint than traditional hyperspectral cameras and weight only 1.4 kg.

Figure 14. Spectrum fingerprint identification

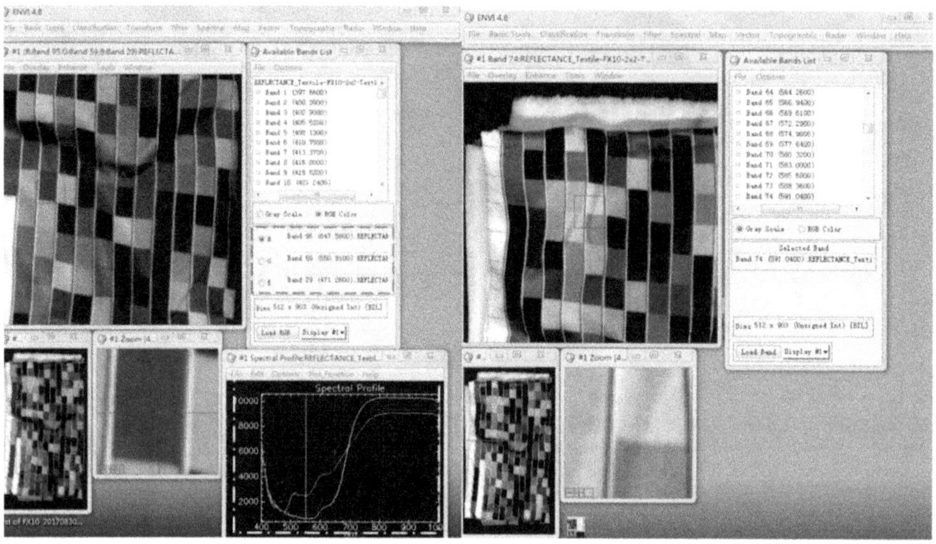

4. **Integration:** FX10e has various options for controlling software, including Lumo ToolKit, Lumo SDK and a separate ASCII protocol. The FX10 is controlled via GigE Interface."

Figure 4C was the experiments result tested by Specim (Specim 2019). Band 1 (397.66000), 397.66000 is central wavelength. For every band has the same spatial information: same rows and same lines. Extracted 1 same pixel of all the bands and make them connected in the experiment. Then, the spectrum of that pixel, which is showed in the yellow dotted box of the Test of FX10-screenshot, with the spectrum, the qualitative and quantitative data can be analysed.

Figure 15. Textile material identification and spectrum detection

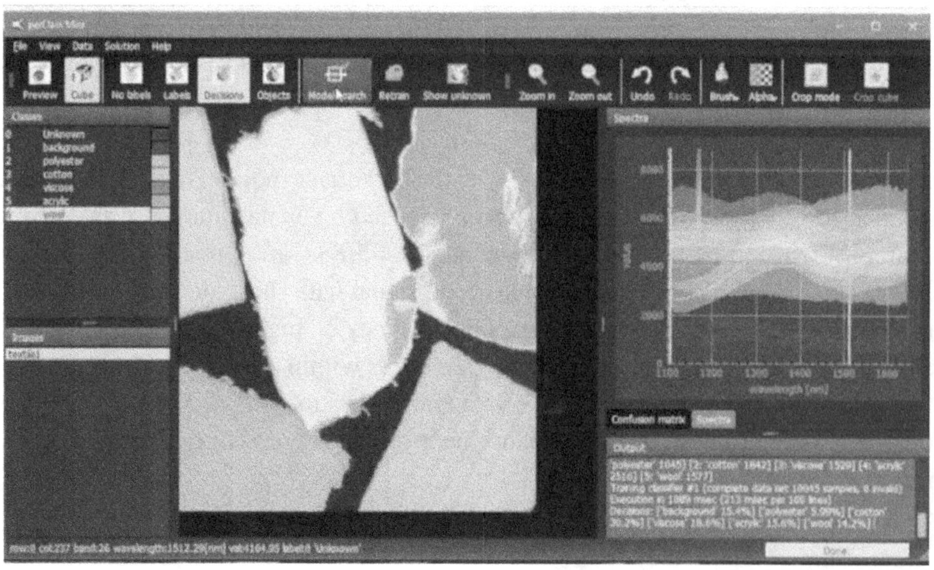

Figure 16. Dataset matrix of all attributes

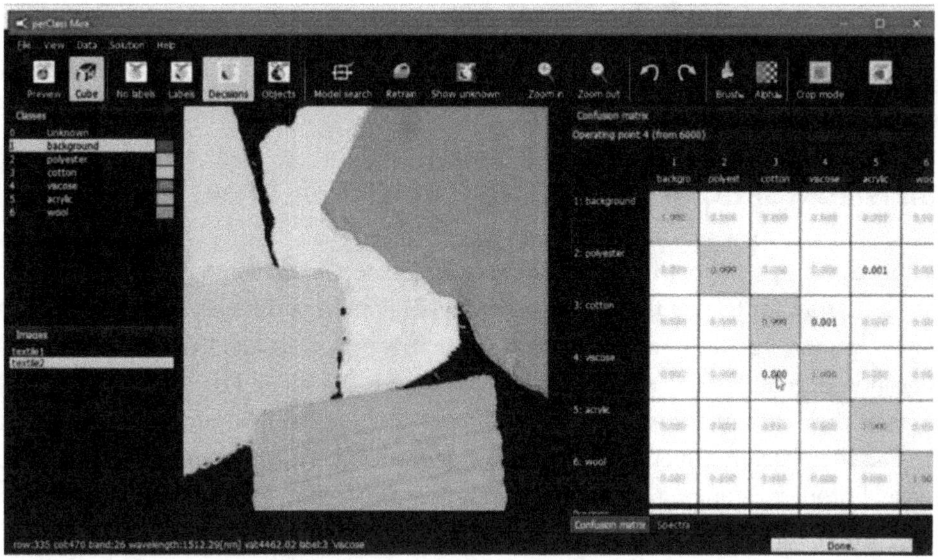

NIR imaging / hyperspectral spectroscopy camera / sensors have been used in the VC and ML material identification system for preliminary experiments testing. The cameras / sensors captured the image and measurement of reflectance relative to the textile materials shown in the Figure 4D. The spectrum

Figure 17. Preliminary system of hyperspectral camera with halogen lamp

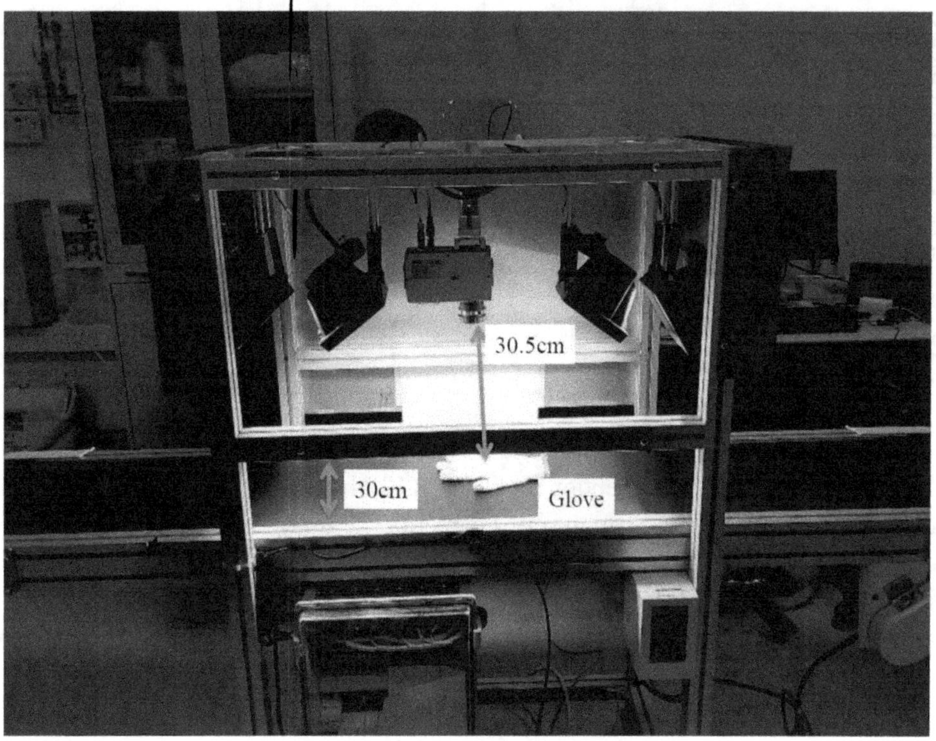

reflection of each material can be identified and all material wavelengths will be showed the system for comparison.

Figure 15 is the example to show the same material and feature matching for comparing whether the same material can be identified.

IN-HOUSE HYPERSPECTRAL IMAGING CAMERA SETTING AND EXPERIMENTS

Base on the above findings, the preliminary system of hyperspectral camera with halogen lamp was established in the laboratory for testing the textile materials. Figure 16 is the system architecture to stimulate the recycle material through the conveyor belt for the textile materials identification. The distance of the object was proposed to 30.5cm, the width of the conveyor as 30cm, and the sample was used a glove.

The setting of the experiments was tested on 14 April 2020 in the laboratory. The dimension is listed on Table 4. The identification method requires a standardized measurement distance, corrections for varying physical properties among the samples and identification tolerances that allow natural spectral variation but restrict impurities. It will cover a broad range of textile materials.

Table 4. Hyperspectral camera system dimension

Model	FX10E	Unit	Remark
Measurement Distance:	305	mm	Camera above the conveyor
Field of view (FOV)	38	degree	
Image line length(L)	206.596568	mm	Width
Spatial Sampling	1024	px	
Spatial Resolution (SR)	0.202	mm/px	
Conveyor speed	24.3	mm/s	
Frame rate	120.443	Hz	Maximum 324 Hz
Exposure time	5	ms	Maximum 8.303 ms
Spectral binning	1	set	1x, 2x, 4x, 8x
Spatial binning	1	set	1x, 2x, 4x, 8x
Sample Length	150	mm	
Sample Imaging Time	6.17283951	s	
Sample Frame Count	743.477985	s	

Figure 16 used the software of perception part and FX10E hyperspectral camera to identify the glove image. This is unique software for Specim series camera. The image of glove and spectral range were showed in the screen. If the graphics card is not powerful enough, the image may not be defected. The warning signal will be showed in red.

Through the RGB color identification for defining each pixel material type of the image in different color spectrum for building up the database. Each pixel color range of the image can be classified in RGB color range for training up the model. Figure 18 showed the pixel image color range, the spectrum range

Figure 18. Hyperspectral camera with perception park software

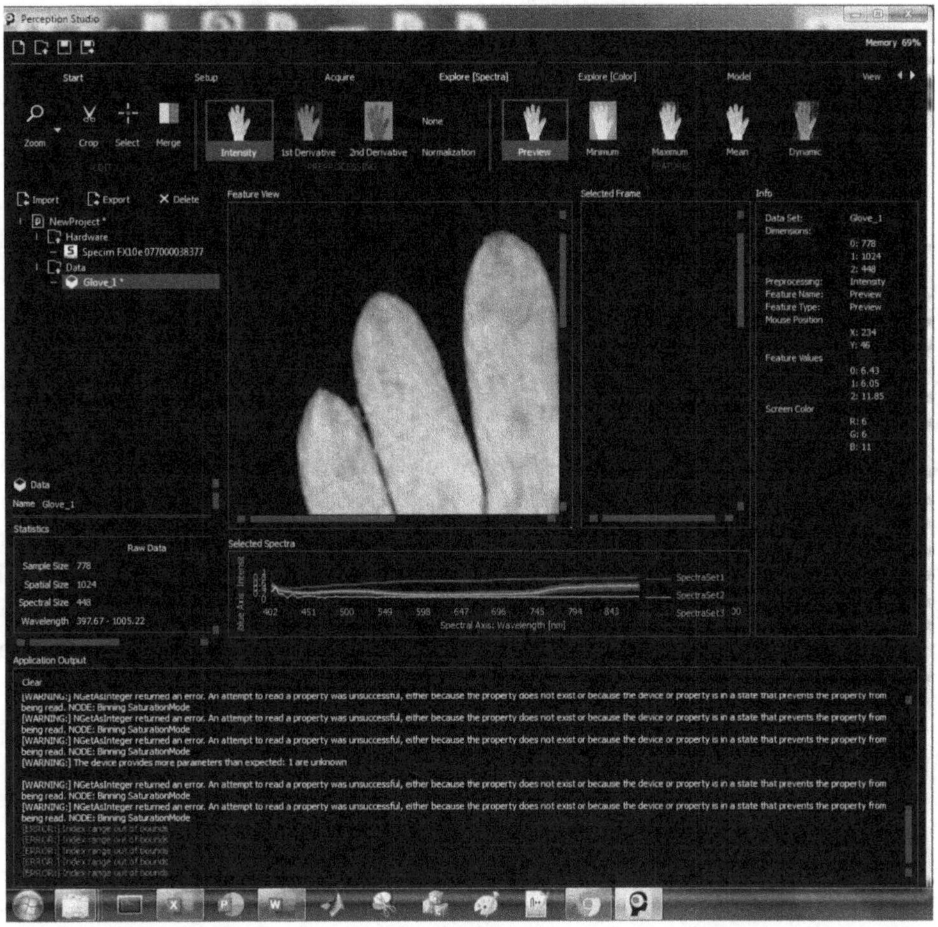

Figure 19. Imaging identification through RGB and spectrum imaging

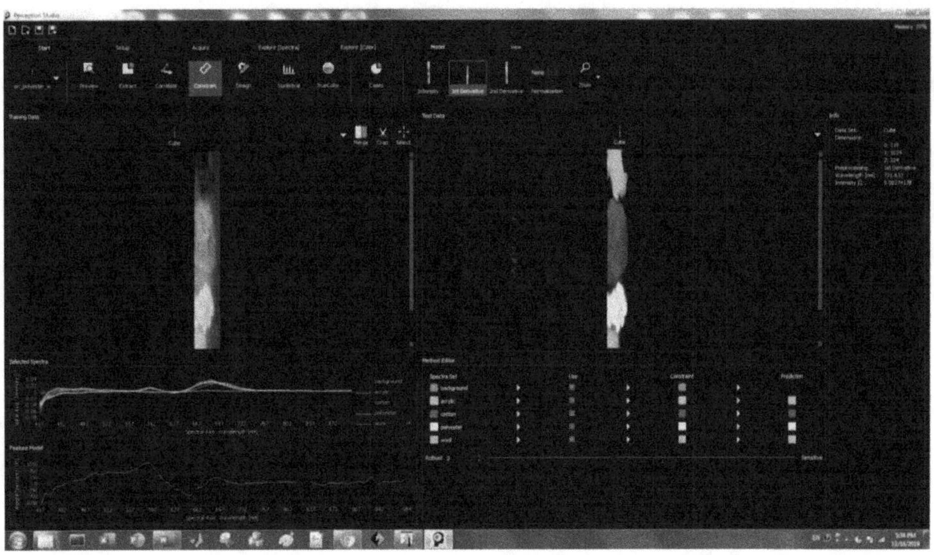

was fluctuated when the object was moving forward. Therefore, the data volume is very large for using hyperspectral imaging for textile material. The material moving speed, measurement distance, imaging time, frame count time and exposure time were critical setting and parameter before the experiment.

CONCLUSION AND FUTURE WORK

In order to solve the material identification issue by using hyperspectral imaging technology, it has many literature and study in this decade for applying the method of hyperspectral imaging solutions in different recycle industry application. These include hyperspectral systems, hyperspectral cameras, a huge number of datasets and spectrographs. The proposed VC and ML combination feasibility study of garment type and materials identification with a comprehensive garment type and textile materials database is used to detect the materials by using complex garment identification and material sorting algorithm for building up a machine learning module for textile recycling. The dataset is combined the data from garment type, fabric content label data and captured image of textile garment and material by using NIR / hyperspectral spectroscopy camera. The database will be built. Base on different garment type and material sorting methodologies and datasets including the data of 1) OCR from garment image and fabric content label, 2) ANN / CNN algorithm, 3) garment image and textile material spectrum imaging from different spectral and light source, 4) Visual identification module by using NIR / hyperspectral camera with different image, wavelength, pixel, frame size.

The proposed technology is novel to the market and is capable to leverage labour shortage, which is common in many sectors of textile and materials recycling industry.

Upon completion of the project, the project deliverables would be disseminated in technical seminars or exhibitions. Relevance with government policies and in line to re-industrialization of textile industry, it helps to address textile waste problem (landfill problem) and recover materials for making new clothing. This will support innovation and technology development in Hong Kong and benefit to social community and industry. In Hong Kong, textile recycling rate is low and the issue is gaining greater and greater concern. This project promotes the further deployment of innovative and advanced technologies to spearhead the development of "re-industrialization" in Hong Kong.

REFERENCES

Alam, F. I., Zhou, J., Liew, A. W.-C., & Jia, X. (2016). CRF Learning with CNN Features for Hyperspectral Image Segmentation. *IEEE Conference Publications*, 6890 – 6893. 10.1109/IGARSS.2016.7730798

Alipourfard, T., Arefi, H., & Mahmoudi, S. (2018). A novel Deep Learning Framework by Combination of Subspace-based Feature Extraction Convolutional Neural Network for Hyperspectral Images Classification. *IEEE Conference Publications*, 4780 – 4783. 10.1109/IGARSS.2018.8518956

Chen, C., Jiang, J., Zhang, B., Yang, W., & Guo, J. (2015). Hyperspectral Image Classification Using Gradient Local Auto-Correlations. *IEEE Conference Publications*, 454 - 458. 10.1109/ACPR.2015.7486544

Chua, R. J., Richard, N., Fernandez-Maloigne, C., & Hardeberg, J. Y. (2019). A Metrological Measurement of Texture in Hyperspectral Images Using Relocated Spectral Difference Occurrence Matrix. *IEEE Conference Publications*, 1-6. 10.1109/ICIP.2019.8803378

Clark, J. D., & Mendenhall, M. J. (2010). Stochastic Feature Selection with Distributed Feature Spacing for Hyperspectral Data. *IEEE Conference Publications*, 1 – 4. 10.1109/WHISPERS.2010.5594951

Feng, J., Wu, X., Chen, J., Zhang, X., & Tang, X. (2019). Joint Multilayer Spatial-Spectral Classification of Hyperspectral Images Based on CNN and CONVLSTM. *IEEE Conference Publications*, 588 - 591.

Hisanaga, F., & Zhong, Z. (2004). Visual inspection of Textile Surfaces with Translation Invarient Wavelet Shrinkage. *IEEE Conference Publications*, 1252 - 1257.

Jia, D. Y., & Ding, T. (2004). Detection of foreign fibres in cotton using NIR optimal wavelength imaging. *IEEE Conference Publications*, 751 – 752.

Karayiannis, Y. A., Stojanovic, R., Mitropoulos, P., Koulamas, C., Stouraitis, T., Koubias, S., & Papadopoulos, G. (1999). Defect Detection and Classification on Web Textile Fabric using Multiresolution Decomposition and Neural Networks. *IEEE Conference Publications*, 765 - 768. 10.1109/ICECS.1999.813221

Li, J., Du, Q., Xi, B., & Li, Y. (2018). Hyperspectral Image Classification via Sample Expansion for Convolutional Neural Network. *IEEE Conference Publications*, 1 - 5. 10.1109/WHISPERS.2018.8747245

Lin, Q., Xu, X., & Zhan, Y. (2011). Extracting Regions of Interest Based on Visual Attention Model. *IEEE Conference Publications*, 313 - 316.

Liu, J., & Edwin, L. (2015). Affective Property Computation of Visual Texture. *IEEE Conference Publications*, 52 - 57.

Liu, J., Zuol, B., Vroman, P., Rabenasolov, B., & Zeng, X. (2009). Visual Quality Recognition of Nonwovens Based on Wavelet Transform and LVQ Neural Network. *IEEE Conference Publications*, 1885 - 1890. 10.1109/ICCIE.2009.5223507

Liu, S., & Zhang, L. (2008). Textile Pattern Generation Technique Based on Quasi-regular Pattern Theory and Their Transform. *IEEE Conference Publications*, 264 – 266.

Markowski, H. (2012). *Textiles for Textiles (T4T).* www.textiles4textiles.eu

Mei, S., Liu, X., Zhang, G., & Du, Q. (2019). Sensor-speciðc Transfer Learning for Hyperspectral Image Processing. *IEEE Conference Publications*, 1-6.

Mei, S., Yuan, X., Ji, J., Wan, S., Hou, J., & Du, Q. (2017). Hyperspectral Image Super-resolution via Convolutional Neural Network. *IEEE Conference Publications*, 4297 – 4301. 10.1109/ICIP.2017.8297093

Middleton. (2020). *Enhanced Machine Vision (AI).* https://www.middletonspectral.com/applications-project-examples/applications/enhanced-machine-vision-ai/

Paoletti, M. E., Haut, J. M., Fernandez-Beltran, R., Plaza, J., Plaza, A. J., & Pla, F. (2019, February). Deep Pyramidal Residual Networks for Spectral–Spatial Hyperspectral Image Classification. *IEEE Journal of Transactions on Geoscience and Remote Sensing*, *57*(2), 740–754. doi:10.1109/TGRS.2018.2860125

Paoletti, M. E., Haut, J. M., Fernandez-Beltran, R., Plaza, J., Plaza, A. J., & Pla, F. (2019, April). Capsule Networks for Hyperspectral Image Classification. *IEEE Journal of Transactions on Geoscience and Remote Sensing*, *57*(4), 2145–2160. doi:10.1109/TGRS.2018.2871782

Pooja, K., Nidamanuri, R. R., & Mishra, D. (2019). Multi-scale Dilated Resuudal Convolutional Neural network for Hyperspectral Image Classification. *IEEE Conference Publications*, 1 – 5.

Priya, Ashok, & Varghese. (2011). A Novel Approach to Fabric Defect Detection Using Digital Image Processing. *IEEE Conference Publications*, 228 - 232. 10.1109/ICSCCN.2011.6024549

Priya, Ashok, & Varghese. (2012). Defect Detection In Wooven Fabric Using Weighted Morphology. *IEEE Conference Publications*, 1 - 6.

Rahaman, Atiqur, Jussi, Hauta-Kasari, & Seyed Hossein. (2017). Enhanced color visualization by spectral imaging: an application in cultural heritage. *IEEE Conference Publications*, 1-6. 10.1109/ICIVPR.2017.7890870

Rao, M., Tang, L., Tang, P., & Zhang, Z. (2019). ES-CNN: An End-to-End Siamese Convolutional Neural Network for Hyperspectral Image Classification. *IEEE Conference Publications*, 1 – 4. 10.1109/JURSE.2019.8808991

Sabeenian, R. S., & Paramasivam, M. E. (2010). Defect detection and Identification in Textile Fabrics using Multi Resolution Combined Statistical and spatial Frequency Method. *IEEE Conference Publications*, 162 - 166. 10.1109/IADCC.2010.5423017

Specim. (2019). *FX-10e VNIR Hyperspectral Camera (GigE Interface)*. https://www.middletonspectral.com/product/fx-10e/

Su, T.-L., Hong, G.-B., Chang, W.-Y., & Kung, F.-C. (2011). Automatic Recognition of Textile Texture Using Back-propagation Neural Network. *IEEE Conference Publications*, 52 – 55.

Wang, X., Georganas, N. D., & Petriu, E. M. (2009). Fiber-level Structure Recognition of Woven Textile. *IEEE Conference Publications*, 1 – 6.

Wedin, H., Gupta, C., Mzikian, P., Englund, F., Hornbuckle, R., Troppenz, V., Kobal, L., Krečič, M., Costi, M., Ellams, D., & Olsson, S. (2017). *Can automated NIR technology be a way to improve the sorting quality of textile waste?* https://static1.squarespace.com/static/5891ce37d2b857f0c58457c1/t/5a0049d26c3194747a7de384/1509968342157/october_2017_rise_report.pdf

Xu, Y., Du, B., & Zhang, L. (2019). Simultaneous Segmentation and Edge Detection for Hyperspectral Image via a Deep Supervised and Boundary-constrained Network. *IEEE Conference Publications*, 3860 – 3863. 10.1109/IGARSS.2019.8899054

Xue, Z., Zeng, X., Koehl, L., & Chen, Y. (2012). Development of a fuzzy inclusion measure for investigating relations between visual and tactile perception of textile products. *IEEE Conference Publications*, 108 – 113. 10.1109/MFI.2012.6342995

Yu, Zhao, Song, Wang, Li, Han, & Chang. (2019). Hyperspectral Image Classiðcation Method Based on CNN Architecture Embedding With Hashing Semantic Feature. *IEEE Journal of Selected Topics in Applied Earth Observation and Remote Sensing, 12*(6), 1866 – 1881.

Zeng, S., Bai, J., Jiang, L., & Kang, Z. (2017). Multiple Kernel Fuzzy Discriminant Analysis for Hyperspectral Imaging Classification. *IEEE Conference Publications*, 1-6. 10.1109/FUZZ-IEEE.2017.8015719

Compilation of References

Acharyya, J. (2009). FDI, growth and the environment: Evidence from India on CO_2 emission during the last two decades. *Journal of Economic Development, 34*(1), 43–58. doi:10.35866/caujed.2009.34.1.003

Adger, W. N. (2000). Social and ecological resilience: Are they related? *Progress in Human Geography, 24*(3), 347–364. doi:10.1191/030913200701540465

Adger, W. N. (2006). Vulnerability. *Global Environmental Change, 16*(3), 268–281. doi:10.1016/j.gloenvcha.2006.02.006

Adger, W. N., Hughes, T. P., Folke, C., Carpenter, S. R., & Rockstrom, J. (2005). Social– ecological resilience to coastal disasters. *Science, 309*(5737), 1036–1039. doi:10.1126cience.1112122 PMID:16099974

Aeschelmann, F., & Carus, M. (2015). Bio-based building blocks and polymers in the world. *Capacities, production and applications: Status quo and trends towards 2020.*

Ahmed, M., & Shimade, K. (2019). The effect of renewable energy consumption on sustainable economic development; Evidence from emerging and developing economies. *Energies, 12*(15), 2954. doi:10.3390/en12152954

Ahuja, D., & Tatsutani, M. (2009). Sustainable energy for developing countries. *SAPIENS, 2*(1). Retrieved from https://journals.openedition.org/sapiens/823

Ajayi, M. T. (1987). Effects of spacing/number of plants per stand and fertilizer placement on performance of two maize varieties intercropped with cowpea in the forest zone of Nigeria. *Ghana Journal of Agricultural Science, 20*, 39–46.

Akbostanci, E., Tunç, G. I., & Türüt-Asik, S. (2004). *Pollution haven hypothesis and the role of dirty industries in Turkey's exports*. ERC Working Papers 0403, ERC-Economic Research Center, Middle East Technical University, revised Feb 2004.

Akers, J. (2013). Making markets: Think tank legislation and private property in Detroit. *Urban Geography, 34*(8), 1070–1095. doi:10.1080/02723638.2013.814272

Alaimo, K., Packnett, E., Miles, R. A., & Kruger, D. J. (2008). Fruit and vegetable intake among urban community gardeners. *Journal of Nutrition Education and Behavior, 40*(2), 94–101. doi:10.1016/j.jneb.2006.12.003 PMID:18314085

Alam, F. I., Zhou, J., Liew, A. W.-C., & Jia, X. (2016). CRF Learning with CNN Features for Hyperspectral Image Segmentation. *IEEE Conference Publications*, 6890 – 6893. 10.1109/IGARSS.2016.7730798

Alam, M. J., Begum, I. A., Buysse, J., Rahman, S., & Huylenbroeck, G. V. (2011). Dynamic modeling of causal relationship between energy consumption, CO_2 emissions and economic growth in India. *Renewable & Sustainable Energy Reviews, 15*(6), 3243–3251. doi:10.1016/j.rser.2011.04.029

Alberti, M. (2006). *Advances in Urban Ecology. Integrating Humans and Ecological Processes in Urban Ecosystems.* Springer.

Aldabas, M., Gstrein, M., & Teufel, S. (2015). Changing Energy Consumption Behaviour: Individuals' Responsibility and Government Role. *Journal of Electronic Science and Technology*, *13*(4), 343–348.

Algert, S., Baameur, A., & Renvall, M. (2014). Vegetable output and cost savings of community gardens in San Jose, California. *Journal of the Academy of Nutrition and Dietetics*, *114*(7), 1072–1076. doi:10.1016/j.jand.2014.02.030 PMID:24751664

Alipourfard, T., Arefi, H., & Mahmoudi, S. (2018). A novel Deep Learning Framework by Combination of Subspace-based Feature Extraction Convolutional Neural Network for Hyperspectral Images Classification. *IEEE Conference Publications*, 4780 – 4783. 10.1109/IGARSS.2018.8518956

Alkon, A., & Agyeman, J. (2011). *Cultivating food justice: Race, class, and sustainability*. MIT Press. doi:10.7551/mitpress/8922.001.0001

Allen, J. O., Alaimo, K., Elam, D., & Perry, E. (2008). Growing vegetables and values: Benefits of neighborhood-based community gardens for youth development and nutrition. *Journal of Hunger & Environmental Nutrition*, *3*(4), 418–439. doi:10.1080/19320240802529169

Allen, P. (2010). Realizing justice in local food systems. *Cambridge Journal of Regions, Economy and Society*, *3*(2), 295–308. doi:10.1093/cjres/rsq015

Almenar, E., Samsudin, H., Auras, R., & Harte, J. (2010). Consumer acceptance of fresh blueberries in bio-based packages. *Journal of the Science of Food and Agriculture*, *90*(7), 1121–1128. doi:10.1002/jsfa.3922 PMID:20393992

Almfraji, M. A., & Almsafir, M. K. (2014). Foreign direct investment and economic growth: Literature review from 1994 to 2012. *Procedia: Social and Behavioral Sciences*, *129*, 206–213. doi:10.1016/j.sbspro.2014.03.668

Al-Mumin, A., Khattab, O., & Sridhar, G. (2003). Occupants' behavior and activity patterns influencing the energy consumption in the Kuwaiti residences. *Energy and Building*, *35*(6), 549–559. doi:10.1016/S0378-7788(02)00167-6

Al-shaleh, Y.M. (2011). An Empirical Insight into the Functionality of Emerging Sustainble Innovation Systems: The Case of Renewable Energy in Oil-Rich Saudi Arabia. *Int. J. Transit. Innov.*, 302–320.

Altieri, M. A., & Nicholls, C. I. (2004). *Biodiversity and Pest Management in Agroecosystems*. Haworth Press. doi:10.1201/9781482277937

Altieri, M. A., & Toledo, V. M. (2011). The agroecological revolution in Latin America: Rescuing nature, ensuring food sovereignty and empowering peasants. *The Journal of Peasant Studies*, *38*(3), 587–612. doi:10.1080/03066150.2011.582947

Altieri, M., & Nelso Companioni, A. (1999). The greening of the "barrios": Urbanagriculture for food security in Cuba. *Agriculture and Human Values*, *16*(2), 131–140. doi:10.1023/A:1007545304561

Ambrose, D. C. (2020). Biodegradable Packaging – An Eco-Friendly Approach. *Current Agriculture Research Journa*, *8*(1), 4-6.

Anderies, J. M., Janssen, M. A., & Ostrom, E. (2004). A framework to analyze the robustness of social-ecological systems from an institutional perspective. *Ecology and Society*, *9*(1), 18. http://www. ecologyandsociety.org/vol9/iss1/art18/

Angelakis, A. N., Koutsoyiannis, D., & Tchobanoglous, G. (2005). Urban wastewater and stormwater technologies in ancient Greece. *Water Research*, *39*(1), 210–220. doi:10.1016/j.watres.2004.08.033 PMID:15607179

Angelofd. (2012). Environmental innovation: in search of a meaning. *World Journal of Entrepreneurship, Management and Sustainable Development*, *8*(2).

Ansuategi, A., Greño, P., & Houlden, V. (2015). *The impact of climate change on the achievement of the post-2015 sustainable development goals*. CDKN & HR Wallingford.

Arendt, H. (1970). *The Human Condition. The University of Chicago Press.*

Armstrong, D. (2000). A survey of community gardens in upstate New York: Implications for health promotion and community development. *Health & Place, 6*(4), 319–327. doi:10.1016/S1353-8292(00)00013-7 PMID:11027957

Arndt, C., Arent, D., Hartley, F., Merven, B., & Mondal, H. (2019). Faster than you think: Renewable energy and developing countries. *Annual Review of Resource Economics, 11*(1), 149–168. doi:10.1146/annurev-resource-100518-093759

Arrow, K., Bolin, B., Costanza, R., Dasgupta, P., Folke, C., Holling, C. S., Jansson, B. O., Levin, S., Mäler, K. G., Perrings, C., & Pimentel, D. (1995). Economic growth, carrying capacity and the environment. *Science, 268*(5210), 520–521. doi:10.1126cience.268.5210.520 PMID:17756719

Asteriou, D., & Hall, S. G. (2007). *Applied econometrics: A modern approach using Eviews and Microfit* (Revised Edition). Palgrave Macmillan.

Averous, L., Fauconnier, N., Moro, L., & Fringant, C. (2000). Blends of thermoplastic starch and polyesteramide: Processing and properties. *Journal of Applied Polymer Science, 76*(7), 1117–1128. doi:10.1002/(SICI)1097-4628(20000516)76:7<1117::AID-APP16>3.0.CO;2-W

Ayieko, Z. (2011). *World Bank Resource.* Available from: http://siteresources.worldbank.org/EXTAFRREGTOPENERGY/ Resources/717305-1327690230600/8397692-1327691237767/Rural_Electrification_in_Kenya_presentation_Final_11thNov2011.pdf

Bâc, P. D. (2008). A history of the concept of sustainable development: literature review. *Annals of the University of Oradea, Economic Science Series, 17*(2), 576-580. Retrieved November 20, 2015, from http://www.cfr.washington.edu/classes.esrm.458/Paul.%202008.pdf

Baek, J., & Won, W. K. (2008). *A dynamic approach to the FDI-environment nexus: The case of China and India.* No 6508, 2008 Annual Meeting, American Agricultural Economics Association (New Name 2008: Agricultural and Applied Economics Association), Orlando, FL.

Bağbaşı, G. (2010). *İstanbul Kent Meydanlarının Peyzaj Mimarlığı İlkeleri Açısından İrdelenmesi: Sultanahmet, Beyazıt, Taksim, Beşiktaş, Ortaköy Meydanı Örneği. Yüksek Lisans Tezi.* Bartın Üniversitesi, Bartın.

Baker, M. N. (1981). The quest for pure water: the history of water purification from the earliest records to the twentieth century. American Water Works Association.

Balmer, K., Gill, J., Kaplinger, H., Miller, J., Peterson, M., Rhoads, A., … Wall, T. (2005). *The diggable city: Making urban agro ecology a planning priority* [Student project]. Retrieved from https://pdxscholar.library.pdx.edu/usp_murp/52/

Basheer, A. A. (2018). New generation nano-adsorbents for the removal of emerging contaminants in water. *Journal of Molecular Liquids, 261*, 583–593. doi:10.1016/j.molliq.2018.04.021

Bashmakov, I., Bernstein, L., Bogner, J. E., Bosch, P., Dave, R., Davidson, O., ... Ribeiro, S. K. (2007). *Technical summary. Climate Change 2007: Mitigation.* Contribution of Working Group III to the Fourth Assessment Report of the Intergovernmental Panel on Climate Change, 70.

Beckerman, W. (1992). Economic growth and the environment: Whose growth? Whose environment? *World Development, 20*(4), 481–496. doi:10.1016/0305-750X(92)90038-W

Beder, S. (1996). The nature of sustainable development. *Metascience, 10*, 304.

Beisner, B. E., Dent, C. L., & Carpenter, S. R. (2003). Variability of lakes on the landscape: Roles of phosphorus, food webs, and dissolved organic carbon. *Ecology*, *84*(6), 1563–1575. doi:10.1890/0012-9658(2003)084[1563:VOLOTL]2.0.CO;2

Bennett, E. M., Carpenter, S. R., Peterson, G. D., Cumming, G. S., Zurek, M., & Pingali, P. (2003). Why global scenarios need ecology. *Frontiers in Ecology and the Environment*, *1*(6), 322–329. doi:10.1890/1540-9295(2003)001[0322:WGSNE]2.0.CO;2

Benoit, M., Rizzo, D., Marraccini, E., Moonen, A. C., Galli, M., Lardon, S., Rapey, H., Thenail, C., & Bonari, E. (2012). Landscape agronomy: A new field for addressing agricultural landscape dynamics. *Landscape Ecology*, *27*(10), 1385–1394. doi:10.100710980-012-9802-8

Bentrup, G., & Kellerman, T. (2003). Agroforestry and GIS: achieving land productivity and environmental protection. *Proc. of the 8th North American Agroforestry Conference*, 15–25.

Bentrup, G., Schoeneberger, M., Josiah, S., & Francis, C. (2001) Ecobelts: reconnecting agriculture and communities–case studies. In *Proceedings of the Ecospheres Conference*. University of Nebraska.

Benyathiar, P., Almenar, E., Auras, R., Ryser, E., Harte, J., & Siddiq, M. (2009). A sustainable package for fresh cut green salad. *United Fresh 2009 Convention*.

Berger, A. (2007). *Drosscape: Wasting Land Urban America*. Princeton Architectural Press.

Berkes, F., Colding, J., & Folke, C. (Eds.). (2003). *Navigating Social–Ecological Systems: Building Resilience for Complexity and Change*. Cambridge University Press.

Berti, C., Binassi, E., Colonna, M., Fiorini, M., Kannan, G., Karanam, S., . . . Vannini, M. (2010). *U.S. Patent Application No. 12/347,423*. Washington, DC: US Patent Office.

Bhadra, J., Abdulkareem, A., & Al-Thani, N. (2019). Nanotechnology in Decontamination. In *Composite Nanoadsorbents* (pp. 119–137). Elsevier., doi:10.1016/B978-0-12-814132-8.00006-X

Bhatti, M., & Church, A. (2000). I never promised you a rose garden: Gender, leisure and home-making. *Leisure Studies*, *19*(3), 183–197. doi:10.1080/02614360050023071

Biggs, R., Westley, F. R., & Carpenter, S. R. (2010). Navigating the back loop: Fostering social innovation and transformation in ecosystem management. *Ecology and Society*, *15*(2), 9. doi:10.5751/ES-03411-150209

Bikales, N. M., Segal, L., Ott, E., Spurlin, H. M., & Grafflin, M. W. (1954). *Cellulose and cellulose derivatives*. Wiley-Interscience.

Bin, S. & Dowlatabad, H. (2009). Consumer lifestyle approach to US energy use and the related CO2 emissions. *Energy Policy*, *33*, 197–208.

Binh, P. T. (2013). *Unit root tests, cointegration, ECM, VECM and causality models*. Topics in Time Series Econometrics.

Blair, R. B. (2001). Birds and butterflies along urban gradients in two ecoregions of theU.S. In J. L. Lockwood & M. L. McKinney (Eds.), *Biotic Homogenization* (pp. 459–486). Kluwer. doi:10.1007/978-1-4615-1261-5_3

Blakistone, B., & Sand, C. (2007, March). Using sustainable packaging technologies to respond to consumer, retailer, and seafood industry needs. In *International Smoked Seafood Conference Proceedings* (pp. 75-79). Academic Press.

Blanco, L., Gonzalez, F., & Ruiz, I. (2011). *The impact of FDI on CO_2 emissions in Latin America*. Pepperdine University, School of Public Policy Working Papers. Paper 28.

Blanco, H., & Lal, R. (2010). *Principles of Soil Conservation and Management.* doi:10.1007/978-1-4020-8709-7

Blay-Palmer, A., Levkoe, C. Z., Mount, P., & Nelson, E. (Eds.). (2017). *Nourishing Communities: From Fractured Food Systems to Transformative Pathways*. Springer.

Blue & Green Tomorrow. (2015). *Renewable energy as a catalyst of economic development in Kenya.* Available from: https://blueandgreentomorrow.com/energy/renewable-energy-as-a-catalyst-of-economic-development-in-kenya/

Boampong, R., & Phillips, M. (2016). *Renewable energy incentives in Kenya: Feed-in tariffs and rural expansion.* National Science Foundation's Sustainable Energy Pathways Initiative.

Bogaert, J. C., & Coszach, P. (2000, March). Poly (lactic acids): A potential solution to plastic waste dilemma. *Macromolecular Symposia, 153*(1), 287–303. doi:10.1002/1521-3900(200003)153:1<287::AID-MASY287>3.0.CO;2-E

Bokpin, G. A. (2017). Foreign direct investment and environmental sustainability in Africa: The role of institutions and governance. *Research in International Business and Finance, 39*, 239–247. doi:10.1016/j.ribaf.2016.07.038

Bonanno, G. A., Galea, S., Bucciarelli, A., & Vlahov, D. (2006). Psychological Resilience after Disaster: New York City in the Aftermath of the September 11th Terrorist Attack. *Psychological Science, 17*(3), 181–186. doi:10.1111/j.1467-9280.2006.01682.x PMID:16507055

Bond, C. A., & Klonsky, K. (2006). Ecological and economic indicators for sustainability. *SAFS Newsletters, 6*(3). http://safs.ucdavis.edu/newsletter/v06n3/page1.htm

Borensztein, E., De Gregorio, J., & Lee, J.-W. (1998). How does foreign direct investment affect economic growth? *Journal of International Economics, 45*(1), 115–135. doi:10.1016/S0022-1996(97)00033-0

Born, B., & Purcell, M. (2006). Avoiding the local trap: Scale and food systems in planning research. *Journal of Planning Education and Research, 26*(2), 195–207. doi:10.1177/0739456X06291389

Borras, S. M. Jr, Edelman, M., & Kay, C. (2008). Transnational agrarian movements: Origins and politics, campaigns and impact. *Journal of Agrarian Change, 8*(2-3), 169–204. doi:10.1111/j.1471-0366.2008.00167.x

Brannen, S. (2011). *Food works: A vision to improve NYC's food system.* The New York City Council.

Brenner, N., & Theodore, N. (2002). Cities and the geographies of 'actually existing neoliberalism'. *Antipode, 34*(3), 349–379. doi:10.1111/1467-8330.00246

Brian Walker and David Salt. (2012). *Resilience Practice: Building Capacity to Absorb Disturbance and Maintain Function.* Island Press.

Broad, G. (2016). *More than just food: Food justice and community change.* University of California Press. doi:10.1525/california/9780520287440.001.0001

Brown, J. H., Ernest, S. K. M., Parody, J. M., & Haskell, J. P. (2001). Regulation of diversity: Maintenance of species richness in changing environments. *Oecologia, 126*(3), 321–332. doi:10.1007004420000536 PMID:28547444

Bryson, J. R., & Lombardi, R. (2009). Balancing Product and Process Sustainability against Business Profitability: Sustainability as a competitive strategy in the Property Development process. *Business Strategy and the Environment, 18*(2), 97–107. doi:10.1002/bse.640

Buchanan, C. M., Gardner, R. M., & Komarek, R. J. (1993). Aerobic biodegradation of cellulose acetate. *Journal of Applied Polymer Science, 47*(10), 1709–1719. doi:10.1002/app.1993.070471001

Bunch, R. (2002). Increasing productivity through agroecological approaches in Central America: experiences from hillside agriculture. In N. Uphoff (Ed.), *Agroecological innovations increasing food production with participatory development* (pp. 162–172). Earthscan.

Burian, S. J., Nix, S. J., Pitt, R. E., & Durrans, S. R. (2000). Urban Wastewater Management in the United States: Past, Present, and Future. *Journal of Urban Technology*, *7*(3), 33–62. doi:10.1080/713684134

Burney. (2010). Greenhouse gas mitigation by agricultural intensification. Proceedings of the National Academy of Sciences.

Byun, Y., & Kim, Y. T. (2014). Bioplastics for food packaging: chemistry and physics. In *Innovations in food packaging* (pp. 353–368). Academic Press. doi:10.1016/B978-0-12-394601-0.00014-X

Cadenasso, M. L., & Pickett, S. T. A. (2008). Urbanprinciples for ecological landscape design and management: Scientiðc fundamentals. *Cities and the Environment*, *1*(2), 4. doi:10.15365/cate.1242008

Cadenasso, M. L., & Pickett, S. T. A. (2013). Three tides: The development and state of the art of urban ecological studies. In S. T. A. Pickett, M. L. Cadenasso, & B. P. McGrath (Eds.), *Resilience in urban ecology and design: Linking theory and practice for sustainable cities* (pp. 29–46). Springer. doi:10.1007/978-94-007-5341-9_2

Cadenasso, M. L., Pickett, S. T. A., & Grove, J. M. (2006a). Dimensions of ecosystem complexity: Heterogeneity, connectivity, and history. *Ecological Complexity*, *3*(1), 1–12. doi:10.1016/j.ecocom.2005.07.002

Cadenasso, M. L., Pickett, S. T. A., & Grove, J. M. (2006b). Integrative approaches to investigating human-natural systems: The Baltimore EcosystemStudy. *Nature Sciences Sociétés*, *14*(1), 1–14. doi:10.1051/nss:2006002

Cadieux, K. V., & Slocum, R. (2015). What does it mean to do food justice? *Journal of Political Ecology*, *22*(1), 1–26. http://jpe.library.arizona.edu/

Calkins, M. (2005). Strategy use and challenges of ecological design in landscapearchitecture. *Landscape and Urban Planning*, *73*(1), 29–48. doi:10.1016/j.landurbplan.2004.06.003

Calza. (2017). Types of green innovations: ways of implementation in a non-Green industry. *Sustainbility, 9*.

Campanioni, N., Rodriguez, A. A., & Carrion, M. (1997). La Agricultura Urbana en Cuba: su participación en la seguridad alimentaria. In *Proceedings III Encuentro Nacional de Agricultura Organica* (pp. 9–13). Villa Clara, Cuba: Academic Press.

Cannarsi, M., Baiano, A., Marino, R., Sinigaglia, M., & Del Nobile, M. A. (2005). Use of biodegradable films for fresh cut beef steaks packaging. *Meat Science*, *70*(2), 259–265. doi:10.1016/j.meatsci.2005.01.006 PMID:22063482

Canton de Genève. (2013). Plan directeur cantonal 2030. A14: Promouvoir de nouvelles formes de jardins fa-miliaux et encourager la création de plantages. Geneva, Switzerland: République et Canton de Genève.

Cardinale, B. J., Ives, A. R., & Inchausti, P. (2004). Effects of speices diversity on the primary productivity of ecosystems: Extending our spatial and temporal scales of inference. *Oikos*, *104*(3), 437–450. doi:10.1111/j.0030-1299.2004.13254.x

Carley, M. (1995). The bigger picture: Organising for sustainable urban regeneration. *Town and Country Planning*, *64*(9), 236–239.

Carpenter, S. R., & Cottingham, K. L. (1997). Resilience and restoration of lakes. *Conservation Ecology*, *1*(1), 2. doi:10.5751/ES-00020-010102

Carpenter, S. R., & Folke, C. (2006). Ecology for transformation. *Trends in Ecology & Evolution*, *21*(6), 309–315. doi:10.1016/j.tree.2006.02.007 PMID:16769430

Carpenter, S. R., Walker, B., Anderies, J. M., & Abel, N. (2001). From metaphor to measurement: Resilience of what to what? *Ecosystems (New York, N.Y.)*, *4*(8), 765–781. doi:10.100710021-001-0045-9

Çelikyay, S., & Karayılmazlar, A. S. (2016). Bartın Kent Merkezindeki Kamusal Alanların Kentsel Ergonomi ve Kent Kimliği Açısından İncelenmesi. *Bartin Orman Fakültesi Dergisi*, *18*(2), 224–238. doi:10.24011/barofd.278907

Center for Clean Air Policy. (2011). *The value of green infrastructure for urban climate adaptation.* Available at www.ccap.org

Chakraborty, D., & Mukherjee, S. (2013). How do trade and investment flows affect environmental sustainability? Evidence from panel data. *Environmental Development*, *6*, 34–47. doi:10.1016/j.envdev.2013.02.005

Chapin, F. S. III, Shaver, G. R., Giblin, A. E., Nadelhoffer, K. J., & Laundre, J. A. (1995). Response of arctic tundra to experimentaland observed changes in climate. *Ecology*, *76*(3), 694–711. doi:10.2307/1939337

Chappell, M.J.; Lavalle, L.A. (2011). Food security and biodiversity: can we have both? *Agro ecology and Human Values*, *28*, 3-26.

Chen, C., Jiang, J., Zhang, B., Yang, W., & Guo, J. (2015). Hyperspectral Image Classification Using Gradient Local Auto-Correlations. *IEEE Conference Publications*, 454 - 458. 10.1109/ACPR.2015.7486544

Chen, G. Q. (2009). A microbial polyhydroxyalkanoates (PHA) based bio-and materials industry. *Chemical Society Reviews*, *38*(8), 2434–2446. doi:10.1039/b812677c PMID:19623359

Chen, Y. S., Lai, S. B., & Wen, C. T. (2006). The influence of green innovation performance on corporate advantage in Taiwan. *Journal of Business Ethics*, *67*(4), 331339. doi:10.100710551-006-9025-5

Childers, D. L., Pickett, S. T. A., Grove, J. M., Ogden, L., & Whitmer, A. (2014). Advancing urban sustainability theory and action: Challenges and opportunities. *Landscape and Urban Planning*, *125*, 320–328. doi:10.1016/j.landurbplan.2014.01.022

Cho, A. (2010). Energy's tricky tradeoffs. *Science*, *329*(5993), 786–787. doi:10.1126cience.329.5993.786 PMID:20705848

Choi, E., Heshmati, A., & Cho, Y. (2010). *An empirical study of the relationships between CO_2 emissions, economic growth and openness.* IZA Discussion Papers, No. 5304, 1-27.

Christaki, E. (2012). Hippophae rhamnoides L. (Sea Buckthorn): A potential source of nutraceuticals. *Food and Public Heath*, *2*(3), 69–72. doi:10.5923/j.fph.20120203.02

Christopherson, R. W. (1997). Geosystems: An Introduction to Physical Geography (3rd ed.). Upper Saddle River, NJ: Prentice Hall Inc.

Chua, R. J., Richard, N., Fernandez-Maloigne, C., & Hardeberg, J. Y. (2019). A Metrological Measurement of Texture in Hyperspectral Images Using Relocated Spectral Difference Occurrence Matrix. *IEEE Conference Publications*, 1-6. 10.1109/ICIP.2019.8803378

City of Portland. (2007). *ENN-1.01: Food Policy Council (Binding city policy).* Retrieved September 24, 2007, from http://www.portlandonline .com/auditor/index.cfm?a=ihci&c=checj

City of Portland. (2007). *ENN-1.01: Food Policy Council (Binding city policy).* Retrieved September 24, 2007, from http://www.portlandonline. com/auditor/index.cfm?a=ihci&c=checj

City of Vancouver. (2003). *Action plan for creating a just and sustainable food system for the City of Vancouver (RTS Number: 3755).* Author.

Clark, J. D., & Mendenhall, M. J. (2010). Stochastic Feature Selection with Distributed Feature Spacing for Hyperspectral Data. *IEEE Conference Publications*, 1 – 4. 10.1109/WHISPERS.2010.5594951

Clement, D. (2013). *The spatial injustice of crisis-driven neoliberal urban restructuring in Detroit (Unpublished master's thesis)*. Miami, FL: University of Miami; Retrieved from http://scholarlyrepository.miami.edu/oa_theses/406/

Cockrall-King, J. (2012). *Food and the City*. Prometheus Books.

Colasanti, K. (2010). Assessing the local food supply capacity of Detroit, Michigan. *Journal of Agro ecology, and Community Development, 1*(2), 41–58. doi:10.5304/jafscd.2010.012.002

Colding, J. (2011). The role of ecosystem services in contemporary urban planning. In Urban Ecology. Patterns, Processes and Applications. Oxford University Press. doi:10.1093/acprof:oso/9780199563562.003.0028

Colding, J. (2007). Ecological land-use complementation for building resilience in urban ecosystems. *Landscape and Urban Planning, 81*(1-2), 46–55. doi:10.1016/j.landurbplan.2006.10.016

Colding, J., & Barthel, S. (2012). *The potential of Urban Green Commons in the resilience building of cities*. Elsevier B.V.

Colding, J., Lundberg, J., & Folke, C. (2006). Incorporating green-area user groups in urbanecosystem management. *Ambio, 35*(5), 237–244. doi:10.1579/05-A-098R.1 PMID:16989508

Collins, E. M., & Kearins, K. (2010). Delivering on Sustainability's Global and Local orientation. *Academy of Management Learning & Education, 9*(3), 499–506.

Collins, J. P., Kinzig, A., Grimm, N. B., Fagan, W. F., Hope, D., Wu, J., & Borer, E. T. (2000). A new urban ecology. *American Scientist, 88*(5), 416–425. doi:10.1511/2000.5.416

Connell & Slatyer. (1977). Mechanisms of succession in natural communities and their role in community stability and organization. *American Naturalist, 111*, 1119–1144.

Convention on the Biological Diversity. (1992). The Earth Summit. Held in Rio de Janeiro on 5 June 1992 and entered into force on 29 December 1993. United Nations, General Assembly.

Conway-Cranos, L. L. (2012). Geographic variation in resilience: An experimental evaluation of four rocky intertidal assemblages. *Marine Ecology Progress Series, 457*, 67–83. doi:10.3354/meps09715

Cooper, W. S. (1926). The fundamentals of vegetational change. *Ecology, 7*, 391–413.

Cork, S. J., Peterson, G. D., Bennett, E. M., Petschel-Held, G., & Zurek, M. (2006). Synthesis of the storylines. *Ecology and Society, 11*(2), 11. http://www.ecologyandsociety.org/vol11/ iss2/art11/

Cosgrove, S. (2001). *Montréal's community gardening program*. Retrieved July 29, 2007, from http://www.cityfarmer.org/Montreal13.html

Costanza, R., & Patten, B. C. (1995). Defining and predicting sustainability. *Ecological Economics, 15*(3), 193–196. doi:10.1016/0921-8009(95)00048-8

Covert, J. (2017). Planning for the Implementation of SDG-14. *Environmental Policy and Law, 47*(1), 6–8. doi:10.3233/EPL-170003

Črnjar, M. & Črnjar, K. (2009). *Menadžment održivog razvoja*. Opatija: Fakultet za turistički i hotelski menadžment Opatija, Sveučilište u Rijeci.

Cumming, G. S., Alcamo, J., Sala, O., Swart, R., Bennett, E. M., & Zurek, M. (2005). Are existing global scenarios consistent with ecological feedbacks? *Ecosystems (New York, N.Y.), 8*(2), 143–152. doi:10.100710021-004-0075-1

Cumming, G. S., & Child, M. F. (2009). Contrasting spatial patterns of taxonomic andfunctional richness offer insights into potential loss of ecosystem services. *Philosophical Transactions of the Royal Society of London. Series B, Biological Sciences*, *364*(1524), 1683–1692. doi:10.1098/rstb.2008.0317 PMID:19451119

Curwell, S., & Cooper, I. (1998). The implications of urban sustainability. *Building Research and Information*, *26*(1), 17–28. doi:10.1080/096132198370074

Curwell, S., Deakin, M., & Symes, M. (Eds.). (2005). *Sustainable urbandevelopment, volume1: The frameworkand protocols for environmental assessment*. Routledge. doi:10.4324/9780203299913

Daftary-Steel, S., Dignity, F., Herrera, H., & Porter, C. M. (2015). The unattainable trifecta of urban agro ecology. *Journal of Agro ecology. Systems, and Community Development*, *6*(61), 19–32. doi:10.5304/jafscd.2015.061.014

Daniel, D. C. (2011). *A Formal Theory of Resilience*. Dissertation. Master of Science in Geoinformatics, University of M'unster, Germany.

Daniel, D., & Ortmann, J. (2011). Disambiguating Resilience. In Geoinformatik 2011 – Geochange (vol. 41, pp. 117–125). Heidelberg, Germany. AKA Verlag.

Daune, L., & Mongé, N. (2011). L'agriculture urbaine, un fondement dans le projet de territoire. *Urbia—Les Cahiers du développement urbain durable, 12*, 85-106.

Davis, L., & Middleton, J. (2012). The perilous road from community activism to public policy: Fifteen years of community agriculture in Sandwell. In Sustainable Food Planning: Evolving Theory and Practice. Wageningen: Wageningen Academic Publishers.

Day, T., & Kurdziel, M. (2018). *The role of renewable energy mini-grids in Kenya's electricity sectors*. New Climate Institute.

de Jong, M., Joss, S., Schraven, D., Zhan, Ch., & Weijnen, M. (2015). Sustainable–Smart–Resilient–Low Carbon–Eco–Knowledge Cities; Making Sense of a Multitude of Concepts Promoting Sustainable Urbanization. *Journal of Cleaner Production*, *109*(1), 25–38. doi:10.1016/j.jclepro.2015.02.004

De Lotto, R., & Di Tolle, M. L. (2013). Elementi di progettazione urbanistica-rigenerazione urbana nella città contemporanea. Maggioli Publisher.

De Lotto, R., Esopi, G., & Sturla, S. (2017). *Sustainable policies to improve urban ecosystem resilience* (Vol. 96). WIT Transactions on State of the Art in Science and Engineering.

De Mello, L. (1999). Foreign direct investment-led growth: Evidence from time series and panel data. *Oxford Economic Papers*, *51*(1), 133–151. doi:10.1093/oep/51.1.133

De Molina, M. G. (2012). Agro ecology and politics. How to get sustainability? About the necessity for a political agroecologist. *Agroecology and Sustainable Food Systems*, *37*, 45–59. doi:10.1080/10440046.2012.705810

De Schutter, O. (2014). *Report of the Special Rapporteur on the right to food. Final report: the transformative potential of the right to food.* Human Right Council, General Assembly, United Nations.

De Schutter, O. (2010). *Report submitted by the special rapporteur on the right to food*. United Nations Human Rights Council.

De Zeeuw, H., Van Veenhuizen, R., & Dubbeling, M. (2011). The role of urban agriculture in building resilient cities in developing countries. *J Agric Sci*, *149*(S1), 153–163. doi:10.1017/S0021859610001279

Declaration of Nyeleni. (2007). Available online: https://nyeleni.org/IMG/pdf/DeclNyeleni-en.pdf

Deelstra, T., Boyd, D., & van den Biggelaar, M. (2006). Multifunctional land use, promoting urban agriculture in Europe. In Sustainable Development: Opportunities for Change. Department of the Environment, Transport and the Regions. Consultation Paper on a Revised UK Strategy, DETR, London.

Deelstra, T., Boyd, D., & van den Biggelaar, M. (2006). *Multifunctional land use, promoting urban agriculture in Europe*. Academic Press.

DETR. (1998). *Sustainable Development: Opportunities for Change.* Department of the Environment, Transport and the Regions. Consultation Paper on a Revised UK Strategy. DETR.

Detroit Works Project. (2014). *Detroit future city: Detroit strategic framework plan.* Retrieved from http://detroitfuturecity.com/wp-content/uploads/2014/12/DFC_Full_2nd.pdf

Deutsch, L., Folke, C., & Skanberg, K. (2003). The critical natural capital of ecosystem performance as insurance for human well-being. *Ecological Economics*, *44*(2-3), 205–217. doi:10.1016/S0921-8009(02)00274-4

DHA (Dogan News Agency). (2020). *Electricity consumption increased in the quarantine period.* Retrieved from https://www.cnnturk.com/ekonomi/karantinada-elektrik-tuketimi-artti

Dietzenbacher, E., & Mukhopadhyay, K. (2007). An empirical examination of the pollution haven hypothesis for India: Towards a green Leontief paradox? *Environmental and Resource Economics*, *36*(4), 427–449. doi:10.100710640-006-9036-9

Dinda, S. (2004). Environmental Kuznets curve hypothesis: A survey. *Ecological Economics*, *49*(4), 431–455. doi:10.1016/j.ecolecon.2004.02.011

Doğan, C., & Altan, O. (2007). Kamusal Alanda Oturma Eylemi ve Ergonomik İlkeler. *Megaron Dergisi*, *2*(3), 159–166.

Domenek, S., Feuilloley, P., Gratraud, J., Morel, M. H., & Guilbert, S. (2004). Biodegradability of wheat gluten based bioplastics. *Chemosphere*, *54*(4), 551–559. doi:10.1016/S0045-6535(03)00760-4 PMID:14581057

Donald, P. F., Sanderson, F. J., Burfield, I. J., & van Bommel, F. P. (2006). Further evidence of continent-wide impacts of agricultural intensification on European farmland birds, 1990–2000. *Agriculture, Ecosystems and Environment*, *116*, 189–196.

Donastorg, A., Renukappa, S., & Suresh, S. (2017) Financing renewable energy projects in developing countries: A critical review. *IOP Conference Series: Earth and Environmental Science*, *83*, 012012. 10.1088/1755-1315/83/1/012012

Douglas, I., Goode, D., Houke, M., & Wang, R. (2011). *The Routledge Handbook of Urban Ecology*. Routledge.

Dow, K. (2000). Social dimensions of gradients in urban ecosystems. *Urban Ecosystems*, *4*(4), 255–275. doi:10.1023/A:1015767231137

Driessen, P. H., & Hillebrand, B. (2002). Adoption and diffusion of green innovations. *Marketing for sustainability: towards transactional policy-making*, 343-355.

Driffield, N., & Hughes, D. (2003). Foreign and domestic investment: Regional development or crowding out? *Regional Studies*, *37*(3), 277–288. doi:10.1080/0034340032000065433

Druckman, A., & Jackson, T. (2008). Household energy consumption in the UK: A highly geographically and socio-economically disaggregated model. *Energy Policy*, *36*(8), 3177–3192. doi:10.1016/j.enpol.2008.03.021

Dunning, J. B., Danielson, B. J., & Pulliam, H. R. (1992). Ecological processes thataffect populations in complex landscapes. *Oikos*, *65*(1), 169–175. doi:10.2307/3544901

Eberhard, A., Gratwick, K., & Kariuki, L. (2018). Kenya's lessons from two decades of experience with independent power producers. *Utilities Policy*, *52*, 37–49. doi:10.1016/j.jup.2018.04.002

Eco-Innovation Action Plan. (2011). *Innovation for a Sustainable Future.* Communication from the Commission to the European Parliament, the Council, the European Economic and Social Committee and the Committee of the Regions. https://eur-lex.europa.eu/legal-content/EN/ALL/;ELX_SESSIONID=JBJhJh7RB5n2T7nTdZLLVLGRlJPCFRnTGM7ZvF8WhH9TJryvnsvL!1821898924?uri=CELEX:52011DC0899

Educational, U. N. Scientific and Cultural Organisation (UNESCO). (2019). *Solar energy: changing rural lives in Kenya.* Available from: https://en.unesco.org/courier/2019-3/solar-energy-changing-rural-lives-kenya

Efe, B., Efe, Ö. F., & Kurt, M. (2018). An integrated approach for ergonomic product design. *Sakarya University Journal of Science*, *22*(2), 192–202. doi:10.16984aufenbilder.306684

Efe, Ö. F., & Efe, B. (2015). Tekstil Sektöründe İş Kazalarının Oluşumuna Ait Ergonomik Risklerin Değerlendirilmesi. *Süleyman Demirel Üniversitesi Mühendislik Bilimleri ve Tasarım Dergisi*, *3*(3), 623–629.

Eizenberg, E. (2012). Actually existing commons: Three moments of space of community gardens in New York City. *Antipode*, *44*(3), 764–782. doi:10.1111/j.1467-8330.2011.00892.x

Elliott, R., & Shimamoto, K. (2008). Are ASEAN countries havens for Japanese pollution-intensive industry? *World Economy*, *31*(2), 236–254. doi:10.1111/j.1467-9701.2007.01088.x

Elmarzougui, E., Larue, B., & Tamini, L. T. (2016). Trade openness, domestic and foreign investments, and the environment. *Modern Economy*, *7*(05), 591–605. doi:10.4236/me.2016.75065

Elmqvist, T., Wall, M., & Berggren, A. L. (2001). Tropical forest reorganization after cyclone and fire disturbance in Samoa: remnant trees as biological legacies. *Cons Ecol*, *5*, 10. http://www.consecol.org/vol5/iss2/art10

Elmqvist, T., Folke, C., Nyström, M., Peterson, G., Bengtsson, J., Walker, B., & Norberg, J. (2003). Response diversity, ecosystem change, and resilience. *Frontiers in Ecology and the Environment*, *1*(9), 488–494. doi:10.1890/1540-9295(2003)001[0488:RDECAR]2.0.CO;2

Emas, R. (2015). *The concept of Sustainable Development: definition and definig principles.* Florida International University.

Energy Information Administration (EIA). (2017). *Levelized cost and levelized avoided cost of new generation resources in the annual energy outlook 2017.* Available from: https://www.eia.gov/outlooks/archive/aeo17/pdf/electricity_generation.pdf

Engle, R., & Granger, C. W. J. (1987). Cointegration and error correction: Representation, estimation and testing. *Econometrica*, *55*(2), 251–276. doi:10.2307/1913236

Ercan, M. N. (1988). *Çalışma Yerlerinin ve Yaşam Ortamlarının Ergonomik Şekillendirilmelerinde Genel Prensipler. İTC Milli Prodüktivite Merkezi 1.* Ulusal Ergonomi Kongresi.

Estes, J. A., & Duggins, D. O. (1995). Sea otters and kelp forests in Alaska: Generality and variation in a community ecological paradigm. *Ecological Monographs*, *65*(1), 75–100. doi:10.2307/2937159

EurActiv. (2004). *Sustainable Development: Introduction.* Author.

Eurek, K., Sullivan, P., Gleason, M., Hettinger, D., Heimiller, D., & Lopez, A. (2017). An improved global wind resource estimate for integrated assessment models. *Energy Economics*, *64*, 552–567. doi:10.1016/j.eneco.2016.11.015

European Commission. (2001). *A Framework for Indicators for the Economic and Social Dimensions of Sustainable Agriculture and Rural Development.* The Agriculture Directorate-General of the European Union. Available online https://ec.europa.eu/agriculture/publi/reports/sustain/index_en.pdf

Falk, B. (2013). The resilient farm and homestead. Chelsea Green Publishing.

Fan, S., & Polman, P. (2014). *An ambitious development goal: Ending hunger and undernutrition by 2025.* Academic Press.

Fanger, P. O. (1970). *Thermal Comfort- Analysis and Applications in Environmental Engineering.* McGraw-Hill.

FAO. (2011). *Food, agro ecology and cities: Challenges of food and nutrition security, agro ecology and ecosystem management in an urbanizing world.* http://www.fao.org/fileadmin/templates/FCIT/PDF/FoodAgriCities_Oct2011.pdf

FAOSTAT. (2016). *Emissions by sector.* Accessed December, 2016. 7, 2016. http://www.fao.org/faostat/en/#data/GT/visualize

Farla, K., De Crombrugghe, D., & Verspagen, B. (2016). Institutions, foreign direct investment, and domestic investment: Crowding out or crowding in? *World Development, 88*(C), 1–9. doi:10.1016/j.worlddev.2014.04.008

Feagan, R. (2007). The place of food: Mapping out the 'local' in local food systems. *Progress in Human Geography, 31*(1), 23–42. doi:10.1177/0309132507073527

Feng, J., Wu, X., Chen, J., Zhang, X., & Tang, X. (2019). Joint Multilayer Spatial-Spectral Classification of Hyperspectral Images Based on CNN and CONVLSTM. *IEEE Conference Publications*, 588 - 591.

Fergus, S., & Zimmerman, M. A. (2005). Adolescent resilience: A framework for understanding healthy development in the face of risk. *Annual Review of Public Health, 26*(1), 399–419. doi:10.1146/annurev.publhealth.26.021304.144357 PMID:15760295

Firebaugh, G. (1992). Growth effects of foreign and domestic investment. *American Journal of Sociology, 98*(1), 105–130. doi:10.1086/229970

Firzli, N. (2018). Greening, Governance and Growth in the Age of Popular Empowerment. *Financial Times.*

Firzli, M., & Nicolas, J. (2016). *Beyond SDGs: Can Fiduciary Capitalism and Bolder, Better Boards Jumpstart Economic Growth?* Analyse Financière.

Folke, C., Carpenter, S. R., Walker, B., Scheffer, M., Chapin, T., & Rockström, J. (2010). Resilience thinking: integrating resilience, adaptability and transformability. *Ecology and Society, 15*(4), 20. http://www.ecologyandsociety.org/vol15/iss4/ art20/

Folke, C., Pritchard, L., Berkes, F., Colding, J., & Svedin, U. (1998). The Problem of Fit Between Ecosystems and Institutions. International Human Dimensions Programme (IHDP). *IHDP Working Paper No 2.* www.uni-bonn.de/IHDP/public.htm

Folke, C., Carpenter, S. R., Elmqvist, T., Gunderson, L., Holling, C. S., & Walker, B. (2002). Resilience and sustainable development: Building adaptive capacity in a world of transformations. *Ambio, 31*(5), 437–440. doi:10.1579/0044-7447-31.5.437 PMID:12374053

Folke, C., Carpenter, S., Walker, B., Scheffer, M., Elmqvist, T., Gunder-son, L., & Holling, C. S. (2004). Regime shifts, resilience, and biodiversityin ecosystem management. *Annual Review of Ecology, Evolution, and Systematics, 35*(1), 557–558. doi:10.1146/annurev.ecolsys.35.021103.105711

Folke, C., Holling, C. S., & Perrings, C. (1996). Biological diversity, ecosystems and the human scale. *Ecological Applications, 6*(4), 1018–1024. doi:10.2307/2269584

Folke, C., Jansson, A., Larsson, J., & Costanza, R. (1997). Ecosystem appropriation by cities. *Ambio*, *26*, 167–172.

Follain, N., Joly, C., Dole, P., & Bliard, C. (2005). Properties of starch based blends. Part 2. Influence of poly vinyl alcohol addition and photocrosslinking on starch based materials mechanical properties. *Carbohydrate Polymers*, *60*(2), 185–192. doi:10.1016/j.carbpol.2004.12.003

Forys, E. A., & Allen, C. R. (2002). Functional group change within and across scales following invasions and extinctions in the Everglades ecosystem. *Ecosystems (New York, N.Y.)*, *5*(4), 339–347. doi:10.100710021-001-0078-0

Francis, C., Lieblein, G., Gliessman, S., Breland, T. A., Creamer, N., Harwood, R., Salomonsson, L., Helenius, J., Rickerl, D., Salvador, R., Wiedenhoeft, M., Simmons, S., Allen, P., Altieri, M., Flora, C., & Poincelot, R. (2003). Agroecology: The ecology of food systems. *Journal of Sustainable Agriculture*, *2003*(22), 99–118. doi:10.1300/J064v22n03_10

Frankfurt School of Finance and Management. (2018). Global trends in renewable energy investment 2018. Frankfurt, Germany: Author.

Frost, T. M., Carpenter, S. R., Ives, A. R., & Kratz, T. K. (1995). Species compensation and complementarity in ecosystem function. In C. G. Jones & J. H. Lawton (Eds.), *Linking species and ecosystems*. Chapman & Hall. doi:10.1007/978-1-4615-1773-3_22

Fujita, K., Ofosu-Budu, K. G., & Ogata, S. (1992). Biological nitrogen fixation in mixed legume-cereal cropping systems. *Plant and Soil*, *141*(1-2), 155–175. doi:10.1007/BF00011315

Funke, U., Bergthaller, W., & Lindhauer, M. G. (1998). Processing and characterization of biodegradable products based on starch. *Polymer Degradation & Stability*, *59*(1-3), 293–296. doi:10.1016/S0141-3910(97)00163-8

Fussler, C., & James, P. (1996). *Eco-innovation: A Breakthrough Discipline for Innovation and Sustainability*. Pitman Publishing.

Gadhave, R. V., Das, A., Mahanwar, P. A., & Gadekar, P. T. (2018). *Starch based bio-plastics: the future of sustainable packaging*. Academic Press.

Gallant, D. J., Bouchet, B., & Baldwin, P. M. (1997). Microscopy of starch: Evidence of a new level of granule organization. *Carbohydrate Polymers*, *32*(3-4), 177–191. doi:10.1016/S0144-8617(97)00008-8

Galliard, T. (Ed.). (1987). *Starch: properties and potential*. John Wiley & Sons Incorporated.

Gandini, A. (2008). Polymers from renewable resources: A challenge for the future of macromolecular materials. *Macromolecules*, *41*(24), 9491–9504. doi:10.1021/ma801735u

Garlotta, D. (2001). A literature review of poly (lactic acid). *Journal of Polymers and the Environment*, *9*(2), 63–84.

Gedon, S., & Fengl, T. (1993). Encyclopedia of Chemical Technology. New York: Wiley.

Geist, H. J., & Lambin, E. F. (2002). Proximate causes and underlying driving forces of tropical deforestation. *Bioscience*, *52*(2), 143–150. doi:10.1641/0006-3568(2002)052[0143:PCAUDF]2.0.CO;2

German Technical Cooperation-GTZ Kenya/ Ministry of Agriculture, Government of Kenya. (2008). *A roadmap for biofuels in Kenya: opportunities and obstacles*. Available from: http://kerea.org/wp-content/uploads/2012/12/A-Roadmap-for-Biofuels-in-Kenya_Opportunities-and-Obstacles.pdf

GESAMP. (2010). IMO/FAO/UNESCOIOC/UNIDO/WMO/IAEA/UN/UNEP Joint Group of Experts on the Scientific Aspects of Marine Environmental Protection. In *Proceedings of the GESAMP International Workshop on plastic particles as a vector in transporting persistent, bio-accumulating and toxic substances in the oceans* (*Vol. 82*, p. 68). Academic Press.

Ghose, R., & Pettygrove, M. (2014). Actors and networks in urban community garden development. *Geoforum*, *53*(May), 93–103. doi:10.1016/j.geoforum.2014.02.009

Gironès, J., López, J. P., Mutjé, P., Carvalho, A. J. F. D., Curvelo, A. A. D. S., & Vilaseca, F. (2012). Natural fiber-reinforced thermoplastic starch composites obtained by melt processing. *Composites Science and Technology*, *72*(7), 858–863. doi:10.1016/j.compscitech.2012.02.019

Gliessman, S. R. (2007). *Agroecology: The ecology of sustainable food systems* (2nd ed.). CRC Press.

Goal 1: No poverty. (2017). *UNDP*.

Goal 12: Responsible consumption, production. (2018). *UNDP*.

Goal 3: Good health and well-being. (2017). *UNDP*.

Goal 7: Affordable and clean energy. (2015). UNDP.

Gohary, A. (2014). Evolution of sanitation and wastewater technologies in Egypt through centuries. *Water Intelligence Online*, *13*. doi:10.2166/9781780404851

Golfinopoulos, A., Kalavrouziotis, I. K., & Aga, V. (2016). Prehistoric and historic hydraulic technologies in stormwater and wastewater management in Greece: A brief review. *Desalination and Water Treatment*, *57*(58), 28015–28024. doi:10.1080/19443994.2016.1179224

Golley, F. B. (1993). *A History of the Ecosystem Concept in Ecology: More Than the Sum of the Parts*. Yale University Press.

Göllü, S. K., & Türkyılmaz, Ç. C. (2019). Kent Meydanlarının Ergonomik Ölçütler Açısından Değerlendirilmesi: Kadıköy Rıhtım Meydanı Örneği. *Ergonomi Dergisi*, *2*(1), 32–48. doi:10.33439/ergonomi.479502

Gómez, L. F., Ríos-Osorio, L., & Eschenhagen, M. L. (2013). Agroecology publications and coloniality of knowledge. *Agronomy for Sustainable Development*, *33*(2), 355–362. doi:10.100713593-012-0109-6

Gontard, N., Guilbert, S., & Cuq, J.-L. (1993). Water and glycerol as plasticizers affect mechanical and water vapor barrier properties of an edible wheat gluten film. *Journal of Food Science*, *58*(1), 206–211. doi:10.1111/j.1365-2621.1993.tb03246.x

Goodland, R., & Daly, H. (1996). *Environmental Sustainability: Universal and Non-negotiable*. Academic Press.

Gopakumar, D. A., Pai, A. R., & Pasquini, D. (2019). Nanomaterials—State of Art, New Challenges, and Opportunities. In Nanoscale Materials in Water Purification (pp. 1–24). Elsevier. doi:10.1016/B978-0-12-813926-4.00001-X

Gormus, A., Nazlioglu, S., & Soytas, U. (2018). High-Yield Bond and Energy Markets. *Energy Economics*, *69*, 101–110. doi:10.1016/j.eneco.2017.10.037

Gottlieb, R., & Joshi, A. (2009). Food Justice. Cambridge, MA: MIT Press.

Gottlieb, R., & Joshi, A. (2010). *Food justice*. MIT Press. doi:10.7551/mitpress/7826.001.0001

Government of Kenya. (2008). *Least Cost Energy Development plan: 2009–2029*. Ministry of Energy.

Graham, S., & Marvin, S. (2001). *Splintering urbanism: Networked infrastructures, technological mobilities and the urban condition*. Routledge.

Grayson, R. (2010). The permaculture papers 3: childhood. *Pac. Edge*. https://pacific-edge.info/2010/09/the-permaculture-papers-2/

Grewal, S. S., & Grewal, P. S. (2012). Can cities become self-reliant in food? *Cities (London, England), 29*(1), 1–11. doi:10.1016/j.cities.2011.06.003

Grimm, N. B., Faeth, S. H., Golubiewski, N. E., Redman, C. L., Wu, J., Bai, X., & Briggs, J. M. (2008). Global change and the ecology of cities. *Science, 319*(5864), 756–760. doi:10.1126cience.1150195 PMID:18258902

Groffman, P. M., Baron, J. S., Blett, T., Gold, A. J., Goodman, I., Gunderson, L. H., Levinson, B. M., Palmer, M. A., Paerl, H. W., Peterson, G. D., Poff, N. L., Rejeski, D. W., Reynolds, J. F., Turner, M. G., Weathers, K. C., & Weins, J. A. (2006). Ecological thresholds: the key to successful environmental management or an important concept with no practical application? *Ecosystems, 9*(1), 1–13. http://springerlink.com/(mw0hzsfu fbpawv2rf24i3szk)/app/home/contribution.asp?referrer=parent&backto=issue,1,12;journal,2,60; linkingpublicationresults,1:101552,1

Grønhøj, A., & Thøgersen, J. (2011). Feedback on household electricity consumption: Learning and social influence processes. *International Journal of Consumer Studies, 35*(2), 138–145. doi:10.1111/j.1470-6431.2010.00967.x

Grossman, G. M., & Krueger, A. B. (1991). *Environmental impacts of a North American free trade agreement.* NBER Working Paper Series, 3914.

Grossman, G. M., & Krueger, A. B. (1995). Economic growth and the environment. *The Quarterly Journal of Economics, 110*(2), 353–377. doi:10.2307/2118443

Grove, J. M. (2009). Cities: Managing densely settled social–ecological systems. In F. S. Chapin III, G. P. Koðnas, & C. Folke (Eds.), *Principles of ecosystem stewardship: Resilience-based natural resource management in a changing world* (pp. 281–294). doi:10.1007/978-0-387-73033-2_13

Gu, J. D., Eberiel, D. T., McCarthy, S. P., & Gross, R. A. (1993). Cellulose acetate biodegradability upon exposure to simulated aerobic composting and anaerobic bioreactor environments. *Journal of Environmental Polymer Degradation, 1*(2), 143–153. doi:10.1007/BF01418207

Gülçubuk, A. (2005). İnsana Uygun İş Tasarımında İklimlendirmeye İlişkin Ergonomik İlkeler ve Optimum Hava Koşulları. *Celal Bayar Üniversitesi Soma Meslek Yüksekokulu Teknik Bilimler Dergisi, 3*(1), 95–100.

Gunderson, L. H., & Holling, C. S. (Eds.). (2002). *Panarchy: Understanding transformations in human and natural systems.* Island Press.

Gunderson, L. H., & Holling, C. S. (Eds.). (2002). *Panarchy: Understanding Transformations in Human and Systems.* Island Press.

Güney, Ş. (2005). *Bürolardaki Mekan-Mobilya Organizasyonundaki Ergonomi Faktörü ve Verimliliğe Etkisi: Bir Banka Örneği. Yüksek Lisans Tezi.* Yıldız Teknik Üniversitesi Fen Bilimleri Enstitüsü.

Gupta, V. K., Agarwal, S., & Saleh, T. A. (2011). Synthesis and characterization of alumina-coated carbon nanotubes and their application for lead removal. *Journal of Hazardous Materials, 185*(1), 17–23. doi:10.1016/j.jhazmat.2010.08.053 PMID:20888691

Haiven, M. (2009). Food, finance, crisis, catalyst: Global capitalism and the revolutionary value of food sovereignty. *Política y Cultura, 2.* https://www.politicsandculture.org/2009/11/03/max-haiven-food-finance-crisis-catalyst-global-capitalism-and-the-revolutionary-value-of-food-sovereignty

Hamel, G., & Valikangas, L. (2003). The Quest for Resilience. *Harvard Business Review, 2003*(September), 52. PMID:12964393

Hansen, A. J., Defries, R., & Turner, W. (2004). Land use change and biodiversity: a syn-thesis of rates and consequences during the period of satellite imagery. In G. Gutman & C. Justice (Eds.), *Land Change Science: Observing, Monitoring, andUnderstanding Trajectories of Change on the Earth's Surface* (pp. 277–299). Springer Verlag.

Harris, J. M. (2000). Basic Principles of Sustainable Development. A Survey of Sustainable Development: Social and Economic Dimensions. Washington, DC: Island Press.

Hasbullah, H., Sabri, N. S. M., Said, N., Rosid, S. M., Roslan, M. I., Ismail, A. F., Jye, L. W., & Yusof, N. (2019). Nanoengineered Materials for Water and Wastewater Treatments. In *Nanotechnology in Water and Wastewater Treatment* (pp. 303–335). Elsevier. doi:10.1016/B978-0-12-813902-8.00016-2

Hassanein, N. (2003). Practicing food democracy: A pragmatic politics of transformation. *Journal of Rural Studies*, *19*(1), 77–86. doi:10.1016/S0743-0167(02)00041-4

Haugaard, V., Weber, C., Danielsen, B., & Bertelsen, G. (2002). Quality changes in orange juice packed in materials based on polylactate. *European Food Research and Technology*, *214*(5), 423–428. doi:10.100700217-001-0474-x

Havens, E., & Roman-Alcalá, A. (2016). *Land and sovereignty land for food justice? AB 551 and structural change.* Food First. Retrieved from https://foodtank.com/news/2017/01/14707/

Havlicek, T. D., & Carpenter, S. R. (2001). Pelagic species size distributions in lakes: Are they discontinuous? *Limnology and Oceanography*, *46*(5), 1021–1033. doi:10.4319/lo.2001.46.5.1021

He, H. A., & Greenberg, S. (2008). *Motivating Sustainable Energy Consumption in the Home. Motivating Sustainable Energy Consumption in the Home*. Technical Report 2008-914-27, Department of Computer Science, University of Calgary, Calgary, AB, Canada.

Hediger, W. (1999). Reconciling weak and strong sustainability. *International Journal of Social Economics*, *26*(7-9), 1120–1143. doi:10.1108/03068299910245859

Hemosilla, C. (2010). Diversity of eco-innovations: Refections from selected case studies. *Journal of Cleaner Production, 10.*

Heshmati, A. (2015). *A Review of the Circular Economy and its Implementation.* Discussion Paper Series, December 2015, IZA DP No. 9611.

Hisanaga, F., & Zhong, Z. (2004). Visual inspection of Textile Surfaces with Translation Invarient Wavelet Shrinkage. *IEEE Conference Publications*, 1252 - 1257.

Hitam, M. B., & Borhan, H. B. (2012). FDI, growth and the environment: Impact on quality of life in Malaysia. *Procedia: Social and Behavioral Sciences*, *50*, 333–342. doi:10.1016/j.sbspro.2012.08.038

Hobbs, R. J. (1993). Can revegetation assist in the conservation of biodiversity inagricultural areas? *Pacific Conservation Biology*, *1*, 29–38. doi:10.1071/PC930029

Hobbs, R. J. (1997, June). Future landscapes and the future of landscape ecology. *Landscape and Urban Planning*, *37*(1–2), 1–9. doi:10.1016/S0169-2046(96)00364-7

Hodgson, K., Caton Campbell, M., & Bailkey, M. (2011). Urban agriculture: growing healthy, sustainable places. Washington, DC: American Planning Association.

Hogg, P. (2001). *Plastics, rubber and composites at Queen Mary*. Academic Press.

Holling, C. S. (1973). Resilience and stability of ecological systems. *Annual Review of Ecology and Systematics*, *4*(1), 1–23. doi:10.1146/annurev.es.04.110173.000245

Holling, C. S. (1996). Engineering resilience versus ecological resilience. In P. C. Schulze (Ed.), *Engineering within ecological constraints*. National Academy Press.

Holmgren, D. (2004). *Permaculture: principles and pathways beyond sustainability*. Holmgren Design, Hepburn.

Hopkins, M. (2004). *Corporate social responsibility: an issues paper*. Policy Integration Department, Working Paper No. 27 World Commission on the Social Dimension of Globalization International Labor Office Geneva.

Hough, M. (1995). *Cities and Natural Process*. Routledge. doi:10.4324/9780203292433

Howe, J., Bohn, K., & Viljoen, A. (2005). Food in time: The history of English open urban space as a European example. In A. Viljoen (Ed.), *Continuous Productive Urban Landscapes: Designing Urb Agriculture for Sustainable Cities* (pp. 95–107). Architectural Press.

Huang, D., & Drescher, M. (2015). Urban crops and livestock: The experiences, challenges, and opportunities of planning for urban agriculture in two Canadian provinces. *Land Use Policy*, *43*(February), 1–14. doi:10.1016/j.landusepol.2014.10.011

Huber, M. (2008). Community for innovations: Developing an integrated concept for open innovation. In *IFIP Working Conference on Open IT-Based Innovation: Moving Towards Cooperative IT Transfer and Knowledge Diffusion* (pp. 503-510). Springer.

Hui, S. C. M. (2011). *Green roof urban farming for buildings in high-density urban cities*. Invited paper for the Hainan China World Green Roof Conference 2011, Hainan, China.

Hui, S. C. M. (2011). Green roof urban farming for buildings in high-density urban cities. *Invited paper for the Hainan China World Green Roof Conference 2011.*

Hulleman, S. H., Janssen, F. H., & Feil, H. (1998). The role of water during plasticization of native starches. *Polymer*, *39*(10), 2043–2048. doi:10.1016/S0032-3861(97)00301-7

Hu, S., Da Yan, D., Guo, S., Cui, Y., & Dong, B. (2017). A survey on energy consumption and energy usage behavior of households and residential building in urban China. *Energy and Building*, *148*, 366–378. doi:10.1016/j.enbuild.2017.03.064

Hynes, H. P. (1996). *A Patch of Eden: America's Inner-city Gardeners*. Chelsea Green Publishing Company.

IAEA. (2015). *Energy Indicators for Sustainable Development: Guidelines and Methodologies*. International Atomic Energy Agency, United Nations Department Of Economic And Social Affairs, International Energy Agency, Eurostat And European Environment Agency, Austria April 2005, STI/PUB/1222. Retrieved from https://www.iaea.org/Publications/index.html

Ibrahim, M. S., Sani, N., Adamu, M., & Abubakar, M. K. (2018). Biodegradable polymers for sustainable environmental and economic development. *MOJ Biorg. Org. Chem*, *2*, 192–194.

IEA. (2018). *Global energy and CO_2 status report 2017*. IEA.

IEA. (2018). *Status of power system transformation 2018: advanced power plant flexibility*. IEA.

Ifezue, C. (2009). *The effect of bio-based films on quality and shelf life of fresh celery*. Clemson University.

IFRC (International Federation of Red Cross and Red Crescent Societies). (2004). *World Disasters Report: Focus on Community Resilience*. International Federation of Red Cross and Red Crescent Societies.

IISD (2009). *What is Sustainable Development?* International Institute for Sustainable Development.

IISD. (2009). *What is Sustainable Development?* International Institute for Sustainable Development.

Imam, S., Glenn, G., Chiou, B. S., Shey, J., Narayan, R., & Orts, W. (2008). Types, production and assessment of bio-based food packaging materials. In *Environmentally compatible food packaging* (pp. 29–62). Woodhead Publishing. doi:10.1533/9781845694784.1.29

Institute for Agro Ecology and Trade Policy. (2012). *Draft principles of food justice*. Minneapolis, MN: Author. Retrieved from https://www.iatp.org/documents/draft-principles-of-food-justice

International Energy Agency (IEA). (2015). *Kenya energy outlook*. Available from: https://www.iea.org/articles/kenya-energy-outlook

International Energy Agency. (2010). *Energy Performance Certification of Buildings: a Policy Tool to Improve Energy Efficiency*. Author.

Isbell, F., Calcagno, V., Hector, A., Connolly, J., Harpole, W. S., Reich, P. B., Scherer-Lorenzen, M., Schmid, B., Tilman, D., van Ruijven, J., Weigelt, A., Wilsey, B. J., Zavaleta, E. S., & Loreau, M. (2011). High plant diversity is needed to maintain ecosystem services. *Nature*, *477*(7363), 199–202. doi:10.1038/nature10282 PMID:21832994

Isenhour, C. (2011). How the Grass Became Greener in the City: On Urban Imaginings and Practices of Sustainable Living in Sweden. *City & Society*, *23*(2), 117–134. doi:10.1111/j.1548-744X.2011.01058.x

Ishizaka, K., & Tanaka, M. (2003). Resolving public conflict in site selection process—A risk communication approach. *Waste Management (New York, N.Y.)*, *23*(5), 385–396. doi:10.1016/S0956-053X(03)00094-1 PMID:12893010

Ismail, T. N. T. (2011). Corporate Social Responsibility: The Influence of the Silver Book. *International Journal of Business And Management Studies*, *3*(2), 371–383.

Jacke, D., & Toensmeier, E. (2005). *Edible forest gardens: ecological design and practice for temperate climate permaculture* (Vol. 2). Chelsea Green Publishing Company.

Jackson, J. B. C., Kirby, M. X., Berger, W. H., Bjorndal, K. A., Botsford, L. W., Bourque, B. J., Bradbury, R. H., Cooke, R., Erlandson, J., Estes, J. A., Hughes, T. P., Kidwell, S., Lange, C. B., Lenihan, H. S., Pandolfi, J. M., Peterson, C. H., Steneck, R. S., Tegner, M. J., & Warner, R. R. (2001). Historical overfishing and the recent collapse of coastal ecosystems. *Science*, *293*(5530), 629–638. doi:10.1126cience.1059199 PMID:11474098

Jain, R. K., Taylor, J. E., & Culligan, P. J. (2013). Investigating the impact eco-feedback information representation has on building occupant energy consumption behavior and savings. *Energy and Building*, *64*, 408–414. doi:10.1016/j.enbuild.2013.05.011

James, P., Magee, L., Scerri, A., & Steger, M. B. (2015). *Urban Sustainability in Theory and Practice*. Routledge.

Jamison, M. S. (1985). The joy of gardening: Collectivist and bureaucratic cultures in conflict. *The Sociological Quarterly*, *26*(4), 473–490. doi:10.1111/j.1533-8525.1985.tb00239.x

Jamshidian, M., Tehrany, E. A., Imran, M., Jacquot, M., & Desobry, S. (2010). Poly-Lactic Acid: Production, applications, nanocomposites, and release studies. *Comprehensive Reviews in Food Science and Food Safety*, *9*(5), 552–571. doi:10.1111/j.1541-4337.2010.00126.x

Jane, J. L., & Wang, S. (1996). *U.S. Patent No. 5,523,293*. Washington, DC: U.S. Patent and Trademark Office.

Jansson, J., Marell, A., & Nordlund, A. (2010). Green consumer behavior: Determinants of curtailment and eco-innovation adoption. *J. Consum. Mark.*

Jayasekara, R., Harding, I., Bowater, I., Christie, G. B. Y., & Lonergan, G. T. (2004). Preparation, surface modification and characterisation of solution cast starch PVA blended films. *Polymer Testing*, *23*(1), 17–27. doi:10.1016/S0142-9418(03)00049-7

Jendrossek, D., & Handrick, R. (2002). Microbial degradation of polyhydroxyalkanoates. *Annual Review of Microbiology*, *56*(1), 403–432. doi:10.1146/annurev.micro.56.012302.160838 PMID:12213937

Jenkins, W. (2009). *Berkshire encyclopaedia of sustainability: the spirit of sustainability* (1st ed., Vol. 1). Berkshire Publishing Group.

Jenks, M., & Jones, C. (Eds.). (2010). Dimensions of the sustainable city. New York: Springer.

Jia, D. Y., & Ding, T. (2004). Detection of foreign fibres in cotton using NIR optimal wavelength imaging. *IEEE Conference Publications*, 751 – 752.

Johansen, S. (1988). Statistical analysis of cointegration vectors. *Journal of Economic Dynamics & Control*, *12*(2), 231–254. doi:10.1016/0165-1889(88)90041-3

Johansen, S., & Juselius, K. (1992). Testing structural hypothesis in a multivariate cointegration analysis of the PPP and the UIP for UK. *Journal of Econometrics*, *53*(1-3), 211–244. doi:10.1016/0304-4076(92)90086-7

John, A., Pecchenino, R., Schmmelpfennig, D., & Schreft, S. (1995). Short- lived agents and the long- lived environment. *Journal of Public Economics*, *5*(1), 12–20. doi:10.1016/0047-2727(94)01459-2

Johnson, D., & Turner, C. (2010). *International business: Themes and issues in the modern global economy* (2nd ed.). Routledge, Taylor & Francis Group. doi:10.4324/9780203861615

Johnson, E. A., & Miyanishi, K. (Eds.). (2007). *Plant disturbance ecology: The process and the response*. Academic Press.

Johnston, D. (2001). *Sustainable energy for future generations*. NEA News 2001 – No. 19. Retrieved from https://www.oecd-nea.org/pub/newsletter/2001/sustainable-energy19-1.pdf

Jones, L. (2005). The explosion of garden visiting in France. In T. Richardson & N. Kingsbury (Eds.), *Vista: The Culture and Politics of Gardens* (pp. 52–65). FrancesLincoln.

Jorgenson, A. K., Dick, C., & Mahutga, M. C. (2007). Foreign investment dependence and the environment: An eco-structural approach. Society for the Study of Social Problems, 54(3), 371-394.

Joshi, S. (2008). Can nanotechnology improve the sustainability of biobased products? The case of layered silicate biopolymer nanocomposites. *Journal of Industrial Ecology*, *12*(3), 474–489. doi:10.1111/j.1530-9290.2008.00039.x

Kahle, L. R., & Gurel-Atay, E. (Eds.). (2014). *Communicating Sustainability for the Green Economy*. M.E. Sharpe.

Kahlen, L., Kurdziel, M., Day, T., & Schiefer, T. (2019). *The role of geothermal and coal in Kenya's electricity sector and implications for sustainable development*. Ambition to Action.

Kale, G., Auras, R., & Singh, S. P. (2007). Comparison of the degradability of poly (lactide) packages in composting and ambient exposure conditions. *Packaging Technology and Science: An International Journal*, *20*(1), 49–70. doi:10.1002/pts.742

Kalin Arroyo, M. T., Zedler, P. H., & Fox, M. D. (Eds.). (1995). *Ecology and biogeography of Mediterranean ecosystems in Chile, California, and Australia*. Springer-Verlag. doi:10.1007/978-1-4612-2490-7

Kammerer, D. (2009). The effects of customer beneðt and regulation on environmental product innovation. Empirical evidence from appliance manufacturers in Germany. *Ecological Economics*, 68.

Kantola, M., & Helén, H. (2001). Quality changes in organic tomatoes packaged in biodegradable plastic films. *Journal of Food Quality*, *24*(2), 167–176. doi:10.1111/j.1745-4557.2001.tb00599.x

Karami, H. (2013). Heavy metal removal from water by magnetite nanorods. *Chemical Engineering Journal*, *219*, 209–216. doi:10.1016/j.cej.2013.01.022

Karayiannis, Y. A., Stojanovic, R., Mitropoulos, P., Koulamas, C., Stouraitis, T., Koubias, S., & Papadopoulos, G. (1999). Defect Detection and Classification on Web Textile Fabric using Multiresolution Decomposition and Neural Networks. *IEEE Conference Publications*, 765 - 768. 10.1109/ICECS.1999.813221

Kariptaş, F. S., Yarael, B., & Ünver, B. (2012). *Ofis Mekanlarında Ergonominin Çalışanlar Üzerindeki Psikolojik Etkileri. 18.* Ulusal Ergonomi Kongresi.

Kasterine, A., & Vanzetti, D. M. (2010). The Effectiveness, Efficiency and Equity of Market Based and Voluntary Measures to Mitigate Greenhouse Gas Emissions from the Agri-Food Sector (February 24, 2010). *UNCTAD Trade and Environment Review, 2010*. Available at SSRN: https://ssrn.com/abstract=1768948

Kates, R., Parris, T., & Leiserowitz, A. (2005). What is Sustainable Development? Goals, Indicators, Values, and practice. *Environment*, *47*(3), 8–21. doi:10.1080/00139157.2005.10524444

Kaufman, R. K., Davidsdottir, B., Garnham, S., & Pauly, P. (1998). The determinants of atmospheric SO_2 concentrations: Reconsidering the environmental Kuznets curve. *Ecological Economics*, *25*(2), 209–220. doi:10.1016/S0921-8009(97)00181-X

Kazimierczuk, A. H. (2019). Wind energy in Kenya: A status and policy framework review. *Renewable & Sustainable Energy Reviews*, *107*, 434–445. doi:10.1016/j.rser.2018.12.061

Kellogg, D. M. (2017). *The Global Sanitation Crisis: A Role for Business. In Beyond the bottom line: integrating sustainability into business and management practice.* Greenleaf Publishing.

Kemp, R., & Arundel, A. (1998). *Survey indicators for environmental innovation*. Academic Press.

Kennedy, C., Pincelt, S., & Bunje, P. (2011). The study of urban metabolism and its applications to urban planning and design. *Environmental Pollution*, *159*(8-9), 1965–1973. doi:10.1016/j.envpol.2010.10.022 PMID:21084139

Kenya Renewable Energy Association (KEREA). (2017). *Wind Energy. Renew Sources*. Available from: http://kerea.org/renewablesources/wind-energy/

Kim, M. K., Choi, K.-M., Yin, C.-R., Lee, K.-Y., Im, W.-T., Lim, J. H., & Lee, S.-T. (2004). Odorous swine wastewater treatment by purple non-sulfur bacteria, Rhodopseudomonas palustris, isolated from eutrophicated ponds. *Biotechnology Letters*, *26*(10), 819–822. doi:10.1023/B:BILE.0000025884.50198.67 PMID:15269554

Kim, S., Park, C. M., Jang, M., Son, A., Her, N., Yu, M., Snyder, S., Kim, D.-H., & Yoon, Y. (2018). Aqueous removal of inorganic and organic contaminants by graphene-based nanoadsorbents: A review. *Chemosphere*, *212*, 1104–1124. doi:10.1016/j.chemosphere.2018.09.033 PMID:30286540

Kimuyu, P., Mutua, J. & Wainaina, J. (2015). *Role of renewable energy in promoting inclusive and sustainable development in Kenya*. European Report on Development.

Kinyoki, D. (2020). Mapping child growth failure across low- and middle-income countries. *Nature*, *577*(7789), 231–234. doi:10.103841586-019-1878-8 PMID:31915393

Kinzig, A. P., Pacala, S. W., & Tilman, D. (Eds.). (2002). *The functional consequences of biodiversity*. Princeton University Press.

Kiplagat, J., Wang, R., & Li, T. (2011). Renewable energy in Kenya: Resource potential and status of exploitation. *Renewable & Sustainable Energy Reviews*, *15*(6), 2960–2973. doi:10.1016/j.rser.2011.03.023

Kizilkaya, O. (2017). The impact of economic growth and foreign direct investment on CO_2 emissions: The case of Turkey. *Turkish Economic Review*, *4*(1), 106–118.

Koestoer, Y. T. (2007). Corporate Social Responsibility in Indonesia Building internal corporate values to address challenges in CSR Implementation. In Good Corporate and Social Governance in Promoting ASEAN's Regional Integration, Jakarta, Indonesia.

Koide, S., & Shi, J. (2007). Microbial and quality evaluation of green peppers stored in biodegradable film packaging. *Food Control*, *18*(9), 1121–1125. doi:10.1016/j.foodcont.2006.07.013

Kolasa, J., & Pickett, S. T. A. (2005). Changing academic perspectives of ecology: A view from within. In M. J. Mappin & E. A. Johnson (Eds.), *Environmental education and advocacy* (pp. 50–71). Cambridge University Press.

Komarek, R. J., Gardner, R. M., Buchanan, C. M., & Gedon, S. (1993). Biodegradation of radiolabeled cellulose acetate and cellulose propionate. *Journal of Applied Polymer Science*, *50*(10), 1739–1746. doi:10.1002/app.1993.070501009

Kostof, S. (1991). *The City Shaped: Urban Patterns and Meanings Through History*. Thames and Hudson.

Kremen, C., Bacon, C., & Iles, A. (2012). Diversified farming systems: An agro-ecological, systems-based alternative to modern industrial agriculture. *Ecology and Society*, *17*(4), 44. doi:10.5751/ES-05103-170444

Kremen, C., & Miles, A. (2012). Ecosystem services in biologically diversified versus conventional farming systems: Benefits, externalities, and trade-offs. *Ecology and Society*, *17*(4), 40. doi:10.5751/ES-05035-170440

Krzan, A., Hemjinda, S., Miertus, S., Corti, A., & Chiellini, E. (2006). Standardization and certification in the area of environmentally degradable plastics. *Polymer Degradation & Stability*, *91*(12), 2819–2833. doi:10.1016/j.polymdegradstab.2006.04.034

Kucukoglu, M., & Ibrahim, P. (2015). Positive influences of green innovation on company Performance. *Social and Behavioral Sciences, 195.*

Kuhn, W. (2003). Semantic reference systems. *International Journal of Geographical Information Science*, *17*(5), 405–409. doi:10.1080/1365881031000114116

Kunduru, K. R., Nazarkovsky, M., Farah, S., Pawar, R. P., Basu, A., & Domb, A. J. (2017). Nanotechnology for water purification: applications of nanotechnology methods in wastewater treatment. In *Water Purification* (pp. 33–74). Elsevier., doi:10.1016/B978-0-12-804300-4.00002-2

Kurozumi, E., & Yamamoto, T. (2000). Modified lag augmented autoregressions. *Econometric Reviews*, *19*(2), 207–231. doi:10.1080/07474930008800468

Kuru, R., & Türkyılmaz, Ç. C. (2019). Kütüphane Yapılarının Mekansal Organizasyonunun Ergonomik Açıdan Değerlendirilmesi: Bahçeşehir Üniversitesi Kütüphane Binası Örneği. *Ergonomi Dergisi*, *2*(3), 153–166.

Kuznets, S. (1995). Economic growth and income inequality. *The American Economic Review*, *45*, 1–28.

Lal, R. (2012). Urban ecosystems and climate change. In R. Lal & B. Augustin (Eds.), *Carbon sequestration in urban ecosystems* (pp. 3–19). Springer Science. doi:10.1007/978-94-007-2366-5_1

Lambin, E. F. (2005). Conditions for sustainability of human-environment systems: Information, motivation, and capacity. *Global Environmental Change*, *15*(3), 177–180. doi:10.1016/j.gloenvcha.2005.06.002

Lansing, J. S. (1991). *Priests and Programmers: Technologies of Power in the Engineered Landscape of Bali*. Princeton University Press.

Laohaprapanon, S., Matahum, J., Tayo, L., & You, S.-J. (2015). Photodegradation of Reactive Black 5 in a ZnO/UV slurry membrane reactor. *Journal of the Taiwan Institute of Chemical Engineers*, *49*, 136–141. doi:10.1016/j.jtice.2014.11.017

Larsen, O. M. (2005). *Governing innovation for sustainable development: integration of environmental nad innovation policies in Norway*. ProSus Working Paper.

Lassoie, J. P., & Buck, L. E. (2000). Development of agroforestry as an integrated land use management strategy. In H. E. Garrett, W. J. Rietveld, & R. F. Fisher (Eds.), *North American agroforestry: an integrated science and practice* (pp. 1–29). American Society of Agronomy.

Lata, S., & Samadder, S. R. (2016). Removal of arsenic from water using nano adsorbents and challenges: A review. *Journal of Environmental Management*, *166*, 387–406. doi:10.1016/j.jenvman.2015.10.039 PMID:26546885

Lazard, M. (2016). *Lazard's levelized cost of energy analysis—version 10.0*. Available from: https://www.lazard.com/media/438038/levelized-cost-of-energy-v100.pdf

LEAD (Leadership for Environment and Development). (2016). *Sustainable Development Goals (SDGs) 2015-2030*. June 2016 Briefing Note.

Lee, Y.-C., & Yang, J.-W. (2012). Self-assembled flower-like TiO2 on exfoliated graphite oxide for heavy metal removal. *Journal of Industrial and Engineering Chemistry*, *18*(3), 1178–1185. doi:10.1016/j.jiec.2012.01.005

Lele, S. M. (1991). Sustainable development: A Critical Review. *World Development*, *19*(6), 607–621. doi:10.1016/0305-750X(91)90197-P

Lemos, P. C., Serafim, L. S., & Reis, M. A. (2006). Synthesis of polyhydroxyalkanoates from different short-chain fatty acids by mixed cultures submitted to aerobic dynamic feeding. *Journal of Biotechnology*, *122*(2), 226–238. doi:10.1016/j.jbiotec.2005.09.006 PMID:16253370

Levin, S. (1999). *Fragile dominion: complexity and the commons*. Perseus Books.

Levkoe, C. Z. (2011). Towards a transformative food politics. *Local Environment*, *16*(7), 687–705. doi:10.1080/13549839.2011.592182

Liam, M., Andy, S., Paul, J., Thom, J. A., Padgham, L., Hickmott, S., Deng, H., & Cahill, F. (2013). Reframing social sustainability reporting: Towards an engaged approach. *Environment, Development and Sustainability*, *15*(1), 225–243. doi:10.100710668-012-9384-2

Liang, F., Wang, Y., & Sun, X. S. (1999). Curing process and mechanical properties of protein-based polymers. *Journal of Polymer Engineering, 19*(6), 383-394.

Liao, H. T., & Wu, C. S. (2009). Preparation and characterization of ternary blends composed of polylactide, poly (ε-caprolactone) and starch. *Materials Science and Engineering A*, *515*(1-2), 207–214. doi:10.1016/j.msea.2009.03.003

Li, J., Du, Q., Xi, B., & Li, Y. (2018). Hyperspectral Image Classification via Sample Expansion for Convolutional Neural Network. *IEEE Conference Publications*, 1 - 5. 10.1109/WHISPERS.2018.8747245

Linh, D. H., & Lin, S. M. (2015). Dynamic causal relationships among CO_2 emissions, energy consumption, economic growth and FDI in the most populous Asian countries. *Advances in Management and Applied Economics*, *5*(1), 69–88.

Linnenluecke, M. K., Griffiths, A., & Winn, M. (2012). Extreme Weather Events and the Critical Importance of Anticipatory Adaptation and Organizational Resilience in Responding to Impacts. *Business Strategy and the Environment, 21*(1), 17–32. doi:10.1002/bse.708

Lin, Q., Xu, X., & Zhan, Y. (2011). Extracting Regions of Interest Based on Visual Attention Model. *IEEE Conference Publications*, 313 - 316.

Liu, J., & Edwin, L. (2015). Affective Property Computation of Visual Texture. *IEEE Conference Publications*, 52 - 57.

Liu, J., Zuol, B., Vroman, P., Rabenasolov, B., & Zeng, X. (2009). Visual Quality Recognition of Nonwovens Based on Wavelet Transform and LVQ Neural Network. *IEEE Conference Publications*, 1885 - 1890. 10.1109/ICCIE.2009.5223507

Liu, L. (2006). Bioplastics in food packaging: Innovative technologies for biodegradable packaging. *San Jose State University Packaging Engineering, 13*, 1348–1368.

Liu, L., Luo, X.-B., Ding, L., & Luo, S.-L. (2019). Application of Nanotechnology in the Removal of Heavy Metal From Water. In *Nanomaterials for the Removal of Pollutants and Resource Reutilization* (pp. 83–147). Elsevier. doi:10.1016/B978-0-12-814837-2.00004-4

Liu, S., & Zhang, L. (2008). Textile Pattern Generation Technique Based on Quasi-regular Pattern Theory and Their Transform. *IEEE Conference Publications*, 264 – 266.

Li, X., & Liu, X. (2005). Foreign direct investment and economic growth: An increasingly endogenous relationship. *World Development, 33*(3), 393–407. doi:10.1016/j.worlddev.2004.11.001

Ljungqvist, J., Barthel, S., Finnveden, G., & Sörlin, S. (2010). The Urban Anthropocene: Lessons for Sustainability from the Environmental History of Constantinople. In P. Sinclair, F. Herschend, C. Isendahl, & G. Nordquist (Eds.), *The Urban Mind, cultural and environmental dynamics Studies in Global Archaeol-ogy* (Vol. 15, pp. 368–394). University Press.

Local Burden of Disease Educational Attainment Collaborators. (2020). Mapping disparities in education across low- and middle-income countries. *Nature, 577*(7789), 235–238. doi:10.103841586-019-1872-1 PMID:31875853

Lofrano, G., & Brown, J. (2010). Wastewater management through the ages: A history of mankind. *The Science of the Total Environment, 408*(22), 5254–5264. doi:10.1016/j.scitotenv.2010.07.062 PMID:20817263

Lopez, R. (1994). The environment as a factor of production: The effects of economic growth and trade liberalisation. *Journal of Environmental Economics and Management, 27*(2), 163–184. doi:10.1006/jeem.1994.1032

Loreau, M., Mouquet, N., & Gonzalez, A. (2003). Biodiversity as spatial insurance in heterogeneous landscapes. *Proceedings of the National Academy of Sciences of the United States of America, 100*(22), 12765–12770. doi:10.1073/pnas.2235465100 PMID:14569008

Loreau, M., Naeem, S., & Inchausti, P. (2001). Biodiversity and ecosystem functioning: Current knowledge and future challenges. *Science, 294*(5543), 804–808. doi:10.1126cience.1064088 PMID:11679658

Lovell, S. (2010). Multifunctional urban agro ecology for sustainable land use planning in the United States. *Sustainability, 2*(8), 2499–2522. doi:10.3390u2082499

Lovell, S. T., DeSantis, S., Nathan, C. A., Olson, M. B., Mendez, V. E., Kominami, H. C., Erickson, D. L., Morris, K. S., & Morris, W. B. (2010). Integrating agroecology and landscape multifunc-tionality in Vermont: An evolving framework to evaluate the design of agroecosystems. *Agricultural Systems, 103*(5), 327–341. doi:10.1016/j.agsy.2010.03.003

Luck, G. W., Daily, G. C., & Ehrlich, P. R. (2003). Population diversity and ecosystem services. *Trends in Ecology & Evolution, 18*(7), 331–336. doi:10.1016/S0169-5347(03)00100-9

Ludwig, D., Brock, W. A., & Carpenter, S. R. (2006). Uncertainty in discount models and environmental accounting. *Ecology and Society*, 10.

Lundberg, J., & Moberg, F. (2003). Mobile link organisms and ecosystem functioning: Implications for ecosystem resilience and management. *Ecosystems (New York, N.Y.)*, *6*(1), 87–98. doi:10.100710021-002-0150-4

Lu, Y., Tighzert, L., Dole, P., & Erre, D. (2005). Preparation and properties of starch thermoplastics modified with waterborne polyurethane from renewable resources. *Polymer*, *46*(23), 9863–9870. doi:10.1016/j.polymer.2005.08.026

MA (Millennium Ecosystem Assessment). (2005). *Ecosystems and Human Well-Being: Synthesis*. Island Press.

Ma, Y. (2018). The sources of green management innovation: Does internal efficiency demand pull or external knowledge supply push? *J. Clean. Prod.*

MacArthur, R. H., & Wilson, E. O. (1967). *The Theory of Island Biogeography*. Princeton University Press.

MacDaniels, L. H., & Lieberman, A. S. (1979). Tree crops: A neglected source of food and forage from marginal lands. *Bioscience*, *29*(3), 173–175. doi:10.2307/1307798

Machlis, G. E., Force, J. E., & Burch, W. R. Jr. (1997). The human ecosystem. 1. The human ecosystem as an organizing concept in ecosystem management. *Society & Natural Resources*, *10*(4), 347–367. doi:10.1080/08941929709381034

Mahmure, O., Selam, A., Firat, S., Kara, M., & Ozel, S. (2015). A comparative analysis of renewable energy use and policies: Global and Turkish Perspectives. *Sustainability*, *7*(12), 16379–16407. doi:10.3390u71215820

Makino, Y., & Hirata, T. (1997). Modified atmosphere packaging of fresh produce with a biodegradable laminate of chitosan-cellulose and polycaprolactone. *Postharvest Biology and Technology*, *10*(3), 247–254. doi:10.1016/S0925-5214(96)01402-0

Mallakpour, S., & Rashidimoghadam, S. (2019). Carbon Nanotubes for Dyes Removal. In *Composite Nanoadsorbents* (pp. 211–243). Elsevier. doi:10.1016/B978-0-12-814132-8.00010-1

Malla, S. (2013). Household energy consumption patterns and its environmental implications: Assessment of energy access and poverty in Nepal. *Energy Policy*, *61*, 990–1002. doi:10.1016/j.enpol.2013.06.023

Mangi, P. (2018). Geothermal development in Kenya-country updates. *Proceedings, 7th African Rift Geothermal Conference.*

Mares, T., & Alkon, A. (2011). Mapping the food movement: Addressing inequality and neoliberalism. *Environment and Society*, *2*(1), 68–86. doi:10.3167/ares.2011.020105

Maria Ribeiro, S., Maria Bógus, C., & Wada Watanabe, H. A. (2015). Agroecological urban agro ecology from the perspective of health promotion. Saúde e Sociedade, 24(2), 730-743.

Markowski, H. (2012). *Textiles for Textiles (T4T)*. www.textiles4textiles.eu

Martinez-Alier, J. (2002). *Environmentalism of the poor: a study of ecological conflicts & valuation.* Edward Elgar. doi:10.4337/9781843765486

Martínez-Torres, M. E., & Rosset, P. M. (2010). La Vía Campesina: The birth and evolution of a transnational social movement. *The Journal of Peasant Studies*, *37*(1), 149–175. doi:10.1080/03066150903498804

Martin, O., Schwach, E., Averous, L., & Couturier, Y. (2001). Properties of biodegradable multilayer films based on plasticized wheat starch. *Stärke*, *53*(8), 372–380. doi:10.1002/1521-379X(200108)53:8<372::AID-STAR372>3.0.CO;2-F

Masolo, C., & Borgo, S. (2005). *Qualities in formal ontology. InFoundational Aspects ofOntologies (FOnt 2005).* Workshop at KI.

Masolo, C., Borgo, S., Gangemi, A., Guarino, N., & Oltramari, A. (2003). *WonderWeb Deliverable D18 Ontology Library*. Final.

Ma, X., Chang, P. R., & Yu, J. (2008). Properties of biodegradable thermoplastic pea starch/carboxymethyl cellulose and pea starch/microcrystalline cellulose composites. *Carbohydrate Polymers*, *72*(3), 369–375. doi:10.1016/j.carbpol.2007.09.002

Ma, X., Yu, J., & Wang, N. (2008). Glycerol plasticized-starch/multiwall carbon nanotube composites for electroactive polymers. *Composites Science and Technology*, *68*(1), 268–273. doi:10.1016/j.compscitech.2007.03.016

May, R. M. (1974). *Stability and Complexity in Model Ecosystems*. Princeton University Press.

Mbaka, J., & Mwaniki, M. (2016). Small hydro-power plants in Kenya: A review of status, challenges and future prospects. *Journal of Renewable Energy and Environment*, *3*(4), 20–26.

McCall, T. (1973). *Legislative address, Governor Tom McCall, Oregon, (1973).* Retrieved July 20, 2008, from http://arcweb.sos.state.or.us/governors/McCall/legis1973.html

McCollum, D., Zhou, W., Bertram, C., Boer, H., & Bosetti, V. (2018). Energy investment needs for fulfilling the Paris agreement and achieving sustainable development goals. *Nature Energy*, *3*(7), 589–599. doi:10.103841560-018-0179-z

McConnell, K. E. (1997). Income and the demand for environmental quality. *Environment and Development Economics*, *2*(4), 383–399. doi:10.1017/S1355770X9700020X

McCrone, A. (2015). *Frankfurt School-UNEP Centre/BNEF. Global Trends in Renewable Energy Investment*. Frankfurt School of Finance & Management.

McDonnell, M. J., & Hahs, A. (2009). Comparative ecology of cities and towns: Past, present and future. In M. J. McDonnell, A. Hahs, & J. Breuste (Eds.), *Ecology of cities and towns: A comparative approach* (pp. 71–89). Cambridge University Press. doi:10.1017/CBO9780511609763.006

McIntyre, & ... (Eds.). (2009). *The International Assessment of Agricultural Knowledge, Science and Technology for Development: Summary for Decision-Makers Report of the Global Report*. Island Press. M. A.

McKinney, M. L. (2002). Urbanization, biodiversity, and conservation. *Biol. Sci.*, *52*, 883–890.

McKinney, M. L. (2002). Urbanization, biodiversity, and conservation. *Bioscience*, *52*(10), 883–890. doi:10.1641/0006-3568(2002)052[0883:UBAC]2.0.CO;2

Mehta, P. S. (2002). The Indian mining sector: Effects on the environment & FDI inflows. *CCNM Global Forum on International Investment, Conference on Foreign Direct investment and the Environment.*

Mei, S., Liu, X., Zhang, G., & Du, Q. (2019). Sensor-speciðc Transfer Learning for Hyperspectral Image Processing. *IEEE Conference Publications*, 1-6.

Mei, S., Yuan, X., Ji, J., Wan, S., Hou, J., & Du, Q. (2017). Hyperspectral Image Super-resolution via Convolutional Neural Network. *IEEE Conference Publications*, 4297 – 4301. 10.1109/ICIP.2017.8297093

Micale, V., Trabacchi, C., & Boni, L. (2015). *Using Public Finance to Attract Private Investment in Geothermal: Olkaria III Case Study, Kenya (2015).* Climate Policy Initiative, Venice, Italy. Available from: http://climatepolicyinitiative.Org/wpcontent/uploads/2015/06/150601_Final_Olkaria_ForWeb.Pdf

Middleton. (2020). *Enhanced Machine Vision (AI).* https://www.middletonspectral.com/applications-project-examples/applications/enhanced-machine-vision-ai/

Millennium Ecosystem Assessment (MA). (2005). *Synthesis Island Press.* Available on the Internet http://www.MAweb.org

Millennium Ecosystem Assessment (MA). (2005). *Synthesis*. Island Press. Available on the Internet http://www.MAweb.org

Millennium Ecosystem Assessment, M. A. (2003). *Ecosystems and their services. Chapter 2 in Ecosystems and human well-being*. Island Press.

Millennium Ecosystem Assessment, M. A. (2005c). *Ecosystems and human well-being: policy responses*. Island Press.

Miller, J. R. (2005). Biodiversity conservation and the extinction of experience. *Trends in Ecology & Evolution*, *20*(8), 430–434. doi:10.1016/j.tree.2005.05.013 PMID:16701413

Minelli, F. (2017). *Achieving energy security And Environmental sustainable development*. ODUMUNC Issue Brief Economic and Social Council (ECOSOC). Retrieved from https://www.odu.edu/content/dam/odu/offices/mun/2017/ib-2017-ecosoc-energy-security.pdf

Ministry of Energy. (2008). *Strategy for the Development of the Biodiesel Industry in Kenya 2008-2012, Final Draft*. Government Printer.

Ministry of Energy. (2016). *Active developers of feed-in tariffs*. Author.

Ministry of Finance. (2011). Physical Infrastructure Sector Medium Term Expenditure Framework (MTEF) Report, 2011/12 – 2013/14. Nairobi: Government Printer.

Mohanty, A. K., Wibowo, A., Misra, M., & Drzal, L. T. (2003). Development of renewable resource–based cellulose acetate bioplastic: Effect of process engineering on the performance of cellulosic plastics. *Polymer Engineering and Science*, *43*(5), 1151–1161. doi:10.1002/pen.10097

Mok, H. F., Williamson, V. G., Grove, J. R., Burry, K., Barker, F., & Hamilton, A. J. (2014). Strawberry fields forever? Urban agriculture in developed countries: A review. *Agronomy for Sustainable Development*, *34*(1), 21–43. doi:10.100713593-013-0156-7

Moksnes, N., Korkovelos, A., Mentis, D., & Howells, M. (2017). Electrification pathways for Kenya-linking spatial electrification analysis and medium to long term energy planning. *Environmental Research Letters*, *12*(9), 095008. doi:10.1088/1748-9326/aa7e18

Mokveld, K., & Eije, S. (2018). Final energy report Kenya. Netherlands Enterprise Agency, RVO.

Mollison, B., & Holmgren, D. (1978). *Permaculture one: a perennial agricultural system for human settlements*. Tagari, Tyalgum.

Mollison, B. (1979). *Permaculture: a designer's manual*. Tagari Publications.

Morgan, K. (2009). Editorial: Feeding the city: The challenge of urban food planning. *International Planning Studies*, *14*(4), 341–348. doi:10.1080/13563471003642852

Morgan, K., & Sonnino, R. (2010). The urban foodscape: World cities and the new food equation, Cambridge Journal of Regions, Economy and Society, Cambridge Political. *Economy and Society*, *3*(2), 209–224.

Mo, X., Sun, X. S., & Wang, Y. (1999). Effects of molding temperature and pressure on properties of soy protein polymers. *Journal of Applied Polymer Science*, *73*(13), 2595–2602. doi:10.1002/(SICI)1097-4628(19990923)73:13<2595::AID-APP6>3.0.CO;2-I

Mukherji, N., & Morales, A. (2010). *Zoning for urban Agriculture*. Zoning Practice 3.

Müller, C. M., Laurindo, J. B., & Yamashita, F. (2011). Effect of nanoclay incorporation method on mechanical and water vapor barrier properties of starch-based films. *Industrial Crops and Products*, *33*(3), 605–610. doi:10.1016/j.indcrop.2010.12.021

Munasinghe, M. (2002). The Sustainomics Trans-disciplinary Meta-framework for making Development more Sustainable: Applications to energy issues. *International Journal of Sustainable Development*, *4*(2), 6–54. doi:10.1504/IJSD.2002.002563

Muok, B., & Palit, D. (2015). *Solar photovoltaic for enhancing electricity access in Kenya: what policies are required?* Policy Brief.

Musacchio, L. R. (2008). Metropolitan landscape ecology: Using translational research to increase sustainability, resilience, and regeneration. *Landscape Journal*, *27*(1), 1–8. doi:10.3368/lj.27.1.1

Nakano, D., & Muniz, J. Jr. (2018). Writing the literature review for an empirical paper. *Production*, *28*(0), e20170086. doi:10.1590/0103-6513.20170086

Nally, D. (2014). Governing precarious lives: Land grabs, geopolitics, and 'food security'. *The Geographical Journal*, *181*(4), 340–349. doi:10.1111/geoj.12063

Narayan, R., & Patel, M. (2003). Review and analysis of bio-based product LCA's. In *Proceedings of the International Workshop Assessing the Sustainability of Bio-based Products*. Institute for Science & Public Policy.

Nasrollahzadeh, M., Sajadi, S. M., Sajjadi, M., & Issaabadi, Z. (2019). Applications of Nanotechnology in Daily Life. In Interface Science and Technology (pp. 113–143). doi:10.1016/B978-0-12-813586-0.00004-3

National Land Use Database. (2000) *Previously Developed Land (PDL): Vacant Sites and Derelict Sites, and Other Previously Developed Land and Buildings that May be Available for Redevelopment, Data Specification*. NLUD – a partnership Rethinking urban capacity.

National Land Use Database. (2000). *Previously Developed Land (PDL): Vacant Sites and Derelict Sites, and Other Previously Developed Land and Buildings that May be Available for Redevelopment, Data Specification*. NLUD – a partnership Rethinking urban capacity.

Naveh, Z. (2000). The total human ecosystem: Integrating ecology and economics. *Bioscience*, *50*, 357–361.

Nazlioglu, S., Gormus, A., & Soytas, U. (2016). Oil Prices and Real Estate Investment Trusts (REITs): Gradual-Shift Causality and Volatility Transmission Analysis. *Energy Economics*, *60*(1), 168–175. doi:10.1016/j.eneco.2016.09.009

Ndegwa, G., Moraa, V., Jamnadass, R., Mowo, J., Nyabenge, M., & Iiyama, M. (2011). *Potential for biofuel feedstock in Kenya*. ICRAF Working Paper No. 139 Nairobi: World Agroforestry Centre.

Nelson, D. R., Adger, W. N., & Brown, K. (2007). Adaptation to environmental change: Contributions of a resilience framework. *Annual Review of Environment and Resources*, *32*, 395–419.

New York City Council. (2010). *FoodWorks: a vision to improve NYC's food system*. The New York City Council, New York City. Obtained from http://ngfn.org/resources/ngfndatabase/knowledge/foodworks_fullreport_11_22_10.pdf

Newell, P., Phillips, J., & Pueyo, A. (2014). *The political economy of low carbon energy in Kenya*. Institute of Development Studies. doi:10.1111/j.2040-0209.2014.00445.x

Ng, L. Y., Mohammad, A. W., Leo, C. P., & Hilal, N. (2013). Polymeric membranes incorporated with metal/metal oxide nanoparticles: A comprehensive review. *Desalination*, *308*, 15–33. doi:10.1016/j.desal.2010.11.033

Nnaji, C. O., Jeevanandam, J., Chan, Y. S., Danquah, M. K., Pan, S., & Barhoum, A. (2018). Engineered nanomaterials for wastewater treatment: current and future trends. In *Fundamentals of Nanoparticles* (pp. 129–168). Elsevier. doi:10.1016/B978-0-323-51255-8.00006-9

Nordahl, D. (2009). *Public Produce: The New Urban Agriculture*. Island Press.

Norris, F. H., Stevens, S. P., Pfefferbaum, B., Wyche, K. F., & Pfefferbaum, R. L. (2008). Community Resilience as a Metaphor, Theory, Set of Capacities, and Strategy for Disaster Readiness. *American Journal of Community Psychology*, *41*(1-2), 127–150. doi:10.100710464-007-9156-6 PMID:18157631

Nyström, M., & Folke, C. (2001). Spatial resilience of coral reefs. *Ecosystems (New York, N.Y.)*, *4*(5), 406–417. doi:10.100710021-001-0019-y

O' Bowman, A. M., & Pagano, M.A. (2004). Terra Incognita: Vacant Land and UrbanStrategies. Georgetown University Press.

O'Bowman, A. M., & Pagano, M.A. (2004). Terra Incognita: Vacant Land and UrbanStrategies. Georgetown University Press.

O'Brien, D. (2010b). Cultivating our garden: David Hume and gardening as therapy. In D. O'Brien (Ed.), *Gardening: Philosophy for Everyone* (pp. 192–203). Wiley-Blackwell. doi:10.1002/9781444324563.ch14

O'Brien, D. (Ed.). (2010a). *Gardening: Philosophy for Everyone*. Wiley-Blackwell. doi:10.1002/9781444324563

O'Rourke, T. (2007). Critical Infrastructure, Interdependencies, and Resilience. *The Bridge*, *37*(1), 22–29.

Odum, E. P. (1969). The strategy of ecosystem development. –. *Science*, *164*(3877), 262–270. doi:10.1126cience.164.3877.262 PMID:5776636

Odum, E. P. (1971). *Fundamentals of Ecology* (3rd ed.). W. B. Saunders.

OECD. (2001). The Application of Biotechnology to Industrial Sustainability. OECD Publishing.

OECD. (2007). OECD Contribution to the United Nations Commission on Sustainable Development 15 Energy for Sustainable Development. OECD Publishing.

Okamoto, K., Ichikawa, T., Yokohara, T., & Yamaguchi, M. (2009). Miscibility, mechanical and thermal properties of poly (lactic acid)/polyester-diol blends. *European Polymer Journal*, *45*(8), 2304–2312. doi:10.1016/j.eurpolymj.2009.05.011

Okvat, H., & Zautra, A. (2011). Community gardening: A parsimonious path to individual, community, and environmental resilience. *American Journal of Community Psychology*, *47*(3), 374–387. doi:10.100710464-010-9404-z PMID:21222153

Olson, R. K., Schoeneberger, M. M., & Aschmann, S. G. (2000). An ecological foundation for temperate agroforestry. In H. E. Garrett, W. J. Rietveld, & R. F. Fisher (Eds.), *North American agroforestry: an integrated science and practice* (pp. 31–61). American Society of Agronomy.

Oltra, V. (2008). *Environmental innovation and industrial dynamics: the contributions of evolutionary economics*. DIME Working Papers on Environmental Innovation.

Opitz, I., Specht, K., Bergers, R., Sieber, T. R., & Piorr, A. (2016): Toward Sustainability: Novelties, Areas of Learning and Innovation. *Urban Agriculture. Sustainability*, *8*, 1–18.

Opiyo, N. (2016). A survey informed PV-based cost-effective electrification options for rural sub-Saharan Africa. *Energy Policy*, *91*, 1–11. doi:10.1016/j.enpol.2015.12.044

Orlandini, S., Marta, A. D., & Mancini, M. (2006). The agroclimatic analysis at farm scale. *Meteorological Applications*, *13*(S1), 87. doi:10.1017/S1350482706002593

Osborne, M. (2015). *Urban farming's growing importance*. TechBerlin. Available at: https://techberlin.com/articles/urban-farmings-growing-importance/

Osterwalder, A., Pigneur, Y., & Clark, T. (2010). *Business model generation: A handbook for visionaries, game changers, and challengers*. Alexander Osterwalder & Yves Pigneur.

Östman, Ö., Ekbom, B., & Bengtsson, J. (2003). Yield increase attributable to aphid predation by ground-living polyphagous natural enemies in spring barley in Sweden. *Ecological Economics*, *45*(1), 149–158. doi:10.1016/S0921-8009(03)00007-7

Otte, A., Simmering, D., & Wolters, V. (2007). Biodiversity at the landscape level: Recent concepts and perspectives for multifunctional land use. *Landscape Ecology*, *22*(5), 639–642. doi:10.100710980-007-9094-6

Owino, H. (2017). *Kenya: solar energy drives socio-economic growth in off-grid regions*. Available from: https://scienceafrica.co.ke/kenya-solar-energy-drives-socio-economic-growth-in-off-grid-regions/

Owino, T., Kamphof, R., Kuneman, E., & Tilburg, X. (2016). Towards a green trajectory of economic growth and energy security in Kenya? Netherlands Institute of International Relations 'Clingendael', The Hague.

Owusu, P. A., & Asumadu-Sarkodie, S. (2016). A review of renewable energy sources, sustainability issues and climate change mitigation. *Cogent Engineering*, *3*(1), 1167990. doi:10.1080/23311916.2016.1167990

Özbek, M. (2004). *Kamusal Alan*. Hil Yayınları.

Öz, O. Z. (1992). Kütüphane Binalarının Mekansal Organizasyonlarında Ergonominin Önemi ve Standardizasyon. *Türk Kütüphaneciliği*, *6*(3), 159–171.

Paetau, I., Chen, C. Z., & Jane, J. L. (1994). Biodegradable plastic made from soybean products. 1. Effect of preparation and processing on mechanical properties and water absorption. *Industrial & Engineering Chemistry Research*, *33*(7), 1821–1827. doi:10.1021/ie00031a023

Pagano, M. A., & Bowman, A. O'M. (2000). Land in Cities: An Urban Resource. Washington, DC: Brookings Institution, Center on Urban and Metropolitan Policy.

Pananond, P., & Cuervo-Cazurra, A. (2018). The complementarity of foreign and domestic investment by emerging market multinationals. In Contemporary Issues in International Business: Institutions, Strategy and Performance. doi:10.1007/978-3-319-70220-9_7

Panayotou, T. (1993*). Empirical tests and policy analysis of environmental degradation at different stages of economic development*. Working Paper WP238, Technology and Employment Programme, International Labour Office (ILO).

Pao, H. T., & Tsai, C. M. (2011). Multivariate Granger causality between CO_2 emissions, energy consumption, FDI (foreign direct investment) and GDP (gross domestic product): Evidence from a panel of BRIC (Brazil, Russian Federation, India, and China) countries. *Energy*, *36*(1), 685–693. doi:10.1016/j.energy.2010.09.041

Paoletti, M. E., Haut, J. M., Fernandez-Beltran, R., Plaza, J., Plaza, A. J., & Pla, F. (2019, April). Capsule Networks for Hyperspectral Image Classification. *IEEE Journal of Transactions on Geoscience and Remote Sensing*, *57*(4), 2145–2160. doi:10.1109/TGRS.2018.2871782

Paoletti, M. E., Haut, J. M., Fernandez-Beltran, R., Plaza, J., Plaza, A. J., & Pla, F. (2019, February). Deep Pyramidal Residual Networks for Spectral–Spatial Hyperspectral Image Classification. *IEEE Journal of Transactions on Geoscience and Remote Sensing*, *57*(2), 740–754. doi:10.1109/TGRS.2018.2860125

Park, S. K., Hettiarachchy, N. S., & Were, L. (2000). Degradation Behavior of Soy Protein– Wheat Gluten Films in Simulated Soil Conditions. *Journal of Agricultural and Food Chemistry*, *48*(7), 3027–3031. doi:10.1021/jf0000272 PMID:10898660

Parvin, F., Rikta, S. Y., & Tareq, S. M. (2018). 8 -Application of Nanomaterials for the Removal of Heavy Metal from Wastewater. In Nanotechnology in Water and Wastewater Treatment: Theory and Applications (pp. 137–157). Elsevier. doi:10.1016/B978-0-12-813902-8.00008-3

Patel, R. (2009). What does food sovereignty like? *The Journal of Peasant Studies*, *36*(3), 663–706. doi:10.1080/03066150903143079

Patnaik, P. R. (2005). Perspectives in the modeling and optimization of PHB production by pure and mixed cultures. *Critical Reviews in Biotechnology*, *25*(3), 153–171. doi:10.1080/07388550500301438 PMID:16294831

Pawar, P. A., & Purwar, A. H. (2013). Biodegradable polymers in food packaging. *American Journal of Engineering Research*, *2*(5), 151–164.

Pearson, C. J. (2010). Guest editorial: Challenging, multidimensional agriculture in cities. *International Journal of Agricultural Sustainability*, *8*(1-2), 3–4. doi:10.3763/ijas.2009.c5008

Pelling, M., & Manuel-Navarrete, D. (2011). From resilience to transformation: The adaptive cycle in two Mexican urban centers. *Ecology and Society*, *16*(2), 11. doi:10.5751/ES-04038-160211

Penke, Y. K., Anantharaman, G., Ramkumar, J., & Kar, K. K. (2019). Redox synergistic Mn-Al-Fe and Cu-Al-Fe ternary metal oxide nano adsorbents for arsenic remediation with environmentally stable As(0) formation. *Journal of Hazardous Materials*, *364*, 519–530. doi:10.1016/j.jhazmat.2018.10.069 PMID:30388635

Peoples Food Sovereignty Network. (2007). *Statement on peoples' food sovereignty*. Available at: https://www2.world-governance.org/article70.html

Peraboni, C. (2010). *Reti ecologiche e infrastrutture verdi*. Maggioli Publisher.

Pesaran, M. H., & Shin, Y. (1999). An autoregressive distributed lag modelling approach to cointegration analysis. In S. Strom (Ed.), *Econometrics and Economic Theory in the 20th Century: The Ragnar Frisch Centennial Symposium*. Cambridge University Press. 10.1017/CCOL521633230.011

Pesaran, M. H., Shin, Y., & Smith, R. J. (2001). Bounds testing approaches to the analysis of level relationships. *Journal of Applied Econometrics*, *16*(3), 289–326. doi:10.1002/jae.616

Peterson, G. D. (2005). Ecological management: control, uncertainty and understanding. In *Ecological paradigms lost: routes of theory change*. Elsevier. https://www.elsevier.com/wps/find/bookdescription.cws_home/705214/description#description

Peterson, G., Allen, C.R., & Holling, C.S. (1998). Ecological resilience, biodiversity, and scale. *Ecosystems, 1*, 6–18.

Peterson, G., Allen, C. R., & Holling, C. S. (1998). Ecological resilience, biodiversity, and scale. *Ecosystems (New York, N.Y.)*, *1*(1), 6–18. doi:10.1007100219900002

Philley, J. (2006). *Collar hazards with a bowtie: Chemical Processing*. Putman Media. http://www.nxtbook.com/nxtbooks/putman/cp0106/index.php?startid=27

Philp, J. C., Bartsev, A., Ritchie, R. J., Baucher, M. A., & Guy, K. (2013). Bioplastics science from a policy vantage point. *New Biotechnology*, *30*(6), 635–646. doi:10.1016/j.nbt.2012.11.021 PMID:23220474

Pickett, S. T. A., Burch, W. R. Jr, Dalton, S., & Foresman, T. (1997). Integrated urban ecosystem research: Themes, needs, and applications. *Urban Ecosystems*, *1*(4), 183–184. doi:10.1023/A:1018579628818

Pickett, S. T. A., & Cadenasso, M. L. (2002). Ecosystem as a multidimensional concept: Meaning, model and metaphor. *Ecosystems (New York, N.Y.)*, *5*(1), 1–10. doi:10.100710021-001-0051-y

Pickett, S. T. A., Cadenasso, M. L., & Grove, J. M. (2004). Resilient cities: Meaning, models, and metaphor for integrating the ecological, socio-economic, and planning realms. *Landscape and Urban Planning*, *69*(4), 369–384. doi:10.1016/j.landurbplan.2003.10.035

Pickett, S. T. A., Groffman, P. M., Brush, G. S., Galvin, M. F., Grove, J. M., ... Troy, A. R. (2008). Exchanges across land-water-scape boundaries in urban systems: Strategies for reducing nitrate pollution. *Annals of the New York Academy of Sciences*, *1134*(1), 213–232. doi:10.1196/annals.1439.012 PMID:18566096

Pillin, I., Montrelay, N., & Grohens, Y. (2006). Thermo-mechanical characterization of plasticized PLA: Is the miscibility the only significant factor? *Polymer*, *47*(13), 4676–4682. doi:10.1016/j.polymer.2006.04.013

Pimbert, M. (2009). *Towards food sovereignty*. London: International Institute for Environment and Development. https://pubs.iied.org/pdfs/14585IIED.pdf

Planete-Energies. (2016, December). *Understanding electricity. Feature Report.* https://www.planete-energies.com/

Poirier, Y. (1999). Green chemistry yields a better plastic. *Nature Biotechnology*, *17*(10), 960–961. doi:10.1038/13652 PMID:10504694

Pooja, K., Nidamanuri, R. R., & Mishra, D. (2019). Multi-scale Dilated Resuudal Convolutional Neural network for Hyperspectral Image Classification. *IEEE Conference Publications*, 1 – 5.

Popa, M., & Belc, N. (2007). Packaging. In *Food Safety* (pp. 68–87). Springer. doi:10.1007/978-0-387-33957-3_4

Pope, S. E., Fahrig, L., & Merriam, H. G. (2000). Landscape complementation andmetapopulation effects on leopard frog populations. *Ecology*, *81*(9), 2498–2508. doi:10.1890/0012-9658(2000)081[2498:LCAMEO]2.0.CO;2

Postel, S. (1999). *Pillar of sand*. W. W. Norton & Company Ltd.

Postel, S., & Carpenter, S. R. (1997). Freshwater ecosystem services. In G. Daily (Ed.), *Nature's services* (pp. 195–214). Island Press.

Pothukuchi, K., & Kaufman, J. (1999). Placing the food system on the urban agenda: The role of municipal institutions in food systems planning. *Agriculture and Human Values*, *16*(2), 213–244. doi:10.1023/A:1007558805953

Pothukuchi, K., & Kaufman, J. (2000). The food system: A stranger to the planning field. *Journal of the American Planning Association*, *66*(2), 112–124. doi:10.1080/01944360008976093

Pretty, J. (2008). Agricultural sustainability: Concepts, principles and evidence. *Philosoph Trans Royal Soc. Botanical Sciences*, *363*(1491), 447–465. PMID:17652074

Priya, Ashok, & Varghese. (2011). A Novel Approach to Fabric Defect Detection Using Digital Image Processing. *IEEE Conference Publications*, 228 - 232. 10.1109/ICSCCN.2011.6024549

Priya, Ashok, & Varghese. (2012). Defect Detection In Wooven Fabric Using Weighted Morphology. *IEEE Conference Publications*, 1 - 6.

Progress of Goal 10 in 2017. (2017). *Sustainable Development Goal Knowledge Platform.*

Progress of Goal 9 in 2017. (2017). *Sustainable Development Knowledge Platform.*

Purcell, M., & Tyman, S. K. (2014). Cultivating food as a right to the city. *Local Environment*, *20*(10), 1132–1147. doi:10.1080/13549839.2014.903236

Queiroz, A. U., & Collares-Queiroz, F. P. (2009). Innovation and industrial trends in bioplastics. Journal of Macromolecular Science®, Part C. *Polymer Reviews (Philadelphia, Pa.)*, *49*(2), 65–78. doi:10.1080/15583720902834759

Quincerot, R., & Weil, M. (2009). *Genève 2020: Plan di-recteur communal de la Ville de Genève*. Ville de Genève.

Rahaman, Atiqur, Jussi, Hauta-Kasari, & Seyed Hossein. (2017). Enhanced color visualization by spectral imaging: an application in cultural heritage. *IEEE Conference Publications*, 1-6. 10.1109/ICIVPR.2017.7890870

Rajani, B., Kumar, R. S., Devi, M. U., & Nayak, J. B. (2016). *Role of Purple Non-Sulfur Bacteria Rhodopseudomonas Palustris Rsou000 and Rhodopseudomonas Thermotolerance Rsou555 in Waste Water Treatment.* doi:10.20959/wjpps20168-7428

Rajan, S. R., & Duncan, C. A. M. (2013). Ecologies of hope: Environment, technology and habitation: Case studies from the intervenient middle. *Journal of Political Ecology*, *20*(20), 73–79. doi:10.2458/v20i1.21758

Raj, S., Raja, S., & Dukes, B. (2016). Beneficial but constrained: Role of urban agro ecology programs in supporting healthy eating among youth. *Journal of Hunger & Environmental Nutrition*. Advance online publication. doi:10.1080/19320248.2015.1128865

Ramus, C. A. (2002). Encouraging innovative environmental actions: What companies and managers must do. *Journal of World Business*, *37*(2), 151–164. doi:10.1016/S1090-9516(02)00074-3

Rana, R., & Sharma, M. (2018). On the causality between electricity generation, energy consumption, investment patterns and CO2 emissions in India. *Proceedings, 2nd European Conference on Electrical Engineering & Computer Science (EECS 2018)*. 10.1109/EECS.2018.00023

Rana, R., & Sharma, M. (2019). Dynamic causality testing for EKC hypothesis, pollution haven hypothesis and international trade in India FDI. *The Journal of International Trade & Economic Development*, *28*(3), 348–364. doi:10.1080/09638199.2018.1542451

Rana, R., & Sharma, M. (2020). Dynamic causality among FDI, economic growth and CO_2 emissions in India: With open markets and technology gap. *International Journal of Asian Business and Information Management*, *11*(3), 3. doi:10.4018/IJABIM.2020070102

Rao, M., Tang, L., Tang, P., & Zhang, Z. (2019). ES-CNN: An End-to-End Siamese Convolutional Neural Network for Hyperspectral Image Classification. *IEEE Conference Publications*, 1 – 4. 10.1109/JURSE.2019.8808991

Rasal, R. M., Janorkar, A. V., & Hirt, D. E. (2010). Poly (lactic acid) modifications. *Progress in Polymer Science*, *35*(3), 338–356. doi:10.1016/j.progpolymsci.2009.12.003

Rasanen, T., Juhani Ruuskanen, J., & Kolehmainen, M. (2008). Reducing energy consumption by using self-organizing maps to create more personalized electricity use information. *Applied Energy*, *85*(9), 830–840. doi:10.1016/j.apenergy.2007.10.012

Raskin, P. D. (2005). Global scenarios: Background review for the Millennium. *Ecosystem Assessment. Ecosystems (New York, N.Y.)*, *8*(2), 133–142. doi:10.100710021-004-0074-2

Rasul, G., & Thapa, G. B. (2003). Shifting cultivation in the mountains of South and Southeast Asia: Regional patterns and factors influencing the change. *Land Degradation & Development*, *14*(5), 495–508. doi:10.1002/ldr.570

Rave, T., Goetzke, F., & Larch, M. (2011). *The determinants of environmental innovations and patenting: Germany reconsidered* (No. 97). Ifo Working Paper.

Reddy, R. L., Reddy, V. S., & Gupta, G. A. (2013). Study of bio-plastics as green and sustainable alternative to plastics. *International Journal of Emerging Technology and Advanced Engineering*, *3*(5), 76–81.

Redman, C., Grove, J. M., & Kuby, L. (2000). *Toward a uniðed understanding of human ecosystems: Integrating social sciences into long-term ecological research.* White Paper of the Long-Term Ecological Research. Albuquerque, NM: LTER Network.

Redman, C., Grove, J. M., & Kuby, L. (2000). *Toward a unified understanding of human ecosystems: Integrating social sciences into long-term ecological research.* White Paper of the Long-Term Ecological Research. Albuquerque, NM: LTER Network.

Redwood, M., & Bahri, A. (2010). *Wastewater Irrigation and Health: Assessing and Mitigation Risks in Low-income Countries*. Earthscan/IDRC/IWMI.

Regan, M. J. (1990). *Sites for our solid waste: a guidebook for effective public involvement.* Office of Solid Waste.

Reich, L. (2010). Why I'm not a permaculturist. *Gard. Rant.* https://gardenrant.com/2010/03/why-im-not-a-permaculturist.html

Reich, P. B., Tilman, D., Isbell, F., Mueller, K., Hobbie, S. E., Flynn, D. F., & Eisenhauer, N. (2012, May 4). Impacts of biodiversity loss escalate through time as redundancy fades. *Science*, *336*(6081), 589–592. doi:10.1126cience.1217909 PMID:22556253

Reiss, P. C., & White, M. W. (2008). What changes energy consumption? Prices and public pressures. *The RAND Journal of Economics*, *39*(3), 636–663. doi:10.1111/j.1756-2171.2008.00032.x

Remenyi, J. (2004). What is Development? In D. Kingsbury, J. Remenyi, J. McKay, & J. Hunt (Eds.), *Key Issues in Development* (pp. 22–44). Palgrave Macmillan.

Rennings, K. (2000). Redefining innovation—Eco-innovation research and the contribution from ecological economics. *Ecological Economics*, *32*(2), 319–332. doi:10.1016/S0921-8009(99)00112-3

REVE. (2018). *Lake Turkana Wind Power in Loyangalani, Marsabit, is the largest wind energy project on the continent. REVE 2018*. Available from: https://www.evwind.es/2018/11/ 11/lake-turkana-wind-power-in-loyangalani-marsabit-is-the-largest-wind-energyproject-on-the-continent/65167

Reynolds, K. (2015). Disparity despite diversity: Social injustice in New York City's urban agro ecology system. *Antipode*, *47*(1), 240–259. doi:10.1111/anti.12098

Reynolds, K., & Cohen, N. (2016). *Beyond the kale: Urban agro ecology and social justice activism in New York City.* University of Georgia Press.

Rhim, J. W., & Kim, J. H. (2009). Properties of poly (lactide)-coated paperboard for the use of 1-way paper cup. *Journal of Food Science*, *74*(2), E105–E111. doi:10.1111/j.1750-3841.2009.01073.x PMID:19323738

Riahi, K., Van Vuuren, D., Akimoto, K., Dellink, R., Andersson, M., Bazilian, M., De Cian, E., Cosgove, B., De Vries, B., Fischer, G., Gielen, D., le Blanc, D., Gruebler, A., Hallegatte, S., Heaps, C., Hellmuth, M., Howells, M., Jiang, K., Kok, M., … Loulou, R. (2013). *Sustainable Development Scenarios for Rio+20 A Component of the SD21 project.* https://sustainabledevelopment.un.org/content/documents/793SD21 scenario report.pdf

Ricketts, T., & Imhoff, M. (2003). Biodiversity, urban areas, and agriculture: Locating prior-ity ecoregions for conservation. *Conservation Ecology*, *8*(2), 1. doi:10.5751/ES-00593-080201

Rincones, J., Zeidler, A. F., Grassi, M. C. B., Carazzolle, M. F., & Pereira, G. A. (2009). The golden bridge for nature: The new biology applied to bioplastics. Journal of Macromolecular Science®, Part C. *Polymer Reviews (Philadelphia, Pa.)*, *49*(2), 85–106. doi:10.1080/15583720902834817

Rivard, C. J., Adney, W. S., Himmel, M. E., Mitchell, D. J., Vinzant, T. B., Grohmann, K., Moens, L., & Chum, H. (1992). Effects of natural polymer acetylation on the anaerobic bioconversion to methane and carbon dioxide. *Applied Biochemistry and Biotechnology*, *34*(1), 725–736. doi:10.1007/BF02920592

Rock, M. T. (1996). Pollution intensity of GDP and trade policy: Can the world be wrong? *World Development*, *24*(3), 471–479. doi:10.1016/0305-750X(95)00152-3

Roh, T. (2019). Open for green innovation: from the perspective of green process and green consumer innovation. *Journal Sustainbility, 11*.

Roman-Alcalá, A. (2015). Broadening the land question in food sovereignty to northern settings: A case study of Occupy the Farm. *Globalizations*, *12*(4), 545–558. doi:10.1080/14747731.2015.1033199

Rosset, P. M., Sosa, B. M., Jaime, A. M. R., & Lozano, D. R. Á. (2011). The campesino-to-campesino agroecology movement of ANAP in Cuba: Social process methodology in the construction of sustainable peasant agriculture and food sovereignty. *The Journal of Peasant Studies*, *38*(1), 161–191. doi:10.1080/03066150.2010.538584 PMID:21284238

Rost, S. (2019). Navigating the ancient Tigris – insights into water management in an early state. *Journal of Anthropological Archaeology*, *54*, 31–47. doi:10.1016/j.jaa.2019.01.005

Rothman, D. S. (1998). Environmental Kuznets curve- Real progress or passing the buck? A case for consumption-based approaches. *Ecological Economics*, *25*(2), 177–194. doi:10.1016/S0921-8009(97)00179-1

Rudd, J. W. M., Kelly, C. A., Schindler, D. W., & Turner, M. A. (1988). Disruption of the nitrogen cycle in acidified lakes. *Science*, *240*(4858), 1515–1517. doi:10.1126cience.240.4858.1515 PMID:17798981

Rusk, D. (1993). *Cities without Suburbs*. The Woodrow.

Sabeenian, R. S., & Paramasivam, M. E. (2010). Defect detection and Identification in Textile Fabrics using Multi Resolution Combined Statistical and spatial Frequency Method. *IEEE Conference Publications*, 162 - 166. 10.1109/IADCC.2010.5423017

Sachs, W. (2010). Environment. In W. Sachs (Ed.), *The Development Dictionary: A guide to knowledge as power* (2nd ed., pp. 24–37). Zed Books.

Safarpour, M., & Khataee, A. (2018). Graphene-Based Materials for Water Purification. In *Nanoscale Materials in Water Purification* (pp. 383–430). Elsevier., doi:10.1016/B978-0-12-813926-4.00021-5

Sage, C. (2014). The transition movement and food sovereignty: From local resilience to global engagement in food system transformation. *Journal of Consumer Culture*, *14*(2), 254–275. doi:10.1177/1469540514526281

Sahakian, M. D. (2011). Understanding household energy consumption patterns: When "West Is Best" in Metro Manila. *Energy Policy*, *39*(2), 596–602. doi:10.1016/j.enpol.2010.10.032

Saharan, B. S., & Ankita, S. D. (2012). Bioplastics-for sustainable development: A review. *Int J Microbial Res Technol*, *1*, 11–23.

Saidmamatov, O., Eshchanov, B., Salaev, S., & Shimin, L. (2014). Renewable energy potential of developing countries; the drivers towards a green economy (a case study from Uzbekistan). *International Journal of Green Economics*, *8*(2), 134–143. doi:10.1504/IJGE.2014.065850

Sala, O. E., Chapin, F. S. I., Armesto, J. J., Berlow, E., Bloomfield, J., Dirzo, R., Huber-Sanwald, E., Huenneke, L. F., Jackson, R. B., Kinzig, A., Leemans, R., Lodge, D. M., Mooney, H. A., Oesterheld, M., Poff, N. L., Sykes, M. T., Walker, B. H., Walker, M., & Wall, D. H. (2000). Global biodiversity scenarios for the year 2100. *Science*, *287*(5459), 1770–1774. doi:10.1126cience.287.5459.1770 PMID:10710299

Saleem, M., & Bachmann, R. T. (2019). A contemporary review on plant-based coagulants for applications in water treatment. *Journal of Industrial and Engineering Chemistry*, *72*, 281–297. doi:10.1016/j.jiec.2018.12.029

Sandstr¨om, U.G., Angelstam, P., & Khakee, A., (2006). Urban comprehensiveplanning—identifying barriers for the maintenance of functional habitatnetworks. *Landsc. Urban Plann., 75*, 43–57.

Santo, R., Palmer, A., & Kim, B. (2016). Vacant lots to vibrant plots: A review of the benefits and limitations of urban agro ecology. Baltimore, MD: Johns Hopkins Center for a Livable Future.

Sapci, O., & Considine, T. (2014). The link between environmental attitudes and energy consumption behavior. *Journal of Behavioral and Experimental Economics*, *52*, 29–34. doi:10.1016/j.socec.2014.06.001

Sawhney, A., & Majumder, P. (2015). *Is India becoming a waste haven of metal scrap? MPRA Paper 67186*. University Library of Munich.

Sawhney, A., & Rastogi, R. (2015). Is India specialising in polluting industries? Evidence from US-India bilateral trade. *World Economy*, *38*(2), 360–378. doi:10.1111/twec.12164

Scheffer, M., Carpenter, S. R., Foley, J. A., Folke, C., & Walker, B. H. (2001). Catastrophic shifts in ecosystems. *Nature*, *413*(6856), 591–596. doi:10.1038/35098000 PMID:11595939

Schembri, P., & Petit, O. (2009). Clean technology transfers and North-South technological gap: An important issue for environmental policies. *Inter Economics*, *120*, 109–130.

Schiederig, T. (2011). *What is green innovation? A quantitative literature review*. The XXII ISPIM Conference.

Schindler, D. W. (1990). Experimental perturbations of whole lakes as tests of hypotheses concerning ecosystem structure and function. *Oikos*, *57*(1), 25–41. doi:10.2307/3565733

SDG16 Data Initiative 2017 Global Report. (2017). *SDG16 Report*.

Seana I., Johnson, L. & Peters, K. (1999). Community gardens and sustainable land use planning: A case-study of the Alex Wilson community garden. *Local Environment*, *4*(1), 33-46. Doi:10.1080/13549839908725579

Seker, F., Ertugrul, H. M., & Cetin, M. (2015). The impact of foreign direct investment on environmental quality: A Bounds testing and causality analysis for Turkey. *Renewable & Sustainable Energy Reviews*, *52*, 347–356. doi:10.1016/j.rser.2015.07.118

Selden, T. M., & Song, D. (1994). Environmental quality and development: Is there a Kuznets curve for air pollution emissions? *Journal of Environmental Economics and Management*, *27*(2), 147–162. doi:10.1006/jeem.1994.1031

Seto, K. C., De Groot, R., Bringezu, S., Erb, K., Graedel, T. E., Ramankutty, N., & Skole, D. L. (2010). Stocks, flows, and prospects of land. In T. E. Graedel & E. van der Voet (Eds.), *Linkages of sustainability* (pp. 72–96). MIT Press.

Seto, K. C., De Groot, R., Bringezu, S., Erb, K., Graedel, T. E., Ramankutty, N., & Skole, D. L. (2010). Stocks, ñows, and prospects of land. In T. E. Graedel & E. van der Voet (Eds.), *Linkages of sustainability* (pp. 72–96). MIT Press.

Shafik, N., & Bandyopadhyay, S. (1992). Economic growth and environmental quality: Time-series and cross-country evidence. World Bank Publications.

Shahbaz, M., Nasreen, S., Abbas, F., & Anis, O. (2015). Does foreign direct investment impede environmental quality in high, middle and low-income countries? *Energy Economics*, *51*, 1–35. doi:10.1016/j.eneco.2015.06.014

Shahriar, A., Sadiq, R., & Tesfamariam, S. (2012). Risk analysis for oil & gas pipelines: A sustainability assessment approach using fuzzy based bowtie analysis. *Journal of Loss Prevention in the Process Industries*, *25*(3), 505–523. doi:10.1016/j.jlp.2011.12.007

Shao, Y. (2018). Does FDI affect carbon intensity? New evidence from dynamic panel analysis. *International Journal of Climate Change Strategies and Management*, *10*(1), 27–42. doi:10.1108/IJCCSM-03-2017-0062

Sharpley, R. (2000). Tourism and Sustainable Development: Exploring the Theoretical Divide. *Journal of Susta Hediger inable. Tourism (Zagreb)*, *8*(1), 1–19. doi:10.1080/09669580008667346

Shepard, M. (2013). *Restoration agriculture: real world permaculture for farmers*. Acres U.S.A.

Šimleša, D. (2003). Podržava li biznis održivi razvoj? *Journal for General Social Issues*, *12*(3-4), 403426. Retrieved September 21, 2015, from https://hrcak.srce.hr/19493

Sims, C. A. (1980). Macroeconomics and reality. *Econometrica*, *48*(1), 1–48. doi:10.2307/1912017

Singh, S., & Yassine, A. (2017). Mining Energy Consumption Behavior Patterns for Households in Smart Grid. *IEEE Transactions on Emerging Topics in Computing*, *7*(3), 404–419. doi:10.1109/TETC.2017.2692098

Siracusa, V., Rocculi, P., Romani, S., & Dalla Rosa, M. (2008). Biodegradable polymers for food packaging: A review. *Trends in Food Science & Technology*, *19*(12), 634–643. doi:10.1016/j.tifs.2008.07.003

SIWI. *Cooperation Over Shared Waters - Stockholm International Water Institute*. (n.d.). SIWI. Retrieved May 18, 2020, from https://www.siwi.org/priority-area/transboundary-water-management/

Slawinska, K., Soltysiak, M., Davies, S., Bénard, L., Finelli, T., & Tomassone, A. (2014). A toolkit On Promoting Sustainable Development Practices To Adults Learners 3R's – RECYCLE, REDUCE, REUSE. Library of Work Pedagogy Publish.

Smit, B., & Wandel, J. (2006). Adaptation, adaptive capacity and vulnerability. *Global Environmental Change*, *16*(3), 282–292. doi:10.1016/j.gloenvcha.2006.03.008

Smith. (2007). *Agriculture, in Climate Change 2007: Mitigation*. Contribution of WG III to the Fourth Assessment Report of the Intergovernmental Panel on Climate Change.

Smith, C. (2002). Learning for the environment: An examination of personal empowerment through learning permaculture. *Post Scr*, *3*, 12–25.

Smith, J. R. (1929). *Tree crops: a permanent agriculture*. New York Brace & Co.

Smith, J. R. (1929). *Tree crops: a permanent agriculture. Quinn and Boden Company, Inc.* Rahway.

Smith, L. (2002). What might we learn from climate forecasts? *Proceedings of the National Academy of Sciences of the United States of America*, *99*(Supplement 1), 2487–2492. doi:10.1073/pnas.012580599 PMID:11875200

Smith, L. (2017). Historical Perspectives on Water Purification. In *Chemistry and Water* (pp. 421–468). Elsevier. doi:10.1016/B978-0-12-809330-6.00012-X

Smit, J., & Nasr, J. (1992). Urban agriculture for sustainable cities: Using wastes and idle land and water bodies as resources. *Environment and Urbanization*, *4*(2), 141–152. doi:10.1177/095624789200400214

Södergård, A., & Stolt, M. (2002). Properties of lactic acid based polymers and their correlation with composition. *Progress in Polymer Science*, *27*(6), 1123–1163. doi:10.1016/S0079-6700(02)00012-6

Solon, O. L. (2014): Vast underground bomb shelter reappropriated by urban farmers. *Wired*. Available at: https://www.wired.co.uk/article/underground-farm-zero-carbon-food

Son, H., Park, G., & Lee, S. (1996). Growth-associated production of poly-β-hydroxybutyrate from glucose or alcoholic distillery wastewater by Actinobacillus sp. EL-9. *Biotechnology Letters*, *18*(11), 1229–1234. doi:10.1007/BF00129945

Sonune, A., & Ghate, R. (2004). Developments in Wastewater Treatment Methods. *Desalination*, *167*, 55–63. doi:10.1016/j.desal.2004.06.113

Specim. (2019). *FX-10e VNIR Hyperspectral Camera (GigE Interface)*. https://www.middletonspectral.com/product/fx-10e/

Stambouli, A. B., Khiat, Z., Flazi, S., & Kitamura, Y. (2012). A review on the renewable energy development in Algeria: Current perspective, energy scenario and sustainability issues. *Renewable & Sustainable Energy Reviews*, *16*(7), 4445–4460. doi:10.1016/j.rser.2012.04.031

Staples, D., & Hermes, R. (2012). Marine biodiversity and resource management – what is the link? *Aquatic Ecosystem Health & Management*, *15*(3), 245–252. doi:10.1080/14634988.2012.709429

Steel, C. (2008). *Hungry City: How Food Shapes Our Lives*. Vintage.

Sterk, S., van Ittersum, M. K., Leeuwis, L., Rossing, R., van Keulen, H., & Van de Ven, G. W. J. (2006). Finding niches for whole-farm design models—Contradictio in terminis? *Agricultural Systems*, *87*(2), 211–228. doi:10.1016/j.agsy.2004.11.008

Stern, D. I. (2003). *The environmental Kuznets curve*. International Society for Ecological Economics Internet Encyclopaedia of Ecological Economics.

Sukopp, H., Hejny, S., & Kowarik, I. (Eds.). (1990). *Urban ecology: Plants and plant communities in urban environments*. SPB Academic Publishing.

Sukopp, H., Trautmann, W., & Korneck, D. (1979). The soilflora and vegetation of Berlin'swastelands. In I. Laurie (Ed.), *Nature in Cities*. John Wiley.

Sun, X. S., Kim, H. R., & Mo, X. (1999). Plastic performance of soybean protein components. *Journal of the American Oil Chemists' Society*, *76*(1), 117–123. doi:10.100711746-999-0057-8

Suri, V., & Chapman, D. (1998). Economic growth, trade and the environment: Implications for the environmental Kuznets curve. *Ecological Economics*, *25*(1), 195–208. doi:10.1016/S0921-8009(97)00180-8

Sustainable Development Goal 11. (2017). Sustainable Development Goals.

Sustainable Development Goal 17. (2017). *Sustainable Development Goals*.

Su, T.-L., Hong, G.-B., Chang, W.-Y., & Kung, F.-C. (2011). Automatic Recognition of Textile Texture Using Back-propagation Neural Network. *IEEE Conference Publications*, 52 – 55.

Suthar, R. G., & Gao, B. (2017). Nanotechnology for drinking water purification. In *Water Purification* (pp. 75–118). Elsevier. doi:10.1016/B978-0-12-804300-4.00003-4

Tacio, H. D. (1993). Sloping agricultural land technology (SALT): A sustainable agroforestry scheme for the uplands. *Agroforestry Systems*, *22*(2), 145–152. doi:10.1007/BF00705143

Tamburrino, A. (2010). Water Technology in Ancient Mesopotamia. In *Ancient Water Technologies* (pp. 29–51). Springer Netherlands., doi:10.1007/978-90-481-8632-7_2

Tansley, A. G. (1935). The use and abuse of vegetational concepts and terms. *Ecology, 16*(3), 284–307. doi:10.2307/1930070

Terborgh, J., Lopez, L., & Nunez, P. (2001). Ecological meltdown in predator-free forest fragments. *Science, 294*(5548), 1923–1926. doi:10.1126cience.1064397 PMID:11729317

Termorshuizen, J. W., & Opdam, P. (2009). Landscape services as a bridge between landscape ecology and sustainable development. *Landscape Ecology, 24*(8), 1037–1052. doi:10.100710980-008-9314-8

The Economics of Ecosystems and Biodiversity (TEEB). (2010). *Mainstreaming the economics of nature: a synthesis of the approach, conclusions and recommendations of TEEB*. TEEB.

The Energy and Natural Resources Ministry. (n.d.a). *Electricity*. Retrieved from https://www.enerji.gov.tr/tr-TR/Sayfalar/Elektrik

The Energy and Natural Resources Ministry. (n.d.b). *Energy Efficiency*. Retrieved from https://www.enerji.gov.tr/tr-TR/Sayfalar/Enerji-Verimliligi

The Ministry of Commerce. (2018). *Turkey Waste Report*. General Directorate Of Consumer Protection And Market Surveillance.

The Ministry of Energy and Natural Resources. (2018). National Energy Efficiency Action Plan 2017-2023. Author.

The Ministry of Environment And Urban, Environmental Impact Assessment, General Directorate Of Authorization And Audit, Environmental Inventory And Information Management Department. (2018). *Environmental Indicators 2016*. Publication No: 38-1.

Theobald, D. M., Hobbs, N. T., Bearly, T., Zack, J. A., Shenk, T., & Riebsame, W. E. (2000). Incorporating biological information in local land-use decisionmaking: Designing a system for conservation planning. *Landscape Ecology, 15*(1), 35–45. doi:10.1023/A:1008165311026

Thibert, J. (2012). Making local planning work for urban agro ecology in the North American context: A view from the ground. *Journal of Planning Education and Research, 32*(3), 349–357. doi:10.1177/0739456X11431692

Thomson, G. (1989). *Planning and Design of Library Buildings*. Butterworth Architecture.

Tian, H., Wu, W., Guo, G., Gaolun, B., Jia, Q., & Xiang, A. (2012). Microstructure and properties of glycerol plasticized soy protein plastics containing castor oil. *Journal of Food Engineering, 109*(3), 496–500. doi:10.1016/j.jfoodeng.2011.10.033

Tidball, G. K., Krasny, E. M., Svendsen, E., Cambell, L., & Helphand, K. (2010). Stewardship, learning and memory in disaster resilience. *Environmental Learning Research, 16*(5–6), 591–609. doi:10.1080/13504622.2010.505437

Tidball, K. G., & Krasny, M. E. (2009). *From risk to resilience: what role for community greening and civic ecology in cities?* Wageningen Academic Publishers.

Tigabu, A., Kingiri, A., Odongo, F., Margrethe, R., Andersen, M. & Lema, R. (2017). *Capability development and collaboration for Kenya's solar and wind technologies: analysis of major energy policy frameworks*. Academic Press.

Timko, J., Le Billon, P., Zerriffi, H., Honey-Rosés, J., de la Roche, I., Gaston, C., Sunderland, T. C. H., & Kozak, R. A. (2018). A policy nexus approach to forests and the SDGs: Tradeoffs and synergies. *Current Opinion in Environmental Sustainability, 34*, 7–12. doi:10.1016/j.cosust.2018.06.004

Timothy, A. (2010). WO Patent Application No. 2010/151346.

Toda, H. Y., & Phillips, P. C. B. (1993). Vector autoregressions and causality. *Econometrica*, *61*(6), 1367–1393. doi:10.2307/2951647

Toda, H. Y., & Yamamoto, T. (1995). Statistical inferences in vector autoregressions with possibly integrated processes. *Journal of Econometrics*, *66*(1-2), 225–250. doi:10.1016/0304-4076(94)01616-8

Tomabechi, K. (2010). Energy resources in the future. *Energies*, *3*(4), 686–695. doi:10.3390/en3040686

Tomislav, K. (2018). The Concept of Sustainable Development: From its Beginning to the Contemporary Issues. *Zagreb International Review of Economics & Business*, *21*(1), 67–94. doi:10.2478/zireb-2018-0005

Tornaghi, C. (2011) *Steal this veg. The Red Pepper*. Available at: https://www.redpepper.org.uk/steal-this-veg

Toronto Public Health Department. (2010). *Annual report*. Toronto Public Health.

Torras, M., & Boyce, J. K. (1998). Income inequality and pollution: A re-assessment of the environmental Kuznets curve. *Ecological Economics*, *25*(2), 147–160. doi:10.1016/S0921-8009(97)00177-8

Tranel, M., & Handlin, L. B. Jr. (2006). Metromorphosis: Documenting change. *Journal of Urban Affairs*, *28*(2), 151–167. doi:10.1111/j.0735-2166.2006.00265.x

Trends, G. (2017). *Challenges and Opportunities in the Implementation of the Sustainable Development Goals*. Academic Press.

TSKB (Turkish Industrial Development Bank). (2018). *Sectoral Outlook*. White Goods February 2018. Retrieved from http://www.tskb.com.tr/i/assets/document/pdf/beyaz-esya-sektorel-gorunum-subat-2018.pdf

TURKBESD (White Goods Manufacturers' Association of Turkey). (n.d.). *TÜRKBESD Last 5 Years Figures, Domestic Sales*. Retrieved from http://www.turkbesd.org/bilgiler.php?P=25

TURKBESD. (2020). *White goods sales increased by 11 percent in the first quarter*. Retrieved from http://www.turkbesd.org/Haberler.php?P=37

TURKSTAT (Turkish Statistical Institute). (2016). *Household Information Technologies Usage Research, 2016*. Retrieved from http://tuik.gov.tr/PreHaberBultenleri.do?id=21779

TURKSTAT. (2018). *The Statistics of Energy Consumption by Sectors, 2017*. Retrieved from http://www.tuik.gov.tr/PreHaberBultenleri.do?id=27667

TURKSTAT. (2020a). *Address Based Population Registration System Results 2019*. Retrieved from http://www.tuik.gov.tr/HbGetirHTML.do?id=33705

TURKSTAT. (2020b). *Housing Sales Statistics, March 2020*. Retrieved from http://www.tuik.gov.tr/PreHaberBultenleri.do?id=33878

TURKSTAT. (n.d.). *Basic Statistics, Population Projections*. Retrieved from http://www.tuik.gov.tr/UstMenu.do?metod=temelist

Twigg, J. (2009). *Characteristics of disaster resilient community–a guidance note*. https://discovery.ucl.ac.uk/1346086/1/1346086.pdf

U.S. Environmental Protection Agency. (2013). *Emerging Technologies for Wastewater Treatment and In-Plant Wet Weather Management EPA 832-R-12-011 March 2013*. Office of Wastewater Management.

Ullah, I., Shah, M., & Khan, F. U. (2014). Domestic investment, foreign direct investment, and economic growth nexus: A case of Pakistan. *Economics Research International*, 1-5.

UN (United Nations). (2015). *Transforming Our World: The 2030 Agenda for Sustainable Development, A/RES/70/1.* Retrieved from https://sustainabledevelopment.un.org/content/documents/21252030%20Agenda%20for%20Sustainable%20Development%20web.pdf

UNEP. (2010). *United Nations Environment Programme Green Economy Report.* http://www.unep.org/greeneconomy/GreenEconomyReport

UNEP. (2012). *The Emissions Gap Report 2012. A UNEP Synthesis Report, United Nations Environment Programme.* UNEP.

UNEP. (2012). *Towards a Green Economy: Pathways to Sustainable Development and Poverty Eradication.* Available online at: http://www.unep.org/greeneconomy/Portals/88/documents/ger/GER_synthesis_en.pdf

UNEP-Sustainable Energy Finance Initiative (SEFI). (2011). Global Trends in Renewable Energy Investment 2011. United Nations Environment Programme (UNEP) Sustainable Energy Finance Initiative (SEFI) and Bloomberg New Energy Finance, Paris.

UNESCO. (2014). Water and Energy Sustainability. *UNESCO.* http://www.unesco.org/open-access/terms-use-ccbysa-en%0Ahttp://www.un.org/waterforlifedecade/pdf/01_2014_sustainability_eng.pdf

UNESCO. (2017). *World Water Scenarios.* Author.

Unger, S., & Wooten, H. (2006). *A food systems assessment for Oakland, CA: Toward a sustainable food plan.* Oakland, CA: Mayor's Office of Sustainability Van der Schans, JW, Wiskerke, JSC (2012) Urb agriculture in developed economies. In A. Viljoen & J. S. C. Wiskerke (Eds.), *Sustainable Food Planning: Evolving Theory and Practice* (pp. 245–258). Wageningen Academic Publishers.

UNICEF. (2015). *Diarrhoeal Disease - UNICEF DATA.* https://data.unicef.org/topic/child-health/diarrhoeal-disease/

UNISDR. (2015). The human cost of weather-related disasters 1995-2015. *UNISDR Publications, 1*, 30. doi:10.1017/CBO9781107415324.004

United Nations Educational, Scientific and Cultural Organization. (n.d.). Retrieved May 19, 2020, from http://www.unesco.org/new/en/unesco/events/prizes-and-celebrations/celebrations/international-days/world-water-day-2017/

United Nations Environmental Program (UNEP). (2011). Global Trends in Renewable Energy' Analysis of Trends and Issues in the Financing Renewable Energy. Investment Report 2011, Nairobi: UNEP.

United Nations: Gender equality and women's empowerment. (2018). United Nations Sustainable Development.

van de Berg, L., & van Veenhuizen, R. (2005). Multiple functions of urban Agriculture. *Urban Agriculture Magazine*, 14.

Van Doorslaer, X., Haylamicheal, I. D., Dewulf, J., Van Langenhove, H., Janssen, C. R., & Demeestere, K. (2015). Heterogeneous photocatalysis of moxifloxacin in water: Chemical transformation and ecotoxicity. *Chemosphere, 119*, S75–S80. doi:10.1016/j.chemosphere.2014.03.048 PMID:24735961

Vaneechoutte, M., Kämpfer, P., De Baere, T., Falsen, E., & Verschraegen, G. (2004). Wautersia gen. nov., a novel genus accommodating the phylogenetic lineage including Ralstonia eutropha and related species, and proposal of Ralstonia [Pseudomonas] syzygii (Roberts et al. 1990) comb. nov. *International Journal of Systematic and Evolutionary Microbiology, 54*(2), 317–327. doi:10.1099/ijs.0.02754-0 PMID:15023939

Vassileva, I. (2012). *Characterization Of Household Energy Consumption In Sweden.* School of Sustainable Development of Society and Technology Mälardalen University Press Dissertations No. 129.

Vassileva, I., Wallin, F. & Dahlquist, E. (2012). Understanding energy consumption behavior for future demand response strategy development. *Energy, 46*, 94-100.

Verbeek, C. J. R., & Bier, J. M. (2011). Synthesis and characterization of thermoplastic agro-polymers. In *Handbook of Applied Biopolymer Technology* (pp. 197–242). RSC Publishing. doi:10.1039/9781849733458-00197

Vermeulen, S. J., Campbell, B.M., & Ingram, J. S. I. (2012). *Climate Change and Food Systems*. Academic Press.

Vink, E. T., Rabago, K. R., Glassner, D. A., & Gruber, P. R. (2003). Applications of life cycle assessment to NatureWorks™ polylactide (PLA) production. *Polymer Degradation & Stability*, *80*(3), 403–419. doi:10.1016/S0141-3910(02)00372-5

Vogus, T. J., & Sutcliffe, K. M. (2007). The impact of safety organizing, trusted leadership, and care pathways on reported medication errors in hospital nursing units. *Medical Care*, *41*(10), 992–1002. doi:10.1097/MLR.0b013e318053674f PMID:17890998

von Haaren, C., & Reich, M. (2006). The German way to greenways and habitatnetworks. *Landscape and Urban Planning*, *76*(1-4), 7–22. doi:10.1016/j.landurbplan.2004.09.041

Wagnild, G. M. (2010): *Discovering Your Resilience* Core. https://www.pobal.ie/Publications/Documents/Discovering_Your_Resilience_Core.pdf

Wakefield, S., Yeudall, F., Taron, C., Reynolds, J., & Skinner, A. (2007). Growing urban health: Community gardening in South-East Toronto. *Health Promotion International*, *22*(2), 92–101. doi:10.1093/heapro/dam001 PMID:17324956

Walker, B. H., Holling, C. S., Carpenter, S. R., & Kinzig, A. P. (2004). Resilience, adaptability and transformability in social–ecological systems. *Ecology and Society, 9*(2), 5. http://www.ecologyandsociety. org/vol9/iss2/art5/

Walker, B. H., Kinzig, A. P., & Langridge, J. (1999). Plant attribute diversity, resilience, and ecosystem function: The nature and significance of dominant and minor species. *Ecosystems (New York, N.Y.)*, *2*(2), 95–113. doi:10.1007100219900062

Walker, B., Holling, C. S., Carpenter, S. R., & Kinzig, A. (2004). Resilience, adaptability and transformability in social–ecological systems. *Ecology and Society*, *9*(2), 5. doi:10.5751/ES-00650-090205

Walker, B., & Salt, D. (2006). *Resilience thinking: Sustaining ecosystems and people in a changing world.* Island Press.

Walker, B., & Salt, D. (2012). *Resilience Practice: Engaging the Sources of Our Sustainability*. Island Press. doi:10.5822/978-1-61091-231-0

Walley, M. P. (1991). *Statistical reasoning with imprecise probabilities*. Chapman and Hall.

Wang, H., Jiang, L., & Fu, L. (2007). Properties of molded soy protein isolate plastics. *Journal of Applied Polymer Science*, *106*(6), 3716–3720. doi:10.1002/app.27064

Wang, Q., Cai, J., Zhang, L., Xu, M., Cheng, H., Han, C. C., Kuga, S., Xiao, J., & Xiao, R. (2013). A bioplastic with high strength constructed from a cellulose hydrogel by changing the aggregated structure. *Journal of Materials Chemistry. A, Materials for Energy and Sustainability*, *1*(22), 6678–6686. doi:10.1039/c3ta11130j

Wang, X. L., Yang, K. K., & Wang, Y. Z. (2003). Properties of starch blends with biodegradable polymers. *Journal of Macromolecular Science, Part C: Polymer Reviews*, *43*(3), 385–409. doi:10.1081/MC-120023911

Wang, X., Georganas, N. D., & Petriu, E. M. (2009). Fiber-level Structure Recognition of Woven Textile. *IEEE Conference Publications*, 1 – 6.

Warnes, S. L., & Keevil, C. W. (2011). Mechanism of Copper Surface Toxicity in Vancomycin-Resistant Enterococci following Wet or Dry Surface Contact. *Applied and Environmental Microbiology*, *77*(17), 6049–6059. doi:10.1128/AEM.00597-11 PMID:21742916

Wedin, H., Gupta, C., Mzikian, P., Englund, F., Hornbuckle, R., Troppenz, V., Kobal, L., Krečič, M., Costi, M., Ellams, D., & Olsson, S. (2017). *Can automated NIR technology be a way to improve the sorting quality of textile waste?* https://static1.squarespace.com/static/5891ce37d2b857f0c58457c1/t/5a0049d26c3194747a7de384/1509968342157/october_2017_rise_report.pdf

Weisenburger, D. D. (1993). Human health effects of agrichemical use. *Human Pathology*, *24*(6), 571–576. doi:10.1016/0046-8177(93)90234-8 PMID:8505035

Wei, Y. M., Liu, L. C., Fan, Y., & Wu, G. (2007). The impact of lifestyle on energy use and CO2 emission: An empirical analysis of China's residents. *Energy Policy*, *35*(1), 247–257. doi:10.1016/j.enpol.2005.11.020

Whelan, A., Wrigley, N., Warm, N., & Canning, S. E. (2002). Life in a 'Food Desert'. *Urban Studies (Edinburgh, Scotland)*, *39*(11), 2083–2100. doi:10.1080/0042098022000011371

WHO & UNICEF. (2017). Progress on Drinking Water, Sanitation and Hygiene: 2017 Update and SDG Baselines. Geneva: World Health Organization (WHO) and the United Nations Children's Fund (UNICEF).

WHO. (n.d.). *Drinking-water.* Retrieved May 19, 2020, from https://www.who.int/news-room/fact-sheets/detail/drinking-water

WHO. (n.d.). *Millenium Development Goals (MDGs)*. Retrieved from https://www.who.int/topics/millennium_development_goals/about/en/

Willis, K. (2005). *Theories and Practices of Development*. Routledge.

Winchester, C. L., & Salji, M. (2016). Writing a literature review. *Journal of Clinical Urology*, *9*(5), 308–312. doi:10.1177/2051415816650133

Witten, I. H., Frank, E., & Hall, M. A. (2011). *Data Mining: Practical Machine Learning Tools and Techniques.* Morgan Kaufmann. doi:10.1007/978-1-4899-3472-7

Wooldridge, J. M. (2013). *Introductory econometrics: A modern approach* (5th ed.). South-Western Cengage Learning.

World Bank. (2009). *Global Economic Prospects 2009: Commodities at the Crossroads*. World Bank. Available online at: http://siteresources.worldbank.org/INTGEP2009/ Resources/10363_WebPDF-w47.pdf

Wu, J., & Wu, T. (2013). Ecological resilience as a foundation for urban design and sustainability. In S. T. A. Pickett, M. L. Cadenasso, & B. McGrath (Eds.), *Resilience in ecology and urban design: Linking theory and practice for sustainable cities* (pp. 211–229). Springer. doi:10.1007/978-94-007-5341-9_10

Xiao, C., & Yang, M. (2006). Controlled preparation of physical cross-linked starch-g-PVA hydrogel. *Carbohydrate Polymers*, *64*(1), 37–40. doi:10.1016/j.carbpol.2005.10.020

Xuan, L., Baotong, D., & Hua, Y. (2011). The Research Based on the 3-R Principle of Agro-circular Economy Model-The Erhai Lake Basin as an Example. *Energy*, *5*, 1399–1404.

Xu, B. (2000). Multinational enterprises, technology diffusion, and host country productivity growth. *Journal of Development Economics*, *62*(2), 477–493. doi:10.1016/S0304-3878(00)00093-6

Xue, Z., Zeng, X., Koehl, L., & Chen, Y. (2012). Development of a fuzzy inclusion measure for investigating relations between visual and tactile perception of textile products. *IEEE Conference Publications*, 108 – 113. 10.1109/MFI.2012.6342995

Xu, Y., Du, B., & Zhang, L. (2019). Simultaneous Segmentation and Edge Detection for Hyperspectral Image via a Deep Supervised and Boundary-constrained Network. *IEEE Conference Publications*, 3860 – 3863. 10.1109/IGARSS.2019.8899054

Yadoo, A., & Cruickshank, H. (2012). The role for low carbon electrification technologies in poverty reduction and climate change strategies: A focus on renewable energy mini-grids with case studies in Nepal, Peru and Kenya. *Energy Policy*, *42*, 591–602. doi:10.1016/j.enpol.2011.12.029

Yıldırım, S., & Yıldırım, D. Ç. (2019). Sürdürülebilir Kalkınma Kapsamında Türkiye'nin Sürdürülebilir Enerji Kaynakları Profili [Energy Resources Profile of Turkey in the context of Sustainable Development]. In İktisadi Teknik ve Strateji Boyutları ile Türkiye'de Enerji Sorunsalı [Economic, Technical And Strategy Dimensions in Turkey by The Energy Problems]. Ekin pub.

Yıldırım, S., & Yıldırım, D. Ç. (2020). Achieving Sustainable Development Through a Green Economy Approach. In Advanced Integrated Approaches to Environmental Economics and Policy: Emerging Research and Opportunities. IGI Global Pub.

Yildirim, D. C., Erdogan, S., Yildirim, S., & Can, H. (2017). The effect of the Trans-Anatolian Natural Gas Pipeline Project (TANAP) on industrial production in Turkey. *International Journal of Energy Sector Management*, *11*(3), 404–415. doi:10.1108/IJESM-10-2016-0005

Yıldırım, D. Ç., Yıldırım, S., & Demirtas, I. (2019). Investigating energy consumption and economic growth for BRICS-T countries, *World Journal of Science. Technology and Sustainable Development*, *16*(4), 184–195.

Yildirim, E. (2014). Energy use, CO_2 emission and foreign direct investment: Is there any inconsistence between causal relations? *Frontiers in Energy*, *8*(3), 269–278. doi:10.100711708-014-0326-6

Yohe, G., & Tol, R. S. J. (2002). Indicators for social and economic coping capacity – moving toward a working deðnition of adaptive capacity. *Global Environmental Change*, *12*(1), 25–40. doi:10.1016/S0959-3780(01)00026-7

Yoshida, T., & Zusman, E. (2015). Achieving the multiple benefits of a sustainable development goal for energy. In Achieving the Sustainable Development Goals: From Agenda to Action. Academic Press.

Yu, Zhao, Song, Wang, Li, Han, & Chang. (2019). Hyperspectral Image Classiðcation Method Based on CNN Architecture Embedding With Hashing Semantic Feature. *IEEE Journal of Selected Topics in Applied Earth Observation and Remote Sensing, 12*(6), 1866 – 1881.

Yücel, G. F. (2006). Kamusal Açık Mekanlarda Donatı Elemanlarının Kullanımı. *Ege Mimarlık Dergisi*, *4*(59), 26–29.

Zanoni, P., & Janssens, M. (2009). Sustainable DiverCities. In M. Janssens, D. Pinelli, D. C. Reyman, & S. Wallman (Eds.), *Sustainable Cities. Diversity, Economic Growthand Social Cohesion* (pp. 3–25). Edward Elgar.

Zeng, S., Bai, J., Jiang, L., & Kang, Z. (2017). Multiple Kernel Fuzzy Discriminant Analysis for Hyperspectral Imaging Classification. *IEEE Conference Publications*, 1-6. 10.1109/FUZZ-IEEE.2017.8015719

Zhai, M., Yoshii, F., & Kume, T. (2003). Radiation modification of starch-based plastic sheets. *Carbohydrate Polymers*, *52*(3), 311–317. doi:10.1016/S0144-8617(02)00292-8

Zhai, M., Yoshii, F., Kume, T., & Hashim, K. (2002). Syntheses of PVA/starch grafted hydrogels by irradiation. *Carbohydrate Polymers*, *50*(3), 295–303. doi:10.1016/S0144-8617(02)00031-0

Zhang, J., & Reineccius, G. A. (2016). Factors controlling the turbidity of submicron emulsions stabilized by food biopolymers and natural surfactant. *Lebensmittel-Wissenschaft + Technologie*, *71*, 162–168. doi:10.1016/j.lwt.2016.03.035

Zhang, S. X., & Babovic, V. (2012). A real options approach to the design and architecture of water supply systems using innovative water technologies under uncertainty (PDF). *Journal of Hydroinformatics*, *488*. www.frontiersinecology.org

Zhang, Z., & Meng, X. (2019). Internet penetration and the environmental Kuznets curve: A cross-national analysis. *Sustainability*, *11*(1358), 1–12. doi:10.3390u11051358

Zhou, K., & Yang, S. (2016). Understanding household energy consumption behavior: The contribution of energy big data analytics. *Renewable & Sustainable Energy Reviews*, *56*, 810–819. doi:10.1016/j.rser.2015.12.001

Zimov, S. A., Chuprynin, V. I., Oreshko, A. P., Chapin, F. S. III, Reynolds, J. F., & Chapin, M. C. (1995). Steppe–tundra transition: An herbivore-driven biome shift at the end of the Pleistocene. *American Naturalist*, *146*(5), 765–794. doi:10.1086/285824

Zott, C. (2011). The Business Model: Recent Developments and Future Research. *Journal of Management*, *37*(4), 1019–1042.

Related References

To continue our tradition of advancing academic research, we have compiled a list of recommended IGI Global readings. These references will provide additional information and guidance to further enrich your knowledge and assist you with your own research and future publications.

Abed, S., Khir, T., & Ben Brahim, A. (2016). Thermodynamic and Energy Study of a Regenerator in Gas Turbine Cycle and Optimization of Performances. *International Journal of Energy Optimization and Engineering*, *5*(2), 25–44. doi:10.4018/IJEOE.2016040102

Abu Bakar, W. A., Abdullah, W. N., Ali, R., & Mokhtar, W. N. (2016). Polymolybdate Supported Nano Catalyst for Desulfurization of Diesel. In T. Saleh (Ed.), *Applying Nanotechnology to the Desulfurization Process in Petroleum Engineering* (pp. 263–280). Hershey, PA: IGI Global. doi:10.4018/978-1-4666-9545-0.ch009

Addo-Tenkorang, R., Helo, P., & Kantola, J. (2016). Engineer-To-Order Product Development: A Communication Network Analysis for Supply-Chain's Sustainable Competitive Advantage. In R. Addo-Tenkorang, J. Kantola, P. Helo, & A. Shamsuzzoha (Eds.), *Supply Chain Strategies and the Engineer-to-Order Approach* (pp. 43–59). Hershey, PA: IGI Global. doi:10.4018/978-1-5225-0021-6.ch003

Adebiyi, I. D., Popoola, P. A., & Pityana, S. (2016). Mitigation of Wear Damage by Laser Surface Alloying Technique. In E. Akinlabi, R. Mahamood, & S. Akinlabi (Eds.), *Advanced Manufacturing Techniques Using Laser Material Processing* (pp. 172–196). Hershey, PA: IGI Global. doi:10.4018/978-1-5225-0329-3.ch007

Ahmad, W. (2016). Sulfur in Petroleum: Petroleum Desulfurization Techniques. In T. Saleh (Ed.), *Applying Nanotechnology to the Desulfurization Process in Petroleum Engineering* (pp. 1–52). Hershey, PA: IGI Global. doi:10.4018/978-1-4666-9545-0.ch001

Ahmed, I., Ahmad, N., Mehmood, I., Haq, I. U., Hassan, M., & Khan, M. U. (2016). Applications of Nanotechnology in Transportation Engineering. In A. Khitab & W. Anwar (Eds.), *Advanced Research on Nanotechnology for Civil Engineering Applications* (pp. 180–207). Hershey, PA: IGI Global. doi:10.4018/978-1-5225-0344-6.ch006

Aikhuele, D. (2018). A Study of Product Development Engineering and Design Reliability Concerns. *International Journal of Applied Industrial Engineering*, *5*(1), 79–89. doi:10.4018/IJAIE.2018010105

Al-Najar, B. T., & Bououdina, M. (2016). Bioinspired Nanoparticles for Efficient Drug Delivery System. In M. Bououdina (Ed.), *Emerging Research on Bioinspired Materials Engineering* (pp. 69–103). Hershey, PA: IGI Global. doi:10.4018/978-1-4666-9811-6.ch003

Al-Shebeeb, O. A., Rangaswamy, S., Gopalakrishan, B., & Devaru, D. G. (2017). Evaluation and Indexing of Process Plans Based on Electrical Demand and Energy Consumption. *International Journal of Manufacturing, Materials, and Mechanical Engineering*, *7*(3), 1–19. doi:10.4018/IJMMME.2017070101

Alexakis, H., & Makris, N. (2016). Validation of the Discrete Element Method for the Limit Stability Analysis of Masonry Arches. In V. Sarhosis, K. Bagi, J. Lemos, & G. Milani (Eds.), *Computational Modeling of Masonry Structures Using the Discrete Element Method* (pp. 292–325). Hershey, PA: IGI Global. doi:10.4018/978-1-5225-0231-9.ch012

AlMegren, H. A., Gonzalez-Cortes, S., Huang, Y., Chen, H., Qian, Y., Alkinany, M., ... Xiao, T. (2016). Preparation of Deep Hydrodesulfurzation Catalysts for Diesel Fuel using Organic Matrix Decomposition Method. In H. Al-Megren & T. Xiao (Eds.), *Petrochemical Catalyst Materials, Processes, and Emerging Technologies* (pp. 216–253). Hershey, PA: IGI Global. doi:10.4018/978-1-4666-9975-5.ch009

Alshammari, A., Kalevaru, V. N., Bagabas, A., & Martin, A. (2016). Production of Ethylene and its Commercial Importance in the Global Market. In H. Al-Megren & T. Xiao (Eds.), *Petrochemical Catalyst Materials, Processes, and Emerging Technologies* (pp. 82–115). Hershey, PA: IGI Global. doi:10.4018/978-1-4666-9975-5.ch004

Amel, M. (2016). Synthesis, Characterizations, and Biological Effects Study of Some Quinoline Family. In M. Bououdina (Ed.), *Emerging Research on Bioinspired Materials Engineering* (pp. 160–196). Hershey, PA: IGI Global. doi:10.4018/978-1-4666-9811-6.ch006

Amna, T., Haasan, M. S., Khil, M., & Hwang, I. (2016). Impact of Electrospun Biomimetic Extracellular Environment on Proliferation and Intercellular Communication of Muscle Precursor Cells: An Overview – Intercellular Communication of Muscle Precursor Cells with Extracellular Environment. In M. Bououdina (Ed.), *Emerging Research on Bioinspired Materials Engineering* (pp. 247–265). Hershey, PA: IGI Global. doi:10.4018/978-1-4666-9811-6.ch009

Amuda, M. O., Lawal, T. F., & Akinlabi, E. T. (2017). Research Progress on Rheological Behavior of AA7075 Aluminum Alloy During Hot Deformation. *International Journal of Materials Forming and Machining Processes*, *4*(1), 53–96. doi:10.4018/IJMFMP.2017010104

An, M., & Qin, Y. (2016). Challenges of Railway Safety Risk Assessment and Maintenance Decision Making. In B. Rai (Ed.), *Handbook of Research on Emerging Innovations in Rail Transportation Engineering* (pp. 173–211). Hershey, PA: IGI Global. doi:10.4018/978-1-5225-0084-1.ch009

Anil, M., Ayyildiz-Tamis, D., Tasdemir, S., Sendemir-Urkmez, A., & Gulce-Iz, S. (2016). Bioinspired Materials and Biocompatibility. In M. Bououdina (Ed.), *Emerging Research on Bioinspired Materials Engineering* (pp. 294–322). Hershey, PA: IGI Global. doi:10.4018/978-1-4666-9811-6.ch011

Armutlu, H. (2018). Intelligent Biomedical Engineering Operations by Cloud Computing Technologies. In U. Kose, G. Guraksin, & O. Deperlioglu (Eds.), *Nature-Inspired Intelligent Techniques for Solving Biomedical Engineering Problems* (pp. 297–317). Hershey, PA: IGI Global. doi:10.4018/978-1-5225-4769-3.ch015

Arokiyaraj, S., Saravanan, M., Bharanidharan, R., Islam, V. I., Bououdina, M., & Vincent, S. (2016). Green Synthesis of Metallic Nanoparticles Using Plant Compounds and Their Applications: Metallic Nanoparticles Synthesis Using Plants. In M. Bououdina (Ed.), *Emerging Research on Bioinspired Materials Engineering* (pp. 1–34). Hershey, PA: IGI Global. doi:10.4018/978-1-4666-9811-6.ch001

Atik, M., Sadek, M., & Shahrour, I. (2017). Single-Run Adaptive Pushover Procedure for Shear Wall Structures. In V. Plevris, G. Kremmyda, & Y. Fahjan (Eds.), *Performance-Based Seismic Design of Concrete Structures and Infrastructures* (pp. 59–83). Hershey, PA: IGI Global. doi:10.4018/978-1-5225-2089-4.ch003

Aydin, A., Akyol, E., Gungor, M., Kaya, A., & Tasdelen, S. (2018). Geophysical Surveys in Engineering Geology Investigations With Field Examples. In N. Ceryan (Ed.), *Handbook of Research on Trends and Digital Advances in Engineering Geology* (pp. 257–280). Hershey, PA: IGI Global. doi:10.4018/978-1-5225-2709-1.ch007

Azevedo, N. M., Lemos, J. V., & Rocha de Almeida, J. (2016). Discrete Element Particle Modelling of Stone Masonry. In V. Sarhosis, K. Bagi, J. Lemos, & G. Milani (Eds.), *Computational Modeling of Masonry Structures Using the Discrete Element Method* (pp. 146–170). Hershey, PA: IGI Global. doi:10.4018/978-1-5225-0231-9.ch007

Bamufleh, H. S., Noureldin, M. M., & El-Halwagi, M. M. (2016). Sustainable Process Integration in the Petrochemical Industries. In H. Al-Megren & T. Xiao (Eds.), *Petrochemical Catalyst Materials, Processes, and Emerging Technologies* (pp. 150–163). Hershey, PA: IGI Global. doi:10.4018/978-1-4666-9975-5.ch006

Banerjee, S., Gautam, R. K., Gautam, P. K., Jaiswal, A., & Chattopadhyaya, M. C. (2016). Recent Trends and Advancement in Nanotechnology for Water and Wastewater Treatment: Nanotechnological Approach for Water Purification. In A. Khitab & W. Anwar (Eds.), *Advanced Research on Nanotechnology for Civil Engineering Applications* (pp. 208–252). Hershey, PA: IGI Global. doi:10.4018/978-1-5225-0344-6.ch007

Bas, T. G. (2017). Nutraceutical Industry with the Collaboration of Biotechnology and Nutrigenomics Engineering: The Significance of Intellectual Property in the Entrepreneurship and Scientific Research Ecosystems. In T. Bas & J. Zhao (Eds.), *Comparative Approaches to Biotechnology Development and Use in Developed and Emerging Nations* (pp. 1–17). Hershey, PA: IGI Global. doi:10.4018/978-1-5225-1040-6.ch001

Beale, R., & André, J. (2017). *Design Solutions and Innovations in Temporary Structures*. Hershey, PA: IGI Global. doi:10.4018/978-1-5225-2199-0

Behnam, B. (2017). Simulating Post-Earthquake Fire Loading in Conventional RC Structures. In P. Samui, S. Chakraborty, & D. Kim (Eds.), *Modeling and Simulation Techniques in Structural Engineering* (pp. 425–444). Hershey, PA: IGI Global. doi:10.4018/978-1-5225-0588-4.ch015

Ben Hamida, I., Salah, S. B., Msahli, F., & Mimouni, M. F. (2018). Distribution Network Reconfiguration Using SPEA2 for Power Loss Minimization and Reliability Improvement. *International Journal of Energy Optimization and Engineering*, *7*(1), 50–65. doi:10.4018/IJEOE.2018010103

Benjamin, S. R., de Lima, F., & Rathoure, A. K. (2016). Genetically Engineered Microorganisms for Bioremediation Processes: GEMs for Bioremediaton. In A. Rathoure & V. Dhatwalia (Eds.), *Toxicity and Waste Management Using Bioremediation* (pp. 113–140). Hershey, PA: IGI Global. doi:10.4018/978-1-4666-9734-8.ch006

Bhaskar, S. V., & Kudal, H. N. (2017). Effect of TiCN and AlCrN Coating on Tribological Behaviour of Plasma-nitrided AISI 4140 Steel. *International Journal of Surface Engineering and Interdisciplinary Materials Science*, *5*(2), 1–17. doi:10.4018/IJSEIMS.2017070101

Bhowmik, S., Sahoo, P., Acharyya, S. K., Dhar, S., & Chattopadhyay, J. (2016). Effect of Microstructure Degradation on Fracture Toughness of 20MnMoNi55 Steel in DBT Region. *International Journal of Manufacturing, Materials, and Mechanical Engineering*, *6*(3), 11–27. doi:10.4018/IJMMME.2016070102

Bhutto, A. W., Abro, R., Abbas, T., Yu, G., & Chen, X. (2016). Desulphurization of Fuel Oils Using Ionic Liquids. In H. Al-Megren & T. Xiao (Eds.), *Petrochemical Catalyst Materials, Processes, and Emerging Technologies* (pp. 254–284). Hershey, PA: IGI Global. doi:10.4018/978-1-4666-9975-5.ch010

Bhuyan, D. (2018). Designing of a Twin Tube Shock Absorber: A Study in Reverse Engineering. In K. Kumar & J. Davim (Eds.), *Design and Optimization of Mechanical Engineering Products* (pp. 83–104). Hershey, PA: IGI Global. doi:10.4018/978-1-5225-3401-3.ch005

Bouloudenine, M., & Bououdina, M. (2016). Toxic Effects of Engineered Nanoparticles on Living Cells. In M. Bououdina (Ed.), *Emerging Research on Bioinspired Materials Engineering* (pp. 35–68). Hershey, PA: IGI Global. doi:10.4018/978-1-4666-9811-6.ch002

Brunetti, A., Sellaro, M., Drioli, E., & Barbieri, G. (2016). Membrane Engineering and its Role in Oil Refining and Petrochemical Industry. In H. Al-Megren & T. Xiao (Eds.), *Petrochemical Catalyst Materials, Processes, and Emerging Technologies* (pp. 116–149). Hershey, PA: IGI Global. doi:10.4018/978-1-4666-9975-5.ch005

Bügler, M., & Borrmann, A. (2016). Simulation Based Construction Project Schedule Optimization: An Overview on the State-of-the-Art. In F. Miranda & C. Abreu (Eds.), *Handbook of Research on Computational Simulation and Modeling in Engineering* (pp. 482–507). Hershey, PA: IGI Global. doi:10.4018/978-1-4666-8823-0.ch016

Calderon, F. A., Giolo, E. G., Frau, C. D., Rengel, M. G., Rodriguez, H., Tornello, M., ... Gallucci, R. (2018). Seismic Microzonation and Site Effects Detection Through Microtremors Measures: A Review. In N. Ceryan (Ed.), *Handbook of Research on Trends and Digital Advances in Engineering Geology* (pp. 326–349). Hershey, PA: IGI Global. doi:10.4018/978-1-5225-2709-1.ch009

Carmona-Murillo, J., & Valenzuela-Valdés, J. F. (2016). Motivation on Problem Based Learning. In D. Fonseca & E. Redondo (Eds.), *Handbook of Research on Applied E-Learning in Engineering and Architecture Education* (pp. 179–203). Hershey, PA: IGI Global. doi:10.4018/978-1-4666-8803-2.ch009

Ceryan, N. (2016). A Review of Soft Computing Methods Application in Rock Mechanic Engineering. In P. Samui (Ed.), *Handbook of Research on Advanced Computational Techniques for Simulation-Based Engineering* (pp. 1–70). Hershey, PA: IGI Global. doi:10.4018/978-1-4666-9479-8.ch001

Ceryan, N., & Can, N. K. (2018). Prediction of The Uniaxial Compressive Strength of Rocks Materials. In N. Ceryan (Ed.), *Handbook of Research on Trends and Digital Advances in Engineering Geology* (pp. 31–96). Hershey, PA: IGI Global. doi:10.4018/978-1-5225-2709-1.ch002

Ceryan, S. (2018). Weathering Indices Used in Evaluation of the Weathering State of Rock Material. In N. Ceryan (Ed.), *Handbook of Research on Trends and Digital Advances in Engineering Geology* (pp. 132–186). Hershey, PA: IGI Global. doi:10.4018/978-1-5225-2709-1.ch004

Chandrasekaran, S., Silva, B., Patil, A., Oo, A. M., & Campbell, M. (2016). Evaluating Engineering Students' Perceptions: The Impact of Team-Based Learning Practices in Engineering Education. *International Journal of Quality Assurance in Engineering and Technology Education, 5*(4), 42–59. doi:10.4018/IJQAETE.2016100103

Chen, H., Padilla, R. V., & Besarati, S. (2017). Supercritical Fluids and Their Applications in Power Generation. In L. Chen & Y. Iwamoto (Eds.), *Advanced Applications of Supercritical Fluids in Energy Systems* (pp. 369–402). Hershey, PA: IGI Global. doi:10.4018/978-1-5225-2047-4.ch012

Chen, L. (2017). Principles, Experiments, and Numerical Studies of Supercritical Fluid Natural Circulation System. In L. Chen & Y. Iwamoto (Eds.), *Advanced Applications of Supercritical Fluids in Energy Systems* (pp. 136–187). Hershey, PA: IGI Global. doi:10.4018/978-1-5225-2047-4.ch005

Clementi, F., Di Sciascio, G., Di Sciascio, S., & Lenci, S. (2017). Influence of the Shear-Bending Interaction on the Global Capacity of Reinforced Concrete Frames: A Brief Overview of the New Perspectives. In V. Plevris, G. Kremmyda, & Y. Fahjan (Eds.), *Performance-Based Seismic Design of Concrete Structures and Infrastructures* (pp. 84–111). Hershey, PA: IGI Global. doi:10.4018/978-1-5225-2089-4.ch004

Cortés-Polo, D., Calle-Cancho, J., Carmona-Murillo, J., & González-Sánchez, J. (2017). Future Trends in Mobile-Fixed Integration for Next Generation Networks: Classification and Analysis. *International Journal of Vehicular Telematics and Infotainment Systems, 1*(1), 33–53. doi:10.4018/IJVTIS.2017010103

Cui, X., Zeng, S., Li, Z., Zheng, Q., Yu, X., & Han, B. (2018). Advanced Composites for Civil Engineering Infrastructures. In K. Kumar & J. Davim (Eds.), *Composites and Advanced Materials for Industrial Applications* (pp. 212–248). Hershey, PA: IGI Global. doi:10.4018/978-1-5225-5216-1.ch010

Dalgıç, S., & Kuşku, İ. (2018). Geological and Geotechnical Investigations in Tunneling. In N. Ceryan (Ed.), *Handbook of Research on Trends and Digital Advances in Engineering Geology* (pp. 482–529). Hershey, PA: IGI Global. doi:10.4018/978-1-5225-2709-1.ch014

de la Varga, D., Soto, M., Arias, C. A., van Oirschot, D., Kilian, R., Pascual, A., & Álvarez, J. A. (2017). Constructed Wetlands for Industrial Wastewater Treatment and Removal of Nutrients. In Á. Val del Río, J. Campos Gómez, & A. Mosquera Corral (Eds.), *Technologies for the Treatment and Recovery of Nutrients from Industrial Wastewater* (pp. 202–230). Hershey, PA: IGI Global. doi:10.4018/978-1-5225-1037-6.ch008 del Valle-Zermeño, R., Chimenos, J. M., & Formosa, J. (2016). Flue Gas Desulfurization: Processes and Technologies. In T. Saleh (Ed.), *Applying Nanotechnology to the Desulfurization Process in Petroleum Engineering* (pp. 337–377). Hershey, PA: IGI Global. doi:10.4018/978-1-4666-9545-0.ch011

Delgado, J. M., Henriques, A. A., & Delgado, R. M. (2016). Structural Non-Linear Models and Simulation Techniques: An Efficient Combination for Safety Evaluation of RC Structures. In F. Miranda & C. Abreu (Eds.), *Handbook of Research on Computational Simulation and Modeling in Engineering* (pp. 540–584). Hershey, PA: IGI Global. doi:10.4018/978-1-4666-8823-0.ch018

Delgado, P. S., Arêde, A., Pouca, N. V., & Costa, A. (2016). Numerical Modeling of RC Bridges for Seismic Risk Analysis. In F. Miranda & C. Abreu (Eds.), *Handbook of Research on Computational Simulation and Modeling in Engineering* (pp. 457–481). Hershey, PA: IGI Global. doi:10.4018/978-1-4666-8823-0.ch015

Deng, Y., & Liu, S. (2016). Catalysis with Room Temperature Ionic Liquids Mediated Metal Nanoparticles. In H. Al-Megren & T. Xiao (Eds.), *Petrochemical Catalyst Materials, Processes, and Emerging Technologies* (pp. 285–329). Hershey, PA: IGI Global. doi:10.4018/978-1-4666-9975-5.ch011

Deperlioglu, O. (2018). Intelligent Techniques Inspired by Nature and Used in Biomedical Engineering. In U. Kose, G. Guraksin, & O. Deperlioglu (Eds.), *Nature-Inspired Intelligent Techniques for Solving Biomedical Engineering Problems* (pp. 51–77). Hershey, PA: IGI Global. doi:10.4018/978-1-5225-4769-3.ch003

Dias, G. L., Magalhães, R. R., Ferreira, D. D., & Vitoriano, F. A. (2016). The Use of a Robotic Arm for Displacement Measurements in a Cantilever beam. *International Journal of Manufacturing, Materials, and Mechanical Engineering*, *6*(3), 45–57. doi:10.4018/IJMMME.2016070104

Dimitratos, N., Villa, A., Chan-Thaw, C. E., Hammond, C., & Prati, L. (2016). Valorisation of Glycerol to Fine Chemicals and Fuels. In H. Al-Megren & T. Xiao (Eds.), *Petrochemical Catalyst Materials, Processes, and Emerging Technologies* (pp. 352–384). Hershey, PA: IGI Global. doi:10.4018/978-1-4666-9975-5.ch013

Dixit, A. (2018). Application of Silica-Gel-Reinforced Aluminium Composite on the Piston of Internal Combustion Engine: Comparative Study of Silica-Gel-Reinforced Aluminium Composite Piston With Aluminium Alloy Piston. In K. Kumar & J. Davim (Eds.), *Composites and Advanced Materials for Industrial Applications* (pp. 63–98). Hershey, PA: IGI Global. doi:10.4018/978-1-5225-5216-1.ch004

Drei, A., Milani, G., & Sincraian, G. (2016). Application of DEM to Historic Masonries, Two Case-Studies in Portugal and Italy: Aguas Livres Aqueduct and Arch-Tympana of a Church. In V. Sarhosis, K. Bagi, J. Lemos, & G. Milani (Eds.), *Computational Modeling of Masonry Structures Using the Discrete Element Method* (pp. 326–366). Hershey, PA: IGI Global. doi:10.4018/978-1-5225-0231-9.ch013

Dutta, S., Roy, P. K., & Nandi, D. (2016). Optimal Allocation of Static Synchronous Series Compensator Controllers using Chemical Reaction Optimization for Reactive Power Dispatch. *International Journal of Energy Optimization and Engineering*, *5*(3), 43–62. doi:10.4018/IJEOE.2016070103

Dutta, S., Roy, P. K., & Nandi, D. (2016). Quasi Oppositional Teaching-Learning based Optimization for Optimal Power Flow Incorporating FACTS. *International Journal of Energy Optimization and Engineering*, *5*(2), 64–84. doi:10.4018/IJEOE.2016040104

Eloy, S., Dias, M. S., Lopes, P. F., & Vilar, E. (2016). Digital Technologies in Architecture and Engineering: Exploring an Engaged Interaction within Curricula. In D. Fonseca & E. Redondo (Eds.), *Handbook of Research on Applied E-Learning in Engineering and Architecture Education* (pp. 368–402). Hershey, PA: IGI Global. doi:10.4018/978-1-4666-8803-2.ch017

Elsayed, A. M., Dakkama, H. J., Mahmoud, S., Al-Dadah, R., & Kaialy, W. (2017). Sustainable Cooling Research Using Activated Carbon Adsorbents and Their Environmental Impact. In T. Kobayashi (Ed.), *Applied Environmental Materials Science for Sustainability* (pp. 186–221). Hershey, PA: IGI Global. doi:10.4018/978-1-5225-1971-3.ch009

Ercanoglu, M., & Sonmez, H. (2018). General Trends and New Perspectives on Landslide Mapping and Assessment Methods. In N. Ceryan (Ed.), *Handbook of Research on Trends and Digital Advances in Engineering Geology* (pp. 350–379). Hershey, PA: IGI Global. doi:10.4018/978-1-5225-2709-1.ch010

Erinosho, M. F., Akinlabi, E. T., & Pityana, S. (2016). Enhancement of Surface Integrity of Titanium Alloy with Copper by Means of Laser Metal Deposition Process. In E. Akinlabi, R. Mahamood, & S. Akinlabi (Eds.), *Advanced Manufacturing Techniques Using Laser Material Processing* (pp. 60–91). Hershey, PA: IGI Global. doi:10.4018/978-1-5225-0329-3.ch004

Farag, H., & Kishida, M. (2016). Kinetic Models for Complex Parallel–Consecutive Reactions Assessment of Reaction Network and Product Selectivity. In H. Al-Megren & T. Xiao (Eds.), *Petrochemical Catalyst Materials, Processes, and Emerging Technologies* (pp. 330–351). Hershey, PA: IGI Global. doi:10.4018/978-1-4666-9975-5.ch012

Faroz, S. A., Pujari, N. N., Rastogi, R., & Ghosh, S. (2017). Risk Analysis of Structural Engineering Systems Using Bayesian Inference. In P. Samui, S. Chakraborty, & D. Kim (Eds.), *Modeling and Simulation Techniques in Structural Engineering* (pp. 390–424). Hershey, PA: IGI Global. doi:10.4018/978-1-5225-0588-4.ch014

Fernando, P. R., Hamigah, T., Disne, S., Wickramasingha, G. G., & Sutharshan, A. (2018). The Evaluation of Engineering Properties of Low Cost Concrete Blocks by Partial Doping of Sand with Sawdust: Low Cost Sawdust Concrete Block. *International Journal of Strategic Engineering*, *1*(2), 26–42. doi:10.4018/IJoSE.2018070103

Fragiadakis, M., Stefanou, I., & Psycharis, I. N. (2016). Vulnerability Assessment of Damaged Classical Multidrum Columns. In V. Sarhosis, K. Bagi, J. Lemos, & G. Milani (Eds.), *Computational Modeling of Masonry Structures Using the Discrete Element Method* (pp. 235–253). Hershey, PA: IGI Global. doi:10.4018/978-1-5225-0231-9.ch010

Gaines, T. W., Williams, K. R., & Wagener, K. B. (2016). ADMET: Functionalized Polyolefins. In H. Al-Megren & T. Xiao (Eds.), *Petrochemical Catalyst Materials, Processes, and Emerging Technologies* (pp. 1–21). Hershey, PA: IGI Global. doi:10.4018/978-1-4666-9975-5.ch001

Garg, H. (2016). Bi-Criteria Optimization for Finding the Optimal Replacement Interval for Maintaining the Performance of the Process Industries. In P. Vasant, G. Weber, & V. Dieu (Eds.), *Handbook of Research on Modern Optimization Algorithms and Applications in Engineering and Economics* (pp. 643–675). Hershey, PA: IGI Global. doi:10.4018/978-1-4666-9644-0.ch025

Gaspar, P. D., Dinho da Silva, P., Gonçalves, J. P., & Carneiro, R. (2016). Computational Modelling and Simulation to Assist the Improvement of Thermal Performance and Energy Efficiency in Industrial Engineering Systems: Application to Cold Stores. In F. Miranda & C. Abreu (Eds.), *Handbook of Research on Computational Simulation and Modeling in Engineering* (pp. 1–68). Hershey, PA: IGI Global. doi:10.4018/978-1-4666-8823-0.ch001

Ge, H., Tang, M., & Wen, X. (2016). Ni/ZnO Nano Sorbent for Reactive Adsorption Desulfurization of Refinery Oil Streams. In T. Saleh (Ed.), *Applying Nanotechnology to the Desulfurization Process in Petroleum Engineering* (pp. 216–239). Hershey, PA: IGI Global. doi:10.4018/978-1-4666-9545-0.ch007

Ghosh, S., Mitra, S., Ghosh, S., & Chakraborty, S. (2017). Seismic Reliability Analysis in the Framework of Metamodelling Based Monte Carlo Simulation. In P. Samui, S. Chakraborty, & D. Kim (Eds.), *Modeling and Simulation Techniques in Structural Engineering* (pp. 192–208). Hershey, PA: IGI Global. doi:10.4018/978-1-5225-0588-4.ch006

Gil, M., & Otero, B. (2017). Learning Engineering Skills through Creativity and Collaboration: A Game-Based Proposal. In R. Alexandre Peixoto de Queirós & M. Pinto (Eds.), *Gamification-Based E-Learning Strategies for Computer Programming Education* (pp. 14–29). Hershey, PA: IGI Global. doi:10.4018/978-1-5225-1034-5.ch002

Gill, J., Ayre, M., & Mills, J. (2017). Revisioning the Engineering Profession: How to Make It Happen! In M. Gray & K. Thomas (Eds.), *Strategies for Increasing Diversity in Engineering Majors and Careers* (pp. 156–175). Hershey, PA: IGI Global. doi:10.4018/978-1-5225-2212-6.ch008

Gopal, S., & Al-Hazmi, M. H. (2016). Advances in Catalytic Technologies for Selective Oxidation of Lower Alkanes. In H. Al-Megren & T. Xiao (Eds.), *Petrochemical Catalyst Materials, Processes, and Emerging Technologies* (pp. 22–52). Hershey, PA: IGI Global. doi:10.4018/978-1-4666-9975-5.ch002

Goyal, N., Ram, M., Bhardwaj, A., & Kumar, A. (2016). Thermal Power Plant Modelling with Fault Coverage Stochastically. *International Journal of Manufacturing, Materials, and Mechanical Engineering*, *6*(3), 28–44. doi:10.4018/IJMMME.2016070103

Goyal, N., Ram, M., & Kumar, P. (2017). Welding Process under Fault Coverage Approach for Reliability and MTTF. In M. Ram & J. Davim (Eds.), *Mathematical Concepts and Applications in Mechanical Engineering and Mechatronics* (pp. 222–245). Hershey, PA: IGI Global. doi:10.4018/978-1-5225-1639-2.ch011

Gray, M., & Lundy, C. (2017). Engineering Study Abroad: High Impact Strategy for Increasing Access. In M. Gray & K. Thomas (Eds.), *Strategies for Increasing Diversity in Engineering Majors and Careers* (pp. 42–59). Hershey, PA: IGI Global. doi:10.4018/978-1-5225-2212-6.ch003

Guha, D., Roy, P. K., & Banerjee, S. (2016). Application of Modified Biogeography Based Optimization in AGC of an Interconnected Multi-Unit Multi-Source AC-DC Linked Power System. *International Journal of Energy Optimization and Engineering*, *5*(3), 1–18. doi:10.4018/IJEOE.2016070101

Guha, D., Roy, P. K., & Banerjee, S. (2016). Grey Wolf Optimization to Solve Load Frequency Control of an Interconnected Power System: GWO Used to Solve LFC Problem. *International Journal of Energy Optimization and Engineering*, *5*(4), 62–83. doi:10.4018/IJEOE.2016100104

Gupta, A. K., Dey, A., & Mukhopadhyay, A. K. (2016). Micromechanical and Finite Element Modeling for Composites. In S. Datta & J. Davim (Eds.), *Computational Approaches to Materials Design: Theoretical and Practical Aspects* (pp. 101–162). Hershey, PA: IGI Global. doi:10.4018/978-1-5225-0290-6.ch005

Guraksin, G. E. (2018). Internet of Things and Nature-Inspired Intelligent Techniques for the Future of Biomedical Engineering. In U. Kose, G. Guraksin, & O. Deperlioglu (Eds.), *Nature-Inspired Intelligent Techniques for Solving Biomedical Engineering Problems* (pp. 263–282). Hershey, PA: IGI Global. doi:10.4018/978-1-5225-4769-3.ch013

Hansman, C. A. (2016). Developing Mentoring Programs in Engineering and Technology Education. *International Journal of Quality Assurance in Engineering and Technology Education*, *5*(2), 1–15. doi:10.4018/IJQAETE.2016040101

Hasan, U., Chegenizadeh, A., & Nikraz, H. (2016). Nanotechnology Future and Present in Construction Industry: Applications in Geotechnical Engineering. In A. Khitab & W. Anwar (Eds.), *Advanced Research on Nanotechnology for Civil Engineering Applications* (pp. 141–179). Hershey, PA: IGI Global. doi:10.4018/978-1-5225-0344-6.ch005

Hejazi, T., & Akbari, L. (2017). A Multiresponse Optimization Model for Statistical Design of Processes with Discrete Variables. In M. Ram & J. Davim (Eds.), *Mathematical Concepts and Applications in Mechanical Engineering and Mechatronics* (pp. 17–37). Hershey, PA: IGI Global. doi:10.4018/978-1-5225-1639-2.ch002

Hejazi, T., & Hejazi, A. (2017). Monte Carlo Simulation for Reliability-Based Design of Automotive Complex Subsystems. In M. Ram & J. Davim (Eds.), *Mathematical Concepts and Applications in Mechanical Engineering and Mechatronics* (pp. 177–200). Hershey, PA: IGI Global. doi:10.4018/978-1-5225-1639-2.ch009

Hejazi, T., & Poursabbagh, H. (2017). Reliability Analysis of Engineering Systems: An Accelerated Life Testing for Boiler Tubes. In M. Ram & J. Davim (Eds.), *Mathematical Concepts and Applications in Mechanical Engineering and Mechatronics* (pp. 154–176). Hershey, PA: IGI Global. doi:10.4018/978-1-5225-1639-2.ch008

Henao, J., & Sotelo, O. (2018). Surface Engineering at High Temperature: Thermal Cycling and Corrosion Resistance. In A. Pakseresht (Ed.), *Production, Properties, and Applications of High Temperature Coatings* (pp. 131–159). Hershey, PA: IGI Global. doi:10.4018/978-1-5225-4194-3.ch006

Huirache-Acuña, R., Alonso-Nuñez, G., Rivera-Muñoz, E. M., Gutierrez, O., & Pawelec, B. (2016). Trimetallic Sulfide Catalysts for Hydrodesulfurization. In T. Saleh (Ed.), *Applying Nanotechnology to the Desulfurization Process in Petroleum Engineering* (pp. 240–262). Hershey, PA: IGI Global. doi:10.4018/978-1-4666-9545-0.ch008

Ilori, O. O., Adetan, D. A., & Umoru, L. E. (2017). Effect of Cutting Parameters on the Surface Residual Stress of Face-Milled Pearlitic Ductile Iron. *International Journal of Materials Forming and Machining Processes*, *4*(1), 38–52. doi:10.4018/IJMFMP.2017010103

Imam, M. H., Tasadduq, I. A., Ahmad, A., Aldosari, F., & Khan, H. (2017). Automated Generation of Course Improvement Plans Using Expert System. *International Journal of Quality Assurance in Engineering and Technology Education*, *6*(1), 1–12. doi:10.4018/IJQAETE.2017010101

Injeti, S. K., & Kumar, T. V. (2018). A WDO Framework for Optimal Deployment of DGs and DSCs in a Radial Distribution System Under Daily Load Pattern to Improve Techno-Economic Benefits. *International Journal of Energy Optimization and Engineering*, *7*(2), 1–38. doi:10.4018/IJEOE.2018040101

Ishii, N., Anami, K., & Knisely, C. W. (2018). *Dynamic Stability of Hydraulic Gates and Engineering for Flood Prevention*. Hershey, PA: IGI Global. doi:10.4018/978-1-5225-3079-4

J., J., Chowdhury, S., Goyal, P., Samui, P., & Dalkiliç, Y. (2016). Determination of Bearing Capacity of Shallow Foundation Using Soft Computing. In P. Saxena, D. Singh, & M. Pant (Eds.), *Problem Solving and Uncertainty Modeling through Optimization and Soft Computing Applications* (pp. 292-328). Hershey, PA: IGI Global. doi:10.4018/978-1-4666-9885-7.ch014

Jagan, J., Gundlapalli, P., & Samui, P. (2016). Utilization of Classification Techniques for the Determination of Liquefaction Susceptibility of Soils. In S. Bhattacharyya, P. Banerjee, D. Majumdar, & P. Dutta (Eds.), *Handbook of Research on Advanced Hybrid Intelligent Techniques and Applications* (pp. 124–160). Hershey, PA: IGI Global. doi:10.4018/978-1-4666-9474-3.ch005

Jayapalan, S. (2018). A Review of Chemical Treatments on Natural Fibers-Based Hybrid Composites for Engineering Applications. In K. Kumar & J. Davim (Eds.), *Composites and Advanced Materials for Industrial Applications* (pp. 16–37). Hershey, PA: IGI Global. doi:10.4018/978-1-5225-5216-1.ch002

Jeet, K., & Dhir, R. (2016). Software Module Clustering Using Bio-Inspired Algorithms. In P. Vasant, G. Weber, & V. Dieu (Eds.), *Handbook of Research on Modern Optimization Algorithms and Applications in Engineering and Economics* (pp. 445–470). Hershey, PA: IGI Global. doi:10.4018/978-1-4666-9644-0.ch017

Joshi, S. D., & Talange, D. B. (2016). Fault Tolerant Control for a Fractional Order AUV System. *International Journal of Energy Optimization and Engineering*, *5*(2), 1–24. doi:10.4018/IJEOE.2016040101

Julião, D., Ribeiro, S., de Castro, B., Cunha-Silva, L., & Balula, S. S. (2016). Polyoxometalates-Based Nanocatalysts for Production of Sulfur-Free Diesel. In T. Saleh (Ed.), *Applying Nanotechnology to the Desulfurization Process in Petroleum Engineering* (pp. 426–458). Hershey, PA: IGI Global. doi:10.4018/978-1-4666-9545-0.ch014

Kamthan, P. (2016). On the Nature of Collaborations in Agile Software Engineering Course Projects. *International Journal of Quality Assurance in Engineering and Technology Education*, *5*(2), 42–59. doi:10.4018/IJQAETE.2016040104

Karaman, O., Celik, C., & Urkmez, A. S. (2016). Self-Assembled Biomimetic Scaffolds for Bone Tissue Engineering. In M. Bououdina (Ed.), *Emerging Research on Bioinspired Materials Engineering* (pp. 104–132). Hershey, PA: IGI Global. doi:10.4018/978-1-4666-9811-6.ch004

Karkalos, N. E., Markopoulos, A. P., & Dossis, M. F. (2017). Optimal Model Parameters of Inverse Kinematics Solution of a 3R Robotic Manipulator Using ANN Models. *International Journal of Manufacturing, Materials, and Mechanical Engineering*, *7*(3), 20–40. doi:10.4018/IJMMME.2017070102

Kesimal, A., Karaman, K., Cihangir, F., & Ercikdi, B. (2018). Excavatability Assessment of Rock Masses for Geotechnical Studies. In N. Ceryan (Ed.), *Handbook of Research on Trends and Digital Advances in Engineering Geology* (pp. 231–256). Hershey, PA: IGI Global. doi:10.4018/978-1-5225-2709-1.ch006

Khanh, D. V., Vasant, P. M., Elamvazuthi, I., & Dieu, V. N. (2016). Multi-Objective Optimization of Two-Stage Thermo-Electric Cooler Using Differential Evolution: MO Optimization of TEC Using DE. In F. Miranda & C. Abreu (Eds.), *Handbook of Research on Computational Simulation and Modeling in Engineering* (pp. 139–170). Hershey, PA: IGI Global. doi:10.4018/978-1-4666-8823-0.ch004

Kim, D., Hassan, M. K., Chang, S., & Bigdeli, Y. (2016). Nonlinear Vibration Control of 3D Irregular Structures Subjected to Seismic Loads. In P. Samui (Ed.), *Handbook of Research on Advanced Computational Techniques for Simulation-Based Engineering* (pp. 103–119). Hershey, PA: IGI Global. doi:10.4018/978-1-4666-9479-8.ch003

Knoflacher, H. (2017). The Role of Engineers and Their Tools in the Transport Sector after Paradigm Change: From Assumptions and Extrapolations to Science. In H. Knoflacher & E. Ocalir-Akunal (Eds.), *Engineering Tools and Solutions for Sustainable Transportation Planning* (pp. 1–29). Hershey, PA: IGI Global. doi:10.4018/978-1-5225-2116-7.ch001

Kose, U. (2018). Towards an Intelligent Biomedical Engineering With Nature-Inspired Artificial Intelligence Techniques. In U. Kose, G. Guraksin, & O. Deperlioglu (Eds.), *Nature-Inspired Intelligent Techniques for Solving Biomedical Engineering Problems* (pp. 1–26). Hershey, PA: IGI Global. doi:10.4018/978-1-5225-4769-3.ch001

Kostić, S. (2018). A Review on Enhanced Stability Analyses of Soil Slopes Using Statistical Design. In N. Ceryan (Ed.), *Handbook of Research on Trends and Digital Advances in Engineering Geology* (pp. 446–481). Hershey, PA: IGI Global. doi:10.4018/978-1-5225-2709-1.ch013

Kumar, A., Patil, P. P., & Prajapati, Y. K. (2018). *Advanced Numerical Simulations in Mechanical Engineering*. Hershey, PA: IGI Global. doi:10.4018/978-1-5225-3722-9

Kumar, G. R., Rajyalakshmi, G., & Manupati, V. K. (2017). Surface Micro Patterning of Aluminium Reinforced Composite through Laser Peening. *International Journal of Manufacturing, Materials, and Mechanical Engineering*, *7*(4), 15–27. doi:10.4018/IJMMME.2017100102

Kumari, N., & Kumar, K. (2018). Fabrication of Orthotic Calipers With Epoxy-Based Green Composite. In K. Kumar & J. Davim (Eds.), *Composites and Advanced Materials for Industrial Applications* (pp. 157–176). Hershey, PA: IGI Global. doi:10.4018/978-1-5225-5216-1.ch008

Kuppusamy, R. R. (2018). Development of Aerospace Composite Structures Through Vacuum-Enhanced Resin Transfer Moulding Technology (VERTMTy): Vacuum-Enhanced Resin Transfer Moulding. In K. Kumar & J. Davim (Eds.), *Composites and Advanced Materials for Industrial Applications* (pp. 99–111). Hershey, PA: IGI Global. doi:10.4018/978-1-5225-5216-1.ch005

Lemos, J. V. (2016). The Basis for Masonry Analysis with UDEC and 3DEC. In V. Sarhosis, K. Bagi, J. Lemos, & G. Milani (Eds.), *Computational Modeling of Masonry Structures Using the Discrete Element Method* (pp. 61–89). Hershey, PA: IGI Global. doi:10.4018/978-1-5225-0231-9.ch003

Loy, J., Howell, S., & Cooper, R. (2017). Engineering Teams: Supporting Diversity in Engineering Education. In M. Gray & K. Thomas (Eds.), *Strategies for Increasing Diversity in Engineering Majors and Careers* (pp. 106–129). Hershey, PA: IGI Global. doi:10.4018/978-1-5225-2212-6.ch006

Macher, G., Armengaud, E., Kreiner, C., Brenner, E., Schmittner, C., Ma, Z., ... Krammer, M. (2018). Integration of Security in the Development Lifecycle of Dependable Automotive CPS. In N. Druml, A. Genser, A. Krieg, M. Menghin, & A. Hoeller (Eds.), *Solutions for Cyber-Physical Systems Ubiquity* (pp. 383–423). Hershey, PA: IGI Global. doi:10.4018/978-1-5225-2845-6.ch015

Maghsoodlou, S., & Poreskandar, S. (2016). Controlling Electrospinning Jet Using Microscopic Model for Ideal Tissue Engineering Scaffolds. *International Journal of Chemoinformatics and Chemical Engineering*, *5*(2), 1–16. doi:10.4018/IJCCE.2016070101

Mahendramani, G., & Lakshmana Swamy, N. (2018). Effect of Weld Groove Area on Distortion of Butt Welded Joints in Submerged Arc Welding. *International Journal of Manufacturing, Materials, and Mechanical Engineering*, *8*(2), 33–44. doi:10.4018/IJMMME.2018040103

Maiti, S. (2016). Engineered Gellan Polysaccharides in the Design of Controlled Drug Delivery Systems. In M. Bououdina (Ed.), *Emerging Research on Bioinspired Materials Engineering* (pp. 266–293). Hershey, PA: IGI Global. doi:10.4018/978-1-4666-9811-6.ch010

Majumdar, J. D., Weisheit, A., & Manna, I. (2016). Laser Surface Processing for Tailoring of Properties by Optimization of Microstructure. In E. Akinlabi, R. Mahamood, & S. Akinlabi (Eds.), *Advanced Manufacturing Techniques Using Laser Material Processing* (pp. 121–171). Hershey, PA: IGI Global. doi:10.4018/978-1-5225-0329-3.ch006

Maldonado-Macías, A. A., García-Alcaraz, J. L., Hernández-Arellano, J. L., & Cortes-Robles, G. (2016). An Ergonomic Compatibility Perspective on the Selection of Advanced Manufacturing Technology: A Case Study for CNC Vertical Machining Centers. In G. Alor-Hernández, C. Sánchez-Ramírez, & J. García-Alcaraz (Eds.), *Handbook of Research on Managerial Strategies for Achieving Optimal Performance in Industrial Processes* (pp. 137–165). Hershey, PA: IGI Global. doi:10.4018/978-1-5225-0130-5.ch008

Mamaghani, I. H. (2016). Application of Discrete Finite Element Method for Analysis of Unreinforced Masonry Structures. In V. Sarhosis, K. Bagi, J. Lemos, & G. Milani (Eds.), *Computational Modeling of Masonry Structures Using the Discrete Element Method* (pp. 440–458). Hershey, PA: IGI Global. doi:10.4018/978-1-5225-0231-9.ch017

Mansor, M. R., Sapuan, S. M., Salim, M. A., Akop, M. Z., Musthafah, M. T., & Shaharuzaman, M. A. (2016). Concurrent Design of Green Composites. In D. Verma, S. Jain, X. Zhang, & P. Gope (Eds.), *Green Approaches to Biocomposite Materials Science and Engineering* (pp. 48–75). Hershey, PA: IGI Global. doi:10.4018/978-1-5225-0424-5.ch003

Mansouri, I., & Esmaeili, E. (2016). Nanotechnology Applications in the Construction Industry. In A. Khitab & W. Anwar (Eds.), *Advanced Research on Nanotechnology for Civil Engineering Applications* (pp. 111–140). Hershey, PA: IGI Global. doi:10.4018/978-1-5225-0344-6.ch004

Manzoor, A. (2016). MOOCs for Enhancing Engineering Education. In D. Fonseca & E. Redondo (Eds.), *Handbook of Research on Applied E-Learning in Engineering and Architecture Education* (pp. 204–223). Hershey, PA: IGI Global. doi:10.4018/978-1-4666-8803-2.ch010

Martin, A., Kalevaru, V. N., & Radnik, J. (2016). Palladium in Heterogeneous Oxidation Catalysis. In H. Al-Megren & T. Xiao (Eds.), *Petrochemical Catalyst Materials, Processes, and Emerging Technologies* (pp. 53–81). Hershey, PA: IGI Global. doi:10.4018/978-1-4666-9975-5.ch003

Melnyczuk, J. M., & Palchoudhury, S. (2016). Introduction to Bio-Inspired Hydrogel and Their Application: Hydrogels. In M. Bououdina (Ed.), *Emerging Research on Bioinspired Materials Engineering* (pp. 133–159). Hershey, PA: IGI Global. doi:10.4018/978-1-4666-9811-6.ch005

Mitra-Kirtley, S., Mullins, O. C., & Pomerantz, A. E. (2016). Sulfur and Nitrogen Chemical Speciation in Crude Oils and Related Carbonaceous Materials. In T. Saleh (Ed.), *Applying Nanotechnology to the Desulfurization Process in Petroleum Engineering* (pp. 53–83). Hershey, PA: IGI Global. doi:10.4018/978-1-4666-9545-0.ch002

Moalosi, R., Uziak, J., & Oladiran, M. T. (2016). Using Blended Learning Approach to Deliver Courses in An Engineering Programme. *International Journal of Quality Assurance in Engineering and Technology Education*, *5*(1), 23–39. doi:10.4018/IJQAETE.2016010103

Mohammadzadeh, S., & Kim, Y. (2017). Nonlinear System Identification of Smart Buildings. In P. Samui, S. Chakraborty, & D. Kim (Eds.), *Modeling and Simulation Techniques in Structural Engineering* (pp. 328–347). Hershey, PA: IGI Global. doi:10.4018/978-1-5225-0588-4.ch011

Mohanty, I., & Bhattacherjee, D. (2016). Artificial Neural Network and Its Application in Steel Industry. In S. Datta & J. Davim (Eds.), *Computational Approaches to Materials Design: Theoretical and Practical Aspects* (pp. 267–300). Hershey, PA: IGI Global. doi:10.4018/978-1-5225-0290-6.ch010

Mohebkhah, A., & Sarhosis, V. (2016). Discrete Element Modeling of Masonry-Infilled Frames. In V. Sarhosis, K. Bagi, J. Lemos, & G. Milani (Eds.), *Computational Modeling of Masonry Structures Using the Discrete Element Method* (pp. 200–234). Hershey, PA: IGI Global. doi:10.4018/978-1-5225-0231-9.ch009

Molina, G. J., Aktaruzzaman, F., Soloiu, V., & Rahman, M. (2017). Design and Testing of a Jet-Impingement Instrument to Study Surface-Modification Effects by Nanofluids. *International Journal of Surface Engineering and Interdisciplinary Materials Science*, *5*(2), 43–61. doi:10.4018/IJSEIMS.2017070104

Montalvan-Sorrosa, D., de los Cobos-Vasconcelos, D., & Gonzalez-Sanchez, A. (2016). Nanotechnology Applied to the Biodesulfurization of Fossil Fuels and Spent Caustic Streams. In T. Saleh (Ed.), *Applying Nanotechnology to the Desulfurization Process in Petroleum Engineering* (pp. 378–389). Hershey, PA: IGI Global. doi:10.4018/978-1-4666-9545-0.ch012

Montillet, J., Yu, K., Bonenberg, L. K., & Roberts, G. W. (2016). Optimization Algorithms in Local and Global Positioning. In P. Vasant, G. Weber, & V. Dieu (Eds.), *Handbook of Research on Modern Optimization Algorithms and Applications in Engineering and Economics* (pp. 1–53). Hershey, PA: IGI Global. doi:10.4018/978-1-4666-9644-0.ch001

Moreira, F., & Ferreira, M. J. (2016). Teaching and Learning Requirements Engineering Based on Mobile Devices and Cloud: A Case Study. In D. Fonseca & E. Redondo (Eds.), *Handbook of Research on Applied E-Learning in Engineering and Architecture Education* (pp. 237–262). Hershey, PA: IGI Global. doi:10.4018/978-1-4666-8803-2.ch012

Mukherjee, A., Saeed, R. A., Dutta, S., & Naskar, M. K. (2017). Fault Tracking Framework for Software-Defined Networking (SDN). In C. Singhal & S. De (Eds.), *Resource Allocation in Next-Generation Broadband Wireless Access Networks* (pp. 247–272). Hershey, PA: IGI Global. doi:10.4018/978-1-5225-2023-8.ch011

Mukhopadhyay, A., Barman, T. K., & Sahoo, P. (2018). Electroless Nickel Coatings for High Temperature Applications. In K. Kumar & J. Davim (Eds.), *Composites and Advanced Materials for Industrial Applications* (pp. 297–331). Hershey, PA: IGI Global. doi:10.4018/978-1-5225-5216-1.ch013

Náprstek, J., & Fischer, C. (2017). Dynamic Stability and Post-Critical Processes of Slender Auto-Parametric Systems. In V. Plevris, G. Kremmyda, & Y. Fahjan (Eds.), *Performance-Based Seismic Design of Concrete Structures and Infrastructures* (pp. 128–171). Hershey, PA: IGI Global. doi:10.4018/978-1-5225-2089-4.ch006

Nautiyal, L., Shivach, P., & Ram, M. (2018). Optimal Designs by Means of Genetic Algorithms. In M. Ram & J. Davim (Eds.), *Soft Computing Techniques and Applications in Mechanical Engineering* (pp. 151–161). Hershey, PA: IGI Global. doi:10.4018/978-1-5225-3035-0.ch007

Nazir, R. (2017). Advanced Nanomaterials for Water Engineering and Treatment: Nano-Metal Oxides and Their Nanocomposites. In T. Saleh (Ed.), *Advanced Nanomaterials for Water Engineering, Treatment, and Hydraulics* (pp. 84–126). Hershey, PA: IGI Global. doi:10.4018/978-1-5225-2136-5.ch005

Nogueira, A. F., Ribeiro, J. C., Fernández de Vega, F., & Zenha-Rela, M. A. (2018). Evolutionary Approaches to Test Data Generation for Object-Oriented Software: Overview of Techniques and Tools. In M. Khosrow-Pour, D.B.A. (Ed.), Incorporating Nature-Inspired Paradigms in Computational Applications (pp. 162-194). Hershey, PA: IGI Global. doi:10.4018/978-1-5225-5020-4.ch006

Nunes, J. F., Moreira, P. M., & Tavares, J. M. (2016). Human Motion Analysis and Simulation Tools: A Survey. In F. Miranda & C. Abreu (Eds.), *Handbook of Research on Computational Simulation and Modeling in Engineering* (pp. 359–388). Hershey, PA: IGI Global. doi:10.4018/978-1-4666-8823-0.ch012

Ogunlaja, A. S., & Tshentu, Z. R. (2016). Molecularly Imprinted Polymer Nanofibers for Adsorptive Desulfurization. In T. Saleh (Ed.), *Applying Nanotechnology to the Desulfurization Process in Petroleum Engineering* (pp. 281–336). Hershey, PA: IGI Global. doi:10.4018/978-1-4666-9545-0.ch010

Ong, P., & Kohshelan, S. (2016). Performances of Adaptive Cuckoo Search Algorithm in Engineering Optimization. In P. Vasant, G. Weber, & V. Dieu (Eds.), *Handbook of Research on Modern Optimization Algorithms and Applications in Engineering and Economics* (pp. 676–699). Hershey, PA: IGI Global. doi:10.4018/978-1-4666-9644-0.ch026

Osho, M. B. (2018). Industrial Enzyme Technology: Potential Applications. In S. Bharati & P. Chaurasia (Eds.), *Research Advancements in Pharmaceutical, Nutritional, and Industrial Enzymology* (pp. 375–394). Hershey, PA: IGI Global. doi:10.4018/978-1-5225-5237-6.ch017

Padmaja, P., & Marutheswar, G. (2017). Certain Investigation on Secured Data Transmission in Wireless Sensor Networks. *International Journal of Mobile Computing and Multimedia Communications*, *8*(1), 48–61. doi:10.4018/IJMCMC.2017010104

Paixão, S. M., Silva, T. P., Arez, B. F., & Alves, L. (2016). Advances in the Reduction of the Costs Inherent to Fossil Fuels' Biodesulfurization towards Its Potential Industrial Application. In T. Saleh (Ed.), *Applying Nanotechnology to the Desulfurization Process in Petroleum Engineering* (pp. 390–425). Hershey, PA: IGI Global. doi:10.4018/978-1-4666-9545-0.ch013

Palmer, S., & Hall, W. (2017). An Evaluation of Group Work in First-Year Engineering Design Education. In R. Tucker (Ed.), *Collaboration and Student Engagement in Design Education* (pp. 145–168). Hershey, PA: IGI Global. doi:10.4018/978-1-5225-0726-0.ch007

Panneer, R. (2017). Effect of Composition of Fibers on Properties of Hybrid Composites. *International Journal of Manufacturing, Materials, and Mechanical Engineering*, *7*(4), 28–43. doi:10.4018/IJMMME.2017100103

Parker, J. (2016). Hubble's Expanding Universe: A Model for Quality in Technology Infused engineering and Technology Education. *International Journal of Quality Assurance in Engineering and Technology Education*, *5*(2), 16–29. doi:10.4018/IJQAETE.2016040102

Paul, S., & Roy, P. (2018). Optimal Design of Power System Stabilizer Using a Novel Evolutionary Algorithm. *International Journal of Energy Optimization and Engineering*, *7*(3), 24–46. doi:10.4018/IJEOE.2018070102

Pavaloiu, A. (2018). Artificial Intelligence Ethics in Biomedical-Engineering-Oriented Problems. In U. Kose, G. Guraksin, & O. Deperlioglu (Eds.), *Nature-Inspired Intelligent Techniques for Solving Biomedical Engineering Problems* (pp. 219–231). Hershey, PA: IGI Global. doi:10.4018/978-1-5225-4769-3.ch010

Peña, F. (2016). A Semi-Discrete Approach for the Numerical Simulation of Freestanding Blocks. In V. Sarhosis, K. Bagi, J. Lemos, & G. Milani (Eds.), *Computational Modeling of Masonry Structures Using the Discrete Element Method* (pp. 416–439). Hershey, PA: IGI Global. doi:10.4018/978-1-5225-0231-9.ch016

Penchovsky, R., & Traykovska, M. (2016). Synthetic Approaches to Biology: Engineering Gene Control Circuits, Synthesizing, and Editing Genomes. In M. Bououdina (Ed.), *Emerging Research on Bioinspired Materials Engineering* (pp. 323–351). Hershey, PA: IGI Global. doi:10.4018/978-1-4666-9811-6.ch012

Pieroni, A., & Iazeolla, G. (2016). Engineering QoS and Energy Saving in the Delivery of ICT Services. In P. Vasant & N. Voropai (Eds.), *Sustaining Power Resources through Energy Optimization and Engineering* (pp. 208–226). Hershey, PA: IGI Global. doi:10.4018/978-1-4666-9755-3.ch009

Pioro, I., Mahdi, M., & Popov, R. (2017). Application of Supercritical Pressures in Power Engineering. In L. Chen & Y. Iwamoto (Eds.), *Advanced Applications of Supercritical Fluids in Energy Systems* (pp. 404–457). Hershey, PA: IGI Global. doi:10.4018/978-1-5225-2047-4.ch013

Plaksina, T., & Gildin, E. (2017). Rigorous Integrated Evolutionary Workflow for Optimal Exploitation of Unconventional Gas Assets. *International Journal of Energy Optimization and Engineering*, *6*(1), 101–122. doi:10.4018/IJEOE.2017010106

Puppala, A. J., Bheemasetti, T. V., Zou, H., Yu, X., Pedarla, A., & Cai, G. (2016). Spatial Variability Analysis of Soil Properties using Geostatistics. In P. Samui (Ed.), *Handbook of Research on Advanced Computational Techniques for Simulation-Based Engineering* (pp. 195–226). Hershey, PA: IGI Global. doi:10.4018/978-1-4666-9479-8.ch008

Ramdani, N., & Azibi, M. (2018). Polymer Composite Materials for Microelectronics Packaging Applications: Composites for Microelectronics Packaging. In K. Kumar & J. Davim (Eds.), *Composites and Advanced Materials for Industrial Applications* (pp. 177–211). Hershey, PA: IGI Global. doi:10.4018/978-1-5225-5216-1.ch009

Ramesh, M., Garg, R., & Subrahmanyam, G. V. (2017). Investigation of Influence of Quenching and Annealing on the Plane Fracture Toughness and Brittle to Ductile Transition Temperature of the Zinc Coated Structural Steel Materials. *International Journal of Surface Engineering and Interdisciplinary Materials Science*, *5*(2), 33–42. doi:10.4018/IJSEIMS.2017070103

Razavi, A. M., & Ahmad, R. (2016). Agile Software Development Challenges in Implementation and Adoption: Focusing on Large and Distributed Settings – Past Experiences, Emergent Topics. In I. Ghani, D. Jawawi, S. Dorairaj, & A. Sidky (Eds.), *Emerging Innovations in Agile Software Development* (pp. 175–207). Hershey, PA: IGI Global. doi:10.4018/978-1-4666-9858-1.ch010

Reccia, E., Cecchi, A., & Milani, G. (2016). FEM/DEM Approach for the Analysis of Masonry Arch Bridges. In V. Sarhosis, K. Bagi, J. Lemos, & G. Milani (Eds.), *Computational Modeling of Masonry Structures Using the Discrete Element Method* (pp. 367–392). Hershey, PA: IGI Global. doi:10.4018/978-1-5225-0231-9.ch014

Ro, H. K., & McIntosh, K. (2016). Constructing Conducive Environment for Women of Color in Engineering Undergraduate Education. In U. Thomas & J. Drake (Eds.), *Critical Research on Sexism and Racism in STEM Fields* (pp. 23–48). Hershey, PA: IGI Global. doi:10.4018/978-1-5225-0174-9.ch002

Rodulfo-Baechler, S. M. (2016). Dual Role of Perovskite Hollow Fiber Membrane in the Methane Oxidation Reactions. In H. Al-Megren & T. Xiao (Eds.), *Petrochemical Catalyst Materials, Processes, and Emerging Technologies* (pp. 385–430). Hershey, PA: IGI Global. doi:10.4018/978-1-4666-9975-5.ch014

Rudolf, S., Biryuk, V. V., & Volov, V. (2018). Vortex Effect, Vortex Power: Technology of Vortex Power Engineering. In V. Kharchenko & P. Vasant (Eds.), *Handbook of Research on Renewable Energy and Electric Resources for Sustainable Rural Development* (pp. 500–533). Hershey, PA: IGI Global. doi:10.4018/978-1-5225-3867-7.ch021

Sah, A., Bhadula, S. J., Dumka, A., & Rawat, S. (2018). A Software Engineering Perspective for Development of Enterprise Applications. In A. Elçi (Ed.), *Handbook of Research on Contemporary Perspectives on Web-Based Systems* (pp. 1–23). Hershey, PA: IGI Global. doi:10.4018/978-1-5225-5384-7.ch001

Sahoo, P., & Roy, S. (2017). Tribological Behavior of Electroless Ni-P, Ni-P-W and Ni-P-Cu Coatings: A Comparison. *International Journal of Surface Engineering and Interdisciplinary Materials Science, 5*(1), 1–15. doi:10.4018/IJSEIMS.2017010101

Sahoo, S. (2018). Laminated Composite Hypar Shells as Roofing Units: Static and Dynamic Behavior. In K. Kumar & J. Davim (Eds.), *Composites and Advanced Materials for Industrial Applications* (pp. 249–269). Hershey, PA: IGI Global. doi:10.4018/978-1-5225-5216-1.ch011

Sahu, H., & Hungyo, M. (2018). Introduction to SDN and NFV. In A. Dumka (Ed.), *Innovations in Software-Defined Networking and Network Functions Virtualization* (pp. 1–25). Hershey, PA: IGI Global. doi:10.4018/978-1-5225-3640-6.ch001

Saikia, P., Bharadwaj, S. K., & Miah, A. T. (2016). Peroxovanadates and Its Bio-Mimicking Relation with Vanadium Haloperoxidases. In M. Bououdina (Ed.), *Emerging Research on Bioinspired Materials Engineering* (pp. 197–219). Hershey, PA: IGI Global. doi:10.4018/978-1-4666-9811-6.ch007

Saladino, R., Botta, G., & Crucianelli, M. (2016). Advances in Nanotechnology Transition Metal Catalysts in Oxidative Desulfurization (ODS) Processes: Nanotechnology Applied to ODS Processing. In T. Saleh (Ed.), *Applying Nanotechnology to the Desulfurization Process in Petroleum Engineering* (pp. 180–215). Hershey, PA: IGI Global. doi:10.4018/978-1-4666-9545-0.ch006

Saleh, T. A., Danmaliki, G. I., & Shuaib, T. D. (2016). Nanocomposites and Hybrid Materials for Adsorptive Desulfurization. In T. Saleh (Ed.), *Applying Nanotechnology to the Desulfurization Process in Petroleum Engineering* (pp. 129–153). Hershey, PA: IGI Global. doi:10.4018/978-1-4666-9545-0.ch004

Saleh, T. A., Shuaib, T. D., Danmaliki, G. I., & Al-Daous, M. A. (2016). Carbon-Based Nanomaterials for Desulfurization: Classification, Preparation, and Evaluation. In T. Saleh (Ed.), *Applying Nanotechnology to the Desulfurization Process in Petroleum Engineering* (pp. 154–179). Hershey, PA: IGI Global. doi:10.4018/978-1-4666-9545-0.ch005

Salem, A. M., & Shmelova, T. (2018). Intelligent Expert Decision Support Systems: Methodologies, Applications, and Challenges. In T. Shmelova, Y. Sikirda, N. Rizun, A. Salem, & Y. Kovalyov (Eds.), *Socio-Technical Decision Support in Air Navigation Systems: Emerging Research and Opportunities* (pp. 215–242). Hershey, PA: IGI Global. doi:10.4018/978-1-5225-3108-1.ch007

Samal, M. (2017). FE Analysis and Experimental Investigation of Cracked and Un-Cracked Thin-Walled Tubular Components to Evaluate Mechanical and Fracture Properties. In P. Samui, S. Chakraborty, & D. Kim (Eds.), *Modeling and Simulation Techniques in Structural Engineering* (pp. 266–293). Hershey, PA: IGI Global. doi:10.4018/978-1-5225-0588-4.ch009

Samal, M., & Balakrishnan, K. (2017). Experiments on a Ring Tension Setup and FE Analysis to Evaluate Transverse Mechanical Properties of Tubular Components. In P. Samui, S. Chakraborty, & D. Kim (Eds.), *Modeling and Simulation Techniques in Structural Engineering* (pp. 91–115). Hershey, PA: IGI Global. doi:10.4018/978-1-5225-0588-4.ch004

Santhanakumar, M., Adalarasan, R., & Rajmohan, M. (2016). An Investigation in Abrasive Waterjet Cutting of Al6061/SiC/Al2O3 Composite Using Principal Component Based Response Surface Methodology. *International Journal of Manufacturing, Materials, and Mechanical Engineering*, *6*(4), 30–47. doi:10.4018/IJMMME.2016100103

Sareen, N., & Bhattacharya, S. (2016). Cleaner Energy Fuels: Hydrodesulfurization and Beyond. In T. Saleh (Ed.), *Applying Nanotechnology to the Desulfurization Process in Petroleum Engineering* (pp. 84–128). Hershey, PA: IGI Global. doi:10.4018/978-1-4666-9545-0.ch003

Sarhosis, V. (2016). Micro-Modeling Options for Masonry. In V. Sarhosis, K. Bagi, J. Lemos, & G. Milani (Eds.), *Computational Modeling of Masonry Structures Using the Discrete Element Method* (pp. 28–60). Hershey, PA: IGI Global. doi:10.4018/978-1-5225-0231-9.ch002

Sarhosis, V., Oliveira, D. V., & Lourenco, P. B. (2016). On the Mechanical Behavior of Masonry. In V. Sarhosis, K. Bagi, J. Lemos, & G. Milani (Eds.), *Computational Modeling of Masonry Structures Using the Discrete Element Method* (pp. 1–27). Hershey, PA: IGI Global. doi:10.4018/978-1-5225-0231-9.ch001

Satyam, N. (2016). Liquefaction Modelling of Granular Soils using Discrete Element Method. In P. Samui (Ed.), *Handbook of Research on Advanced Computational Techniques for Simulation-Based Engineering* (pp. 381–441). Hershey, PA: IGI Global. doi:10.4018/978-1-4666-9479-8.ch015

Sawant, S. (2018). Deep Learning and Biomedical Engineering. In U. Kose, G. Guraksin, & O. Deperlioglu (Eds.), *Nature-Inspired Intelligent Techniques for Solving Biomedical Engineering Problems* (pp. 283–296). Hershey, PA: IGI Global. doi:10.4018/978-1-5225-4769-3.ch014

Sezgin, H., & Berkalp, O. B. (2018). Textile-Reinforced Composites for the Automotive Industry. In K. Kumar & J. Davim (Eds.), *Composites and Advanced Materials for Industrial Applications* (pp. 129–156). Hershey, PA: IGI Global. doi:10.4018/978-1-5225-5216-1.ch007

Shah, M. Z., Gazder, U., Bhatti, M. S., & Hussain, M. (2018). Comparative Performance Evaluation of Effects of Modifier in Asphaltic Concrete Mix. *International Journal of Strategic Engineering*, *1*(2), 13–25. doi:10.4018/IJoSE.2018070102

Shah, V. S., Shah, H. R., & Samui, P. (2016). Application of Meta-Models (MPMR and ELM) for Determining OMC, MDD and Soaked CBR Value of Soil. In S. Bhattacharyya, P. Banerjee, D. Majumdar, & P. Dutta (Eds.), *Handbook of Research on Advanced Hybrid Intelligent Techniques and Applications* (pp. 454–482). Hershey, PA: IGI Global. doi:10.4018/978-1-4666-9474-3.ch015

Sharma, N., & Kumar, K. (2018). Fabrication of Porous NiTi Alloy Using Organic Binders. In K. Kumar & J. Davim (Eds.), *Composites and Advanced Materials for Industrial Applications* (pp. 38–62). Hershey, PA: IGI Global. doi:10.4018/978-1-5225-5216-1.ch003

Sharma, T. K. (2016). Application of Shuffled Frog Leaping Algorithm in Software Project Scheduling. In P. Saxena, D. Singh, & M. Pant (Eds.), *Problem Solving and Uncertainty Modeling through Optimization and Soft Computing Applications* (pp. 225–238). Hershey, PA: IGI Global. doi:10.4018/978-1-4666-9885-7.ch011

Shivach, P., Nautiyal, L., & Ram, M. (2018). Applying Multi-Objective Optimization Algorithms to Mechanical Engineering. In M. Ram & J. Davim (Eds.), *Soft Computing Techniques and Applications in Mechanical Engineering* (pp. 287–301). Hershey, PA: IGI Global. doi:10.4018/978-1-5225-3035-0.ch014

Shmelova, T. (2018). Stochastic Methods for Estimation and Problem Solving in Engineering: Stochastic Methods of Decision Making in Aviation. In S. Kadry (Ed.), *Stochastic Methods for Estimation and Problem Solving in Engineering* (pp. 139–160). Hershey, PA: IGI Global. doi:10.4018/978-1-5225-5045-7.ch006

Shukla, R., Anapagaddi, R., Singh, A. K., Allen, J. K., Panchal, J. H., & Mistree, F. (2016). Integrated Computational Materials Engineering for Determining the Set Points of Unit Operations for Production of a Steel Product Mix. In S. Datta & J. Davim (Eds.), *Computational Approaches to Materials Design: Theoretical and Practical Aspects* (pp. 163–191). Hershey, PA: IGI Global. doi:10.4018/978-1-5225-0290-6.ch006

Siero González, L. R., & Romo Vázquez, A. (2017). Didactic Sequences Teaching Mathematics for Engineers With Focus on Differential Equations. In M. Ramírez-Montoya (Ed.), *Handbook of Research on Driving STEM Learning With Educational Technologies* (pp. 129–151). Hershey, PA: IGI Global. doi:10.4018/978-1-5225-2026-9.ch007

Singh, R., & Dutta, S. (2018). Visible Light Active Nanocomposites for Photocatalytic Applications. In K. Kumar & J. Davim (Eds.), *Composites and Advanced Materials for Industrial Applications* (pp. 270–296). Hershey, PA: IGI Global. doi:10.4018/978-1-5225-5216-1.ch012

Singh, R., & Lou, H. H. (2016). Safety and Efficiency Enhancement in LNG Terminals. In H. Al-Megren & T. Xiao (Eds.), *Petrochemical Catalyst Materials, Processes, and Emerging Technologies* (pp. 164–176). Hershey, PA: IGI Global. doi:10.4018/978-1-4666-9975-5.ch007

Sözbilir, H., Özkaymak, Ç., Uzel, B., & Sümer, Ö. (2018). Criteria for Surface Rupture Microzonation of Active Faults for Earthquake Hazards in Urban Areas. In N. Ceryan (Ed.), *Handbook of Research on Trends and Digital Advances in Engineering Geology* (pp. 187–230). Hershey, PA: IGI Global. doi:10.4018/978-1-5225-2709-1.ch005

Stanciu, I. (2018). Stochastic Methods in Microsystems Engineering. In S. Kadry (Ed.), *Stochastic Methods for Estimation and Problem Solving in Engineering* (pp. 161–176). Hershey, PA: IGI Global. doi:10.4018/978-1-5225-5045-7.ch007

Strebkov, D., Nekrasov, A., Trubnikov, V., & Nekrasov, A. (2018). Single-Wire Resonant Electric Power Systems for Renewable-Based Electric Grid. In V. Kharchenko & P. Vasant (Eds.), *Handbook of Research on Renewable Energy and Electric Resources for Sustainable Rural Development* (pp. 449–474). Hershey, PA: IGI Global. doi:10.4018/978-1-5225-3867-7.ch019

Subburaman, D., Jagan, J., Dalkiliç, Y., & Samui, P. (2016). Reliability Analysis of Slope Using MPMR, GRNN and GPR. In F. Miranda & C. Abreu (Eds.), *Handbook of Research on Computational Simulation and Modeling in Engineering* (pp. 208–224). Hershey, PA: IGI Global. doi:10.4018/978-1-4666-8823-0.ch007

Sun, J., Wan, S., Lin, J., & Wang, Y. (2016). Advances in Catalytic Conversion of Syngas to Ethanol and Higher Alcohols. In H. Al-Megren & T. Xiao (Eds.), *Petrochemical Catalyst Materials, Processes, and Emerging Technologies* (pp. 177–215). Hershey, PA: IGI Global. doi:10.4018/978-1-4666-9975-5.ch008

Tüdeş, Ş., Kumlu, K. B., & Ceryan, S. (2018). Integration Between Urban Planning and Natural Hazards For Resilient City. In N. Ceryan (Ed.), *Handbook of Research on Trends and Digital Advances in Engineering Geology* (pp. 591–630). Hershey, PA: IGI Global. doi:10.4018/978-1-5225-2709-1.ch017

Tyukhov, I., Rezk, H., & Vasant, P. (2016). Modern Optimization Algorithms and Applications in Solar Photovoltaic Engineering. In P. Vasant & N. Voropai (Eds.), *Sustaining Power Resources through Energy Optimization and Engineering* (pp. 390–445). Hershey, PA: IGI Global. doi:10.4018/978-1-4666-9755-3.ch016

Ulamis, K. (2018). Soil Liquefaction Assessment by Anisotropic Cyclic Triaxial Test. In N. Ceryan (Ed.), *Handbook of Research on Trends and Digital Advances in Engineering Geology* (pp. 631–664). Hershey, PA: IGI Global. doi:10.4018/978-1-5225-2709-1.ch018

Umar, M. A., Tenuche, S. S., Yusuf, S. A., Abdulsalami, A. O., & Kufena, A. M. (2016). Usability Engineering in Agile Software Development Processes. In I. Ghani, D. Jawawi, S. Dorairaj, & A. Sidky (Eds.), *Emerging Innovations in Agile Software Development* (pp. 208–221). Hershey, PA: IGI Global. doi:10.4018/978-1-4666-9858-1.ch011

Üzüm, O., & Çakır, Ö. A. (2016). A Bio-Inspired Phenomena in Cementitious Materials: Self-Healing. In M. Bououdina (Ed.), *Emerging Research on Bioinspired Materials Engineering* (pp. 220–246). Hershey, PA: IGI Global. doi:10.4018/978-1-4666-9811-6.ch008

Valente, M., & Milani, G. (2017). Seismic Assessment and Retrofitting of an Under-Designed RC Frame Through a Displacement-Based Approach. In V. Plevris, G. Kremmyda, & Y. Fahjan (Eds.), *Performance-Based Seismic Design of Concrete Structures and Infrastructures* (pp. 36–58). Hershey, PA: IGI Global. doi:10.4018/978-1-5225-2089-4.ch002

Vasant, P. (2018). A General Medical Diagnosis System Formed by Artificial Neural Networks and Swarm Intelligence Techniques. In U. Kose, G. Guraksin, & O. Deperlioglu (Eds.), *Nature-Inspired Intelligent Techniques for Solving Biomedical Engineering Problems* (pp. 130–145). Hershey, PA: IGI Global. doi:10.4018/978-1-5225-4769-3.ch006

Vergara, D., Lorenzo, M., & Rubio, M. (2016). On the Use of Virtual Environments in Engineering Education. *International Journal of Quality Assurance in Engineering and Technology Education*, *5*(2), 30–41. doi:10.4018/IJQAETE.2016040103

Verrollot, J., Tolonen, A., Harkonen, J., & Haapasalo, H. J. (2018). Challenges and Enablers for Rapid Product Development. *International Journal of Applied Industrial Engineering*, *5*(1), 25–49. doi:10.4018/IJAIE.2018010102

Wagner, C., & Ryan, C. (2016). Physical and Digital Integration Strategies of Electronic Device Supply Chains and Their Applicability to ETO Supply Chains. In R. Addo-Tenkorang, J. Kantola, P. Helo, & A. Shamsuzzoha (Eds.), *Supply Chain Strategies and the Engineer-to-Order Approach* (pp. 224–245). Hershey, PA: IGI Global. doi:10.4018/978-1-5225-0021-6.ch011

Wang, Z., Wu, P., Lan, L., & Ji, S. (2016). Preparation, Characterization and Desulfurization of the Supported Nickel Phosphide Catalysts. In H. Al-Megren & T. Xiao (Eds.), *Petrochemical Catalyst Materials, Processes, and Emerging Technologies* (pp. 431–458). Hershey, PA: IGI Global. doi:10.4018/978-1-4666-9975-5.ch015

Yardimci, A. G., & Karpuz, C. (2018). Fuzzy Rock Mass Rating: Soft-Computing-Aided Preliminary Stability Analysis of Weak Rock Slopes. In N. Ceryan (Ed.), *Handbook of Research on Trends and Digital Advances in Engineering Geology* (pp. 97–131). Hershey, PA: IGI Global. doi:10.4018/978-1-5225-2709-1.ch003

Zhang, L., Ding, S., Sun, S., Han, B., Yu, X., & Ou, J. (2016). Nano-Scale Behavior and Nano-Modification of Cement and Concrete Materials. In A. Khitab & W. Anwar (Eds.), *Advanced Research on Nanotechnology for Civil Engineering Applications* (pp. 28–79). Hershey, PA: IGI Global. doi:10.4018/978-1-5225-0344-6.ch002

Zindani, D., & Kumar, K. (2018). Industrial Applications of Polymer Composite Materials. In K. Kumar & J. Davim (Eds.), *Composites and Advanced Materials for Industrial Applications* (pp. 1–15). Hershey, PA: IGI Global. doi:10.4018/978-1-5225-5216-1.ch001

Zindani, D., Maity, S. R., & Bhowmik, S. (2018). A Decision-Making Approach for Material Selection of Polymeric Composite Bumper Beam. In K. Kumar & J. Davim (Eds.), *Composites and Advanced Materials for Industrial Applications* (pp. 112–128). Hershey, PA: IGI Global. doi:10.4018/978-1-5225-5216-1.ch006

About the Contributors

Mir Sayed Shah Danish earned double master degrees, in Energy & Electrical Engineering (MSc. 2015), Business Administration (MBA, 2016), and PhD. in Sustainable Energy (2018) from India and Japan respectively. Since 2004, he has been actively involved in the industry of Energy & Electrical Engineering in practical works, academic researches, and teaching. Besides, other professional duties, he has delivered various projects at different scales as team lead and project manager. Currently, he is doing as Assistant Professor (Specially Appointed Teaching Associate) in the field of Energy Engineering with University of the Ryukyus, Japan. He has remained an active member with several academic societies (namely: IEEE Senior Member, IET, IAES, IRED, PMI).

Tomonobu Senjyu was born in Saga, Japan, in 1963. He received the B.S. and M.S. degrees from the University of the Ryukyus, Okinawa, Japan, in 1986 and 1988, respectively, and the Ph.D. degree from Nagoya University, Aichi, Japan, in 1994, all in electrical engineering. Since 1988, he has been with the Department of Electrical and Electronics Engineering, Faculty of Engineering, University of the Ryukyus, where he is currently working as a Professor. His research interests include stability of ac machines, advanced control of electrical machines, renewable energy, and power electronics. Prof. Senjyu is a member of the Institute of Electrical Engineers of Japan.

* * *

Mohammed El Amine Abdelli is a researcher at University of Salamanca (Spain). Mohammed El Amine's research interests are in Corporate Social Responsibility (CSR), Entrepreneurship, Governance, Innovation, Sustainable Development, and Management. He is currently working on the projects related to Entrepreneurship and Corporate Governance in Mena countries. Mohammed El Amine is a member of IEEE, Expert in the Virtus Interpress (Ukraine)and Ambassador & Research Fellow Communication Institute of Greece(Greece) .He is Advisory Committee Member of 03 indexing conferences (International Conference on E-Commerce and Internet Technology (ECIT2020,China). International Conference on Intelligent Engineering And Management (London-UK). International Conference On Research, Innovation, Knowledge Management, and Technology Application For Business Sustainability (India)) .He is an Editorial board member of 03 indexing Journals (The Journal of Entrepreneurship Education (Q2, SCOPUS), the Academy of Entrepreneurship Journal (Q3, SCOPUS)and the Journal of the International Academy for Case Studies (Q4, SCOPUS)).He is reviewer of 13 indexing Journals in SCOPUS and Web of Science (the International Journal of Information and communication technology Education (IJICTE),the International Journal of Electronic Government Research (IJEGR),the International Journal of Cyber Warfare and Terrorism(IJCWT),the International Journal of Knowledge Management (IJKM), the International Journal of Business Analytics, the Information Resources Management Journal (IRMJ), the International Journal of Technology and Human Interaction (IJTHI),the International Journal of Public Administration in the digital age, the International Journal of E-Entrepreneurship and innovation (IJEEI), the Afro-Asian Journal of Finance and Accounting, the Journal of Organizational and End User Computing(JOEUC) and the International Journal of Knowledge and Systems Science (IJKSS). He has also eight book chapters and 03 indexing papers will be published in 2020.

Mazia Ahmed is a Research Scholar, Centre of Food Technology, University of Allahabad.

Mohit dev Awasthi is a Research Scholar in the Department of Mechanical Engineering, ITM University, Gwalior, Madhya Pradesh, India. He has 2 Journal Paper and 1 Book Chapter publication to his credit. His areas of interest are Composite Materials, Biomechanical Engineering, Nanotechnology, and Aerospace.

Jiannong Cao is currently a Chair Professor of Department of Computing at The Hong Kong Polytechnic University, Hong Kong. He is also the director of the Internet and Mobile Computing Lab in the department and the director of University's Research Facility in Big Data Analytics. His research interests include parallel and distributed computing, wireless sensing and networks, pervasive and mobile computing, and big data and cloud computing. He has co-authored 5 books, co-edited 9 books, and published over 600 papers in major international journals and conference proceedings. He received Best Paper Awards from conferences including, IEEE Trans. Industrial Informatics 2018, IEEE DSAA 2017, IEEE SMARTCOMP 2016, IEEE/IFIP EUC 2016, IEEE ISPA 2013, IEEE WCNC 2011, etc. Professor Cao has directed and participated in over 90 research and development projects and, as a principal investigator, obtained over HK$43 million grants from funding agencies such as National Natural Science Foundation of China (NSFC), Ministry of Science and Technology of P. R. China (MOST), Hong Kong Research Grant Council (RGC), Hong Kong Innovation and Technology Commission (ITC), Hong Kong Construction Industry Council, The Society of Hong Kong Scholars, and industries like Alibaba, Huawei, IBM and Nokia. Professor Cao has also involved in the development of 13 patents. Professor Cao served the Chair of the Technical Committee on Distributed Computing of IEEE Computer Society 2012-2014, a member of IEEE Fellows Evaluation Committee of the Computer Society and the Reliability Society, a member of IEEE Computer Society Education Awards Selection Committee, a member of IEEE Communications Society Awards Committee, and a member of Steering Committee of IEEE Transactions on Mobile Computing. Professor Cao has also served as chairs and members of organizing and technical committees of many international conferences, and as associate editor and member of the editorial boards of many international journals. Professor Cao is a fellow of IEEE and ACM distinguished member. In 2017, he received the Overseas Outstanding Contribution Award from China Computer Federation.

Trilok Chauhan is currently working as an Assitant Professor in the Department of Mechanical Engineering at ITM University, Gwalior. He has more than 6 years of teaching and research experience.

Burak Efe graduated from Gazi University, Department of Industrial Engineering in 2011. He received the M.Sc. degree in Industrial Engineering from Gazi University. He received Ph.D. degree in

Industrial Engineering from Gazi University. His research interests are Multi-criteria decision making, fuzzy logic, ergonomics, occupational health and safety. He has been working as an Assistant Professor in Necmettin Erbakan University.

Ömer Faruk Efe graduated from Selçuk University, Department of Industrial Engineering in 2008. He received the M.Sc. degree in Industrial Engineering from Selçuk University. He received Ph.D. degree in Industrial Engineering from Sakarya University. His research interests are Multi-criteria decision making, fuzzy logic, quality management, lean production, ergonomics, occupational health and safety. He has been working as an Assistant Professor in Afyon Kocatepe University.

Nadeem Faisal is currently working as an Assitant Professor, in the Department of Mechanical Engineering, ITM University, Gwalior, India. He has more than 3 years of teaching and research experience. Prior to switching career to academics, he worked as an Assistant Engineer at Voltas Ltd. He has published more than 18 research papers in journals of National & International repute many of which have been indexed in Scopus, SCI and WoS. He has also authored 2 books and has contributed more than 12 Book Chapters in Edited Volumes published by International Publishers like Springer, Wiley, IGI Global and DeGruyter. He is also a reviewer of 5 International Journals and member of 4 Professional bodies. He has been felicitated with many awards and honours including the Best Teacher award at ITM University recently. His areas of interests are Optimization, Materials Engineering, Biomedical Engineering, Rapid Prototyping, Virtual and Digital Manufacturing.

Siu Cheung Ho received the Bachelor's Degree in Logistics and Supply Chain Management from University of Northumbria at Newcastle, Master's Degree in Global Supply Chain Management and Master's Degree in Integrated Engineering from The Hong Kong Polytechnic University, Engineering Doctorate's Degree from City University of Hong Kong. Before he started his Philosophy of Doctor's Degree studies in The Hong Kong Polytechnic University, he is a Senior Research Fellow in The Hong Kong Research Institute of Textile and Apparel (HKRITA). His current research interests include machine learning, system engineering, project management, vision-based analytics, robotic, big data and pervasive computing. He is a committee member in Education and Training Committee & Logistics Policy Committee and Chartered Member of The Chartered Institute of Logistics and Transport in Hong Kong (CILTHK). He is Technical Advisor in The Employees Retraining Board (ERB).

Unaiza Iqbal is a Research Scholar, Centre of Food Technology, University of Allahabad.

Dheeraj Kumar is a Research Scholar in the Department of Mechanical Engineering at National Institute of Technology, Durgapur, India (An Institute of National Importance in India). He completed his Masters in Engineering from the prestigious Birla Institute of Technology, Mesra, India. He has published 3 research papers in journals of International repute. He also has 8 conference publications and 4 book chapters to his credit. His area of research interests are Solar energy and materials, Solar Air Heater and Solar Desalination.

Joan Mwihaki Nyika, Msc is an Environmental Science PhD student at the University of South Africa and a Tutorial Fellow at the Technical University of Kenya, Geoscience and the Environment Department. Her PhD work revolves around solid waste management and the effects of landfill leachate on soil and groundwater. She has interests in sustainable environmental management, water management and ecosystem management. Joan has published more than 15 articles and 5 book chapters in peer reviewed journals and books, respectively. She has and is mentoring both undergraduate and diploma students and has more than 5 years experience teaching and training in environmental science.

Mukesh Kumar Pandey is currently working as a Professor in the Department of Civil Engineering at ITM University, Gwalior, India. He has more than 23 years of teaching and research experience and has published more than 40 research papers. His area of interest is Construction Technology Management.

Ritu Rana is currently enrolled as a PhD Scholar in the Department of Humanities and Social Sciences, National Institute of Technology (NIT) Hamirpur, Himachal Pradesh, India since August, 2016. She received her MBA (Finance and Marketing) and B.Com. (Foreign Trade) from Kurukshetra University, Kurukshetra, Haryana (India) in 2006 and 2003, respectively. She has also worked with Indian Institute of Foreign Trade (IIFT), New Delhi in 2013 as a Programme Associate. She writes and presents widely on issues of Foreign Direct Investment and its impacts on the economic growth and the environment of the host country. Her current research interests are in International Business, Economics and Sustainable Development.

Pinki Saini is presently working as Assistant Professor at Centre of Food Technology, University of Allahabad. She completed degree of MSc and Ph.D. in Food Science and Technology from the G. B. Pant University of Agriculture and Technology, Pantnagar in 2005 and 2008 respectively. She has 11 years of teaching and 14 years of research experience. Published several research papers in journals and conferences. Completed 2 sponsored research projects, filed one patent and authored 10 book chapters. Her research interests include bioprocessing of foods, fermentations studies, biosorption of heavy metals, biofilm formation and development of edible films from plant and microbial sources.

Manoj Sharma is currently working as Assistant Professor in the Department of Humanities and Social Sciences, National Institute of Technology (NIT) Hamirpur, Himachal Pradesh, India. He received his PhD, M.Phil. (Econ.) and MBA from Himachal Pradesh University, Shimla, India. During his fourteen years of teaching and research, he has published more than twenty research papers in journals. He has also served various universities and institutions in different capacities and chaired technical sessions in national and international conferences. His areas of interest are International Business, Applied Economics, Environmental Economics, Service marketing and Transportation Management.

Anchal Singh is a Research Scholar, Centre of Food Technology, University of Allahabad, Allahabad. ORCID Id:0000-0003-1191-2139

Vartika Singh, Assistant Professor, AIGWES She holds Ph.D. degree in Hydrogeology and Remote Sensing, M.Phil. degree in Hydrogeology from Vikram University and Master degree in App. Geology from Kurukshtera University. She has GATE Qualified all India rank 47. She has 7 years of research experience with Defence Terrain Research Laboratory (DTRL), DRDO, Delhi. Where she is associated with various important projects of Defence Research & Development Organization, DTRL lab like GIM, Thar and PISTA. Her specialization of research is in Hydrogeology, Remote Sensing, Environmental Geology & Artificial Intelligence. She has more than 30 publications in various national, international journals. She is one of the founder member of Computer Intelligence Research Group and Magazine. She is in Editorial Board of several national and international journals.

José G. Vargas-Hernández, M.B.A., Ph.D., Member of the National System of Researchers of Mexico and a research professor at University Center for Economic and Managerial Sciences, University of Guadalajara. Professor Vargas-Hernández has a Ph. D. in Public Administration and a Ph.D. in Organizational Economics. He has undertaken studies in Organisational Behaviour and has a Master of Business Administration, published four books and more than 200 papers in international journals and reviews (some translated to English, French, German, Portuguese, Farsi, Chinese, etc.) and more than 300 essays in national journals and reviews. He has obtained several international Awards and recognition.

Seda Yıldırım was born in 1983. She graduated from Kocaeli University, Department of Business Administration in 2006. She completed her postgraduate and doctoral education at the Department of Business Administration, Kocaeli University, Production Management and Marketing. She worked as a research assistant in Kocaeli University between 2010-2013 and as an assistant professor in Recep Tayyip Erdoğan University between 2013-2016. Since 2017, she has been working as an Associate Professor at Namık Kemal University. Her current research interests include sustainable development, green marketing, sustainable consumption, Islamic economics, marketing culture, and values. She works as an editorial board member and reviewer in various journals. She also studies as a member of the organizing committee of various international congresses (İCOMEP, ENSCON, İSEFE).

Arabeche Zina is assistant professor at Ahmed Zabana University of Relizane, Algeria.

Index

T

U

V

W

www.ingramcontent.com/pod-product-compliance
Lightning Source LLC
LaVergne TN
LVHW060416300926
849162LV00006B/1447